574.09 M196h
Magner, Lois N., 1943-
A history of the life sciences
WITHDRAWN

D0056521

A
History
of the
Life
Sciences

Second Edition

Lois N. Magner

Department of History
Purdue University
West Lafayette, Indiana

Marcel Dekker, Inc.   New York•Basel•Hong Kong

Library of Congress Cataloging-in-Publication Data

Magner, Lois N.
   A history of the life sciences / Lois N. Magner. -- 2nd ed.
     p. cm.
   Includes bibliographical references and index.
   ISBN 0-8247-8942-3 (acid-free paper)
   1. Life sciences--History. 2. Biology--History. I. Title.
QH305.M22 1993
574'.09--dc20   93-34946
                                             CIP

The publisher offers discounts on this book when ordered in bulk quantities. For more information, write to Special Sales/Professional Marketing at the address below.

This book is printed on acid-free paper.

Copyright © 1994 by Marcel Dekker, Inc. All Rights Reserved

Neither this book nor any part may be reproduced or transmitted in any form or by any means, electronic or mechanical, including photocopying, microfilming, and recording, or by any information storage and retrieval system, without permission in writing from the publisher.

Marcel Dekker, Inc.
270 Madison Avenue, New York, New York 10016

Current printing (last digit):
10 9 8 7 6 5 4 3 2 1

PRINTED IN THE UNITED STATES OF AMERICA

To Ki-Han and Oliver,
as always

FOREWORD

This new and expanded edition of Lois Magner's *A History of the Life Sciences* is a brilliant and welcome addition to the much neglected field of the history of the life sciences. Of particular importance is the author's success in carrying the story from the beginning of human civilization through the intellectual concepts of the early Mediterranean civilizations and into the present era of molecular biology.

Dr. Magner has neatly woven the thread of biological history from the early knowledge of hunters, butchers, cooks, herdsmen, and primitive farmers on to the more philosophical cultures of Egypt, Greece, and Rome. She skillfully shows how the philosopher's search for "eternal knowledge" drew upon the "common people's" basic, practical knowledge useful for everyday life.

The author moves rapidly from the understanding of the early major civilizations through the Middle Ages to the Renaissance and the Scientific Revolution, where biology—with some additional input, mostly negative, from natural magic and alchemy—begins to have a profound impact on both art and anatomy.

The sixteenth and seventeenth centuries show the growth of scientific academies and the profound improvement and use of instruments—the primitive microscopes, thermometers, barometers, and the analytical balance, in particular. Indeed, Chapters 3 and 4 are particularly useful in tracing the evolution of effective science in all areas of intellectual progress. Although the scientific community was still very small, the growth of scientific communication and the improvement of instrumentation in the period stimulated the study of a large variety of scientific problems.

The text smoothly shifts in the last six chapters to an analysis of biological problems involving cell theory, physiology, microbiology, evolution, genetics, and molecular biology. These chapters reflect a conversion from a generalist treatment to a specific study of the special segments of understanding of life in its most special rudiments. The author has been highly successful in her transition from the generalist view of the past to the specialist view of the present.

Aaron J. Ihde
Emeritus Professor of Chemistry
and History of Science
University of Wisconsin—Madison

PREFACE

Given the remarkable advances in the life sciences and the explosive growth of studies in the history of science that have occurred since the first edition was written, it is virtually impossible to imagine squeezing even a bare outline of new information into a single volume. My primary purpose in writing this book has been to provide an updated introduction to a vast, complex, and fascinating story that can be used in a one-semester introductory course. I hope that this new edition will also be of interest to the general reader, and to teachers who are trying to bring more science or history into their own courses. Primarily, I have tried to call attention to the main themes of biology—the continuity and conflicts of theories and methodologies, the interactions between various disciplines, the diverse attitudes and ideologies with which scientists have approached the phenomena of life, and the interplay between science and society. Thus, this book emphasizes questions, puzzles, and conflicts that seem to reveal and reflect the fundamental themes at the core of the life sciences; it also examines some problems and individuals more closely than others in a way that may not accord with prevailing standards.

In a survey it is generally more important to paint a panoramic view of the whole forest than to enumerate the trees, but it is also important to stop and examine some of the individual trees, both those that might be considered typical and those that appear to be truly exceptional. Some problems that have consumed the energy and attention of natural scientists during particular periods, such as taxonomy, or the discovery of specific organs, glands, secretions, compounds, enzymes, and metabolic pathways, may therefore seem rather slighted, as are various factors that have engaged the attention of the new social historians of science. Based on my experiences as a teacher during the past 20 years, I have found that many of the topics that are of greatest interest to professional historians of biology, and are the subject of excellent recent scholarship, often fail to meet the needs of readers who are encountering the subject for the first time.

Perhaps even a brief introduction to the history of the life sciences can provoke a desire for a fuller and deeper understanding of the questions at the heart of the life sciences and make it possible to recognize the inherent limitations and liabilities of current paradigms in both science and the history of science.

Lois N. Magner
Department of History
Purdue University

INTRODUCTION

In the past 50 years the history of science has experienced phenomenal growth as a professional discipline, acquiring along the way the full panoply of professional insignia, societies, journals, annual meetings, dictionaries, encyclopedias, canonical reference works, professors, departments, graduate students, undergraduate courses, and some 4000 new publications per year, most of which tend to be highly specialized. The history of science can be seen as evolving, like its subject matter, but in directions quite different from that taken by its ostensible substrate. During the past half-century, the history of science has undergone changes quite as profound as the changes in science itself. From a rigorous focus on the evolution of "great ideas" and the thought processes of "great scientists," it has expanded to include new questions about the social, cultural, economic, and ideological context in which science and scientists are embedded. Profoundly influenced by the concepts and techniques borrowed from sociology, demography, anthropology, and psychology, the new social historians of science emphasize factors such as class, race, gender, institutional affiliations, economics, and so forth.

Scientists and some "old-fashioned" historians of science have complained that the history of science has lost its science. Of course there are few who would argue that scientists operate objectively, neutrally, in a world free of normal human motivations, distractions, and prejudices. While scien-

tists consider experimental results the key medium of exchange in their transactions, some historians of science see claims of experimental verification
and falsification as simply a battle of words for power in defining the discipline and monopolizing the academic marketplace. In its most extreme form,
the new social history denigrates "old-fashioned" studies of ideas, facts,
theories, discoveries, experiments, and the possibility of discovering a preexisting objective world of reality. In other words, there is no "there" out
there. Accordingly, scientists do not *discover* facts, theories, or new laws in
a real world accessed by scientific inquiries, but rather they *invent* such
theories to satisfy their own cultural values and preconceptions, influence
public opinion, enhance their power and resource base, and promote their
own interests at the expense of scientists holding competing views. The new
history of science is more likely to present scientists as entrepreneurs rather
than bench workers or experimentalists trying to study nature; presumably
there is nothing to study because all scientific theories are really socially
constructed excuses to advance cultural prejudices and preconceptions. Battles
between scientists are presented as if all experimental methodologies and attempts to test and validate theories about nature had nothing to do with the
special work of science. The fact that scientists are human beings engaged
in the same kinds of battles fought by entrepreneurial agents in other fields
is presented as the core of the issue.

   The fact that scientists are human beings embedded in a network of
relationships hardly seems a major revelation. Indeed, we could predict that
people in any human endeavor, from molecular biology to the making and
marketing of widgets, will fight for more money, more territory, more
power, a room with a view, and a golden parachute. The kind of finding
that we absolutely could not predict was that nitric oxide, NO, would
emerge as the "molecule of the year" in 1992 because of discoveries about
its previously unknown, unsuspected, multifaceted biological properties as a
factor involved in regulating diverse physiological, biological, and cellular
phenomena. Scientists generally see themselves as engaged in a special form
of work in which the core values include craftsmanship, ingenuity, originality, simplicity, and elegance. What scientists look for and admire is very
different from what is currently fashionable in the history of science.

   In moments of weakness, many scientists and even some historians
of science might join me in confessing to the pleasure and inspiration they
derived in their youth from reading currently unfashionable panegyrics to
science and scientists, such as *Microbe Hunters*, *Arrowsmith*, *Crucibles*, and
Madame Curie. It is highly unlikely (probably unimaginable) that young
readers would decide to follow a career in science after immersing them-

selves in the writings of the new historians of science. Unfortunately, the only historical scientific memoir students are likely to read is James Watson's *Double Helix*. The only lesson they seem to get out of that essentially amoral work is that Watson and Crick got the Nobel Prize without learning the structural formulas of the purines and pyrimidines. Scientists like the biochemist Joseph Fruton (coauthor of the classic biochemistry text used when I was a chemistry major) who have seriously reflected on the nature of science and its history generally argue that there is no one scientific method; thus, when historians and philosophers try to impose their own abstract philosophical constructs in analyzing the work of scientists and scientist–historians they are seen as a self-interested group engaged in a bizarre and irrelevant activity. According to Fruton, few if any researchers in biochemistry have sought guidance from contemporary philosophers with respect to scientific problems. In contrast, philosophers such as Aristotle, Bacon, and Descartes have often been the "silent partners" of scientists; William Harvey, the seventeenth-century master of experimental philosophy, for example, respectfully referred to the ancient philosophers as his guides and lights along his path to discovery. Integrating these increasingly divergent images of the history of biology into a comprehensible, one-volume introductory text is probably an impossible mission.

In the past, the term "scientific revolution" conjured up advances generally limited to the physical sciences, but today we are living in the midst of a biological revolution that some find profoundly disturbing. Nevertheless, until recently biology has been less a subject of historical analysis than the physical sciences, perhaps because biology might be thought of as both the oldest and the youngest of the sciences. Although biological problems were presumably among the first issues to interest human beings, the theories and instruments that are at present used to analyze them are of relatively recent origin. Moreover, the pace of progress in the life sciences has accelerated so rapidly that events prior to 1953, when the structure of DNA was determined, are seen as ancient history by most scientists and students. In view of the rapid advances in the life sciences that have occurred in the second half of the twentieth century, it is not surprising that many students assume that even nineteenth-century biology bears little resemblance to the basic scientific principles they have been taught. If forced to contend with apparently archaic, obsolete, and discarded biological theories, modern scientists might generously allow that previous generations of biologists struggled with questions that could not be answered with the tools then available. Thinking about the history of science can, however, provoke a more creative means of exercising critical analysis and imagination.

Although science and technology are of fundamental importance in the complex modern world, there is a serious imbalance in our approach to a general education. Our curriculum and our culture allow and excuse widespread scientific illiteracy, exacerbate differences between the sciences and the humanities, and encourage overspecialization within all disciplines. Attitudes toward science seemed to be in a tense transition phase in the 1970s when the first edition of this book was written. Fear of science and the unknown tended to overshadow the obvious need to understand and effectively exploit the ideas and inventions of twentieth-century science and technology. After a boom period, when science and technology were accorded great respect, and significant sums of money, skepticism set in and the prestige and authority of science was, to say the least, called into question. Many modern problems, including pollution, resource depletion, threats to biodiversity, overpopulation, changing patterns of morbidity and mortality, and even the emergence of virulent new diseases and antibiotic-resistant pathogens, are regarded as the outcome of scientific and technological advances. There are demands that we either stop further progress in science or limit the practice of science to topics that can provide an immediate payoff in practical terms, such as a cure for cancer and AIDS. The probable remedies for such biomedical problems, however, will almost certainly involve the use of techniques, such as genetic engineering and recombinant DNA, that are still regarded with some degree of suspicion and even hostility. Some critics of science focus on highly improbable threats associated with new techniques and others condemn aspects of science that conflict with religious beliefs, or scientific theories that presumably impoverish spiritual and aesthetic values by replacing the mysteries of nature with a soulless, mechanistic universe. On closer inspection, such criticism often turns out to be more properly directed against archaic mechanical concepts that are quite unrelated to the ideas that stimulate modern scientists.

Studies of the history of science might do much to bridge the gap between scientists and nonscientists, and encourage the scientific literacy that will be needed to deal with the increasingly powerful tools of molecular biology. Yet this opportunity has been largely neglected by both historians and scientists, who tend to both engage in and value highly specialized researches. Although science, as a human invention and mode of thought, deserves a prominent place in the drama of history, it is generally absent from general history courses. Science students tend to find history courses boring and irrelevant, merely a chronicle of wars, treaties made and treaties broken, kings, queens, politicians, and more wars. Scientists whose work saved more lives from the ravages of disease than all our wars have been able to eliminate are accorded little or no place in general history books. But the history of modern civilizations cannot be fully understood without attention to the influence of science and technology. As a creative endeavor, history is clearly different from chronology, but historians, especially historians of science, hold many different ideas about what is most fundamental to their discipline. Of course, history is not so much the record of what happened, but rather the story of what survived. This is as true for the history of science as for any other branch of history. This distortion is not simply dependent on the passage of time, but is also quite apparent upon critical examination of recent textbook accounts of science. When the actual doubts and loose ends, ambiguities, and implicit assumptions are glossed over, science falsely emerges as a static thing, with knowledge seemingly handed down from some omniscient celestial experimentalist. More properly, we should see science as a dynamic concept, encompassing both a body of knowledge and a means of obtaining and using more knowledge.

Science today is a vast and productive human endeavor. Its scope is such that no one individual can properly understand its content and the implications of its development on society in general. Scientists, like physicians, tend to be specialists. Whereas science as a whole is constantly changing, some of its goals and methods have a permanence and spiritual continuity worthy of study on their own merits. Here, the history of science can give us some understanding of the things that are fundamental to science and scientists. Although some scientists have claimed that all the interesting questions have been answered, a broader, historical perspective seems to support the opposite view, that science is indeed an "endless frontier."

I would like to thank Marcel Dekker, Inc., for inviting me to revise and update my *History of the Life Sciences*. On the time scale of modern science a book written in the 1970s might be considered a product of the dark ages. The tendency toward ultra-specialization in writing and teaching and the desire of experts to speak only to one another create a problem in reaching the audience that would benefit from a general synthesis in the form of new books and courses. Historians and scientists are unlikely to reach a general audience unless they are willing to transcend narrow professional boundaries. I am very pleased that Marcel Dekker, Inc., has been so willing to encourage the publication of general surveys of the history of science and medicine. I would especially like to thank Andrew Berin, my editor, for his help and encouragement in the completion of this book and my *History of Medicine* (1992). Once again, I should like to express my deep appreciation to Dr. John Parascandola and Dr. Vernard Foley for their invaluable advice and criticism during the preparation of the first edition of this book. Of course, all remaining errors of omission and commission are my own. I am especially indebted to the staff of the Department of History, Purdue University, for their help, support and patience. Without the assistance and co-operation of the staff of the Inter-Library Loan Division of the HSSE Library, Purdue University, it would have been impossible to complete this project. I would also like to acknowledge the History of Medicine Division, National Library of Medicine, for providing the illustrations used in this book. I would especially like to thank Lucy Keister and John Parascandola, History of Medicine Division, National Library of Medicine, for their help and advice in obtaining the illustrations.

CONTENTS

Contents

1

THE ORIGINS OF THE LIFE SCIENCES

Modern biology encompasses both the oldest and newest scientific disciplines, but the term *biology* was not coined until the beginning of the nineteenth century in order to denote a dynamic departure from the conventions of *natural philosophy*. Most simply defined as the science of living things, biology includes development and differentiation, cytology, genetics, molecular biology, evolution, and ecology. Like all the sciences, biology has roots reaching back into prehistory. Indeed, the most important lessons of all to human survival are those that fall within the purview of biology—agriculture, animal husbandry, and the art of healing. Innovations in these areas among prehistoric peoples would involve the most ephemeral products of human endeavor. Stone tools, weapons, pottery, glass, and metals leave a more permanent record than a new understanding of the cycles of nature, or the relationship between the breath, or the beat of the heart, to life itself. Yet it is ultimately on such biological wisdom that human survival and cultural evolution depended.

Driven by necessity, the earliest human beings must have been keen observers of nature. Anthropologists and ethnobotanists have discovered that so-called primitive peoples establish fairly sophisticated classification schemes as a means of understanding and adapting to their environment. Objects may be divided into categories such as the raw and the cooked, the wet and the dry, but the subdivisions within such schemes may reflect keen insights into the medicinal properties of plants and animal products. Even such a seemingly simple division as edible and inedible requires considerable experimentation. Primitive people also learned from experience that a series of operations could transform things from one class to another. For example, in the preparation of manioc both a food and a poison are produced.

Biology as the study of living things probably began with the emergence of *Homo sapiens sapiens* some 50,000 years ago. Ancient human beings living as hunter-gatherers during the Paleolithic Era or Old Stone Age manufactured crude tools made of chipped stones. Presumably these ancient people also produced useful inventions that were fully biodegradable and left no trace in the fossil record, such as digging sticks, bags, and baskets for carrying and storing food. But the inventions of the utmost importance in the separation of human beings from the ways of their ape-like ancestors were fire, speech, abstract thought, religion, and magic. At this stage, conscious of themselves as special entities, human beings could begin to confront the fundamental problems of existence—birth and death, health and disease, pain and hunger—in new ways. Since the transition from "prepeople" to fully modern human status, cultural evolution, a phenomenon unique to humankind, has taken precedence over biological evolution, a process shared with the rest of the organic world. Another great transition took place about 10,000 years ago when agriculture was invented. The transition from the hunter-gatherer mode of food production to farming and animal husbandry is known as the Neolithic Revolution. In the process of domesticating plants and animals, human beings, too, became domesticated and enmeshed in ways of living, thinking, and planning that revolved around the natural forces controlling their new enterprises. Scientists and historians were once most interested in the *when* and *where* of the emergence of agriculture, but given the fact that hunter-gatherers may enjoy a better diet and more leisure than agriculturalists, the greater puzzle is actually the *how* and *why*. In other words, when *progress* is seen as a concept to be analyzed rather than the inexorable march of human history, the Neolithic transformation can no longer be seen as an early and inevitable step along the road to modernity.

Anatomy, both human and animal, was one of the earliest components of biological knowledge. Among ancient peoples, fear and respect for the dead led to elaborate funerary rites involving the manipulation, and perhaps mutilation, of the corpse. The body might be cremated, buried, or preserved after removal of certain organs. Burial customs may reflect compassion and respect, as well as attempts to placate the spirits of the dead, and make it impossible for them to return, as they sometimes seemed to do in dreams, and harm the living. Once death could be anticipated and remembered, attempts could be made to ward it off by both magical and rational means. Primitive attempts at wound management and surgery would provide knowledge of human anatomy and the location of vulnerable sites on and in the body. Observations made while caring for the sick and the dying would have provided important physiological information, such as the importance of the breath, heartbeat, pulse, blood, and body heat.

Knowledge of animal anatomy and behavior is important to the hunter, herdsman, butcher, cook, and shaman. Following the natural migrations of animals forces the herdsman to study their natural behavior, but the domestication of animals can lead to more detailed observations, and ultimately to attempts to select and breed the best of the herd. Animals have served as totems of tribes or clans, and animal organs and behavior have been used as omens. Since divination and prophecy depended on the appearance and behavior of various animals, any peculiarity of action, shape, or color could decide the fortunes of the tribe and the fortuneteller.

BIOLOGY AND ANCIENT CIVILIZATIONS

The invention of science is generally credited to the Greek natural philosophers who lived during the sixth century B.C. This probably reflects our own cultural bias and ignorance about other cultures rather than a lack of achievements in them. Moreover, it is not clear how much of what is thought of as the Greek invention of science was borrowed from or influenced by the ideas and knowledge of other cultures. Great civilizations developed along the valleys of the Nile, Tigris-Euphrates, Yellow, and Indus rivers as much as 5000 years ago. Although these cultures left their mark in terms of technological advances, complex social organization, and written records, only a small portion of what they achieved was recorded and preserved. Because of the paucity of sources, we undoubtedly have a greatly distorted view of their real accomplishments.

The emphasis on the traditional Western heritage from Greek culture has resulted in the neglect of earlier civilizations and other traditions. India and China have been virtually excluded, except for the notice of certain exotic medicines and the invention of printing and gunpowder. Many important discoveries in pharmacology and nutrition could be listed as contributions to the history of biology, but to merely pick out a few items that have found acceptance in the West is a totally inadequate treatment for complex civilizations that, over the course of thousands of years, have developed unique approaches to fundamental questions about the nature of human beings and their relationship to nature. To understand the history of biology in China or India requires a thorough exploration of science and philosophy in the context of those complex civilizations. Such studies have only recently been attempted. Science and civilization have taken such different paths in China and the West that some scholars suggest that no such entity as biology appeared in classical Chinese science or medicine. It could also be said that in the West there was no biology until the nineteenth century. Certainly observations and theories concerning biological phenomena—human health and disease, plants and animals, the natural world in general—can be found in ancient Chinese writings, but they remained dispersed throughout a very different scholarly tradition. Ideas about vital phenomena and chemical processes can, for example, be found in the writings of ancient Chinese alchemists, physicians, and philosophers. Chinese alchemy, especially the alchemical ideas of the Taoists, could be considered part of the science of life. Alchemists are often thought of as mystics engaged in the impossible quest for the philosopher's stone that would allow them to transform lead into gold. Taoist alchemists, however, were more concerned with the search for the great elixirs of life, drugs that would provide health, longevity, and even immortality. While the writings of the alchemists were often obscure, and much of the work of the ancient sages has been imperfectly preserved, the knowledge accumulated by illiterate craftsmen, herbalists, folk healers, and fortunetellers generally remained outside the scholarly tradition and has been lost to history.

Within Chinese classical science and philosophy, the body of ideas and observations belonging to medicine would come closest to providing knowledge corresponding to the life sciences. Classical scholars explained the unity of nature, life, health, disease, and medicine in terms of the theory of *yin* and *yang* and the *five phases*. The five phases are sometimes called the "five elements," but this is generally considered a mistranslation based on a false analogy with the four elements of ancient Greek science. The Chinese term actually implies transition, movement, or passage rather than the stable, homogeneous chemical constituent implied by the term element. Accepting the impossibility of translating the terms yin and yang, scholars simply retain the terms as representations of all the pairs of opposites that express the dualism of the cosmos. Thus, yang corresponds to that which is masculine, light, warm, firm, heaven, full, and so forth, while yin is characterized as female, dark, cold, soft, earth, empty. Rather than simple pairs of opposites, yin and yang represent relational concepts; that is, members of pairs would not be hot and cold *per se*, but only in comparison to other entities or states. Applying these concepts to the human body, the inside is relatively

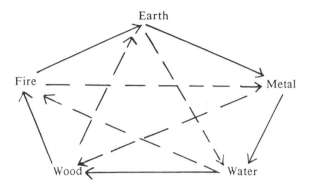

The five phases. As individual names or labels for the finer ramifications of yin and yang, the five phases represent aspects in the cycle of changes. The five phases are linked by relationships of generation and destruction. Patterns of destruction may be summarized as follows: water puts out fire; fire melts metal; a metal ax will cut wood; a wooden plow will turn the earth; an earthen dam will stop the flow of water. The cycle of generation proceeds as water produces the wood of trees; wood produces fire; fire creates ash, or earth; earth is the source of metal; when metals are heated they flow like water.

yin, whereas the outside in relatively yang, and specific internal organs are associated with yang or yin.

According to scholarly tradition, the principle of yin-yang is the basis of everything in creation, the cause of all transformations, and the origin of life and death. The five phases—water, fire, metal, wood, and earth—may be thought of as finer textured aspects of yin and yang. The five phases represent aspects in the cycle of changes. Classical Chinese medical practice was based on the theory of systematic correspondences which linked the philosophy of yin and yang and the five phases with a complex system that purportedly explained the functions of the constituents of the human body. Thus, the goal of classical Chinese anatomy was to explain the body's functional systems and the Western distinction between anatomy and physiology was essentially irrelevant. Within this philosophical system, Chinese scholars were able to accept the relationship between the heart and pulse and ceaseless circulation of the river of blood within the human body. It was not until the seventeenth century that these concepts were incorporated into Western science through the experimental work of William Harvey (1578–1657).

Many aspects of the history of the Indian subcontinent before the fourth century B.C. are obscure, but archaeologists have discovered evidence of the complex Indus River civilization that flourished from about 2700 to 1500 B.C. The sages of ancient India had little interest in chronology or the separation of mythology and history, but traces of their ideas about the nature of the universe and the human condition can be found in the *Vedas*, sacred books of divinely inspired knowledge. Brahma, the First Teacher of the Universe, one of the major gods of Hindu religion, was said to be the author of a great epic entitled *Ayurveda*, or *The Science of Life*. After oral transmission through a long line of divine sages, fragments of the sacred *Ayurveda* were incorporated into the written texts that now provide glimpses into ancient Indian concepts of life, disease, anatomy, physiology, psychology, embryology, alchemy, and medical lore. A striking difference between Indian religions and those of the West is the Hindu concept of a universe of immense size and antiquity undergoing a continuous process of development and decay. Indian scholars excelled in astronomy and mathematics, especially the ability to manipulate the large numbers associated with mythic traditions concerning the duration of the cosmic cycle.

Health, illness, and physiological functions were explained in terms of three primary humors, which are usually translated as *wind*, *bile*, and *phlegm*, in combination with *blood*. The concept of health as a harmonious balance of the primary humors is similar to fundamental Greek concepts, but the Ayurvedic system provided further complications. The body was composed of a combination of the five elements—earth, water, fire, wind, empty space—and the seven basic tissues. Five separate *winds*, plus the *vital soul*, and the *inmost soul* also mediated vital functions. Ayurvedic writers also expressed considerable interest in the fundamental question of conception and development. According to the Ayurvedic classics, living creatures fall into four categories as determined by their method of birth: from the womb, from eggs, from heat and moisture, and from seeds. Within the womb, conception occurred through the union of material provided by the male and female parents and the *spirit*, or external self. The development of the fetus from a shapeless jelly-like mass to the formed infant was described in considerable detail in association with the signs and symptoms characteristic of the stages of pregnancy.

The Ayurvedic surgical tradition presents an interesting challenge to the Western assumption that progress in surgery follows the development of systematic human dissection and animal vivisection. Religious precepts prohibited the use of the knife on human cadavers, but Ayurvedic surgeons performed major operations such as couching the cataract, amputation, trepanation, lithotomy, cesarean section, tonsillectomy, and plastic surgery. The study of human anatomy, the *science of being*, was justified as a form of knowledge that helped explicate the relationship between human beings and the gods. Despite barriers to systematic dissection of human cadavers, Indian scholars developed a remarkably detailed map of the body including a complex system of *vital points*, or *marmas*. Injuries at such points could result in hemorrhage, permanent weakness, paralysis, chronic pain, deformity, loss of speech, change of voice, loss of the sense of taste, or death. Obviously knowledge of such points was critical to ancient physicians and surgeons, who had to take the system into account when performing operations, treating wounds, or prescribing bloodletting and cauterization. Knowledge of the system is also of interest to certain contemporary healers, such as those who use the points as a guide to therapeutic massage. Indeed, Ayurvedic medicine is a living tradition that brings comfort to millions of people, while the ancient classics provide valuable medical insights and inspiration for modern physicians and scientists.

MESOPOTAMIA AND EGYPT

Along the great river valleys of Mesopotamia and Egypt, productivity, life, and death depended on cycles of the flooding and retreat of the rivers. Complex agricultural and engineering activities had to be planned, directed, and supervised by a centralized system of administration. The priests who directed this system dominated the way their subject populations thought about nature. The ancient myths of the separation of chaos into dry land and water and the development of living things from the mud are in essence the life story of these first civilizations. These myths served powerful central governments in dealing with the mundane problems of agriculture—draining marshes, land surveying, irrigation and flood control, storage of food for bad years—and in explaining the capriciousness of nature, which they coped with through divination and ritual tributes to the appropriate gods and demons. For the privileged members of the priesthood, knowledge of biological and astronomical phenomena was a valuable professional monopoly. A world that appeared to be dominated by capricious demons who could be propitiated only through the arcane rituals of the priesthood was not conducive to the free inquiry into nature essential to the development of science.

Mesopotamians and Egyptians had similar ideas about the structure of the universe. Waters were prominent in Mesopotamian cosmology, as might be expected from the experience of dwellers in river valleys subject to rather unpredictable flooding. Early Mesopotamian myths described the earth and heavens as flat disks supported by water. Somewhat later the heavens were described as a hemispherical vault which rested on the waters that surrounded the earth. The heavenly bodies were gods who dwelt beyond the waters above the heavens and came out of their dwelling places for their daily journey through the heavens. Since the gods controlled events on earth, the motions of the heavenly bodies were closely studied to reveal the intentions and designs of the gods.

In Egypt the world was viewed as essentially a rectangular box. The earth was at the bottom, making that part of the box somewhat concave. The sky at the top was supported by four mountain peaks at the corners of the earth, while the Nile was a branch of a universal river that flowed around the earth. This river carried the boat of the sungod on his journey across the sky. Seasonal changes and the annual Nile flood were explained by deviations in the path taken by the boat.

Creation myths explained how the world came into existence from a primeval chaos of waters. Matings of male and female gods in the time of

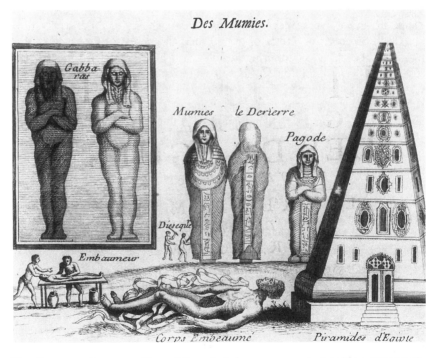

Egyptian mummies, pyramids, and the embalming process as depicted in a seventeenth-century French engraving

chaos brought forth the heavens, earth, air, and the natural forces and objects that were personified by various gods. The gods organized the universe out of the original chaos, separating the land from the waters much as the first settlers had reclaimed the earth from the waters. Egyptian and Mesopotamian civilizations, in their political organization, scientific interests, and religions, reflect the differences in the patterns of the great rivers. The Nile floods were predictable, beneficent, and eagerly anticipated. Egyptian dynasties were generally long-lived, rigid, and complacent. Great efforts were expended in creating appropriate eternal tombs for the mummified remains of the Pharaohs of ancient Egypt. The floods of the Tigris and Euphrates, in contrast, were unpredictable and frightening events. Governments in Mesopotamia were more transient and the future always uncertain. Astrology and divination were the tools and weapons with which Mesopotamian priests wrestled with the gods and the chaotic forces of nature.

Despite great accomplishments during the high Bronze Age, these civilizations remained essentially static while great innovations were made by cultures at the fringes of Bronze Age civilization. The smelting of iron and the use of a simple alphabetic script created a new cultural revolution. Sometime in the second millennium B.C. a tribe in the Armenian mountains developed an efficient method of smelting iron. Originally quite secret, the method became generally known after about 1000 B.C. Iron was used in weapons and agricultural implements, allowing the exploitation of the lands to the north. With iron weapons, barbarians like the Greek tribes were able to conquer rich Bronze Age cultures.

Long after the civilizations of Mesopotamia, Egypt, and India had reached their peak and entered into periods of decline or chaos, Greek civilization seems to have emerged with remarkable suddenness, like Athena from the head of Zeus. This impression is certainly false, but it is difficult to correct because of the paucity of materials from the earliest stages of Greek history. What might be called the prehistory of Greek civilization can be divided into the Mycenaean period, from about 1500 to the collapse of Mycenaean civilization about 1100 B.C., and the so-called Dark Ages from about 1100 to 800 B.C. Fragments of the history of this era have survived in the great epic poems known as the *Iliad* and the *Odyssey*, traditionally attributed to the ninth-century poet known as Homer. Ancient concepts of life and death, the relationship between the gods and human beings, and the structure and vital functions of the parts of the body are encoded in these myths and legends. Later Greek history is conventionally divided into three periods: Archaic (before 480 B.C.), Classical (480–323 B.C.), and Hellenistic (after 323 B.C.). The Greek city-state was quite different from the centralized forms of organization that had developed in previous civilizations. Moreover, unlike previous high cultures, that of the Greeks was not completely based on agriculture. Trade and shipping were important, and since Greece was relatively overpopulated in relation to cultivatable land, colonization and industry were encouraged.

THE PRE-SOCRATIC PHILOSOPHERS

As early as the sixth century B.C., Greek natural philosophers were involved in establishing a secular tradition of enquiry into the natural world and the search for explanations of how and why the world and human beings came to be formed and organized as they were. Such questions are very old, but the Greek philosophers took them out of the realm of religious and mythical explanations into the domains now occupied by science and philosophy. In place of gods and magic, these philosophers attempted to explain the workings of the universe in terms of everyday experience and by analogies with craft processes. Natural philosophy became a significant part of the quest for a reliable means of understanding and controlling nature. Many of the earliest philosophers are known only through a few fragments of their work, but what has survived suggests that their ingenious questions and answers served as the seed crystals that were to stimulate the subsequent development of Western biology, medicine, astronomy, and physics.

Natural science first developed not in the Athens of Socrates, Plato, and Aristotle, but on the Aegean fringes of the mainland of Asia Minor known as Ionia. Miletus, home of the first of the natural philosophers, was the mother city of numerous colonies on the Black Sea and had commercial interests throughout the Mediterranean world as well as contacts with the ancient civilizations of Mesopotamia and Egypt. The first philosophers seem to have had a keen interest in the opportunities for involvement in trade, travel, and politics that characterized these dynamic commercial centers. Scholars in the ancient world were generally highly respected, as shown by the fact that various city-states would ask them to serve as lawgivers, rulers, or tutors to the sons of rulers. On the other hand, in times of political turmoil scholars might find themselves persecuted by hostile factions. Thus, while Greek philosophers had more potential freedom of inquiry than the priests and officials who formed the intellectual elite of other cultures, they lacked the protection that the intermediaries of the gods always claimed.

The Ionian philosophers were preoccupied with finding the natural regularities they believed existed beneath the apparent changes in the world; that is, the common *primary cause* that initiated the entire chain of cause and effects and the common *element* from which everything originated. According to the Egyptians and Babylonians, the primary material constituents of the world were water, earth, and air. Chinese philosophers believed the world was made up of yin-yang and the five phases. Greek philosophers constructed a world of earth, air, water, and fire.

While the world of the early Greek philosophers was certainly still full of gods, their gods were very different from those who so frequently intervened in human affairs in the great epics of Homer. In essence, Greek philosophers from Thales to Aristotle posed the questions that still confound and intrigue philosophers and scientists. What is the grand design of the universe? Are there transcendent laws that explain why the universe is the way it is? Do such laws explain how and why matter and energy became formed, shaped, and organized and ultimately generated conscious beings? Is the universe the way it is *of necessity*, that is, does it contain the reason for its existence within itself, or is it what philosophers call *contingent*, that is, dependent on something beyond itself? The answers presented in the surviving fragments might best be thought of in terms of a theme and its numerous variations. For example, Thales of Miletus (ca. 625 to ca. 547 B.C.) chose *water* as his primary element.

Thales, called the founder of Ionian natural philosophy by no less an authority than Aristotle, flourished at about the same time as Buddha, Confucius, Zoroaster, and Hippocrates. Although Thales apparently distinguished himself as a statesman, merchant, engineer, mathematician, astronomer, and a lover of knowledge, there was considerable uncertainty as to his written work even in antiquity. Some authorities said that he left no writings at all, but others claimed that he was the author of "The Nautical Star Guide." Many legends grew up about his accomplishments and his travels to the older civilizations that represented timeless wisdom to the Greeks of his era. According to one such story, Thales invented a way of calculating the distance of ships at sea by means of the doctrine of similar triangles after becoming familiar with Egyptian geometry during a business trip. Then, having learned Phoenician astronomy while in Mesopotamia, Thales is said to have predicted an eclipse of the sun in 585 B.C., transforming what had been a very terrifying and mysterious phenomenon into a natural, predictable event.

Probably the most popular story associated with Thales demonstrates how astronomy and meteorology can be put to practical use. Through his study of the heavenly bodies, the philosopher predicted an unusually good olive crop. Before the harvest season, he rented all the olive presses in Miletus very cheaply and was then able to make a great profit when the demand for the presses was greatest. Thus, he proved to those who said that philosophy was useless that those who understood nature could easily become rich, though it was not in the nature of philosophers to seek wealth. Aristotle suggested that this story was not literally true, but that it had became attached to Thales because of his reputation for wisdom. It is, thus, rather like the story of young George Washington and the cherry tree, which serves as an emblem of his honesty. On the other hand, Thales was also depicted as the prototype of the absent-minded professor. Plato recounted a story of how Thales was mocked for falling into a well while he was observing the stars and gazing upward, ignoring the hazards just below his feet.

Thales attempted to create a complete cosmological theory in which water served as the primary material basis of all things. The earth was a disk which floated on water and was covered by the rarefied form of water that constituted the sky. Experience tells us that this is not unreasonable; land masses are bordered by oceans and rivers, wells bring us water from below the earth, and water falls from the sky in the form of rain. Whereas Homer had said that the god Poseidon caused earthquakes, Thales argued that just as a ship at sea is rocked by the waves, perturbations in the waters beneath the earth rock the surface. While Thales allowed that "all things are full of gods," supernatural agencies were not to be invoked as explanations for natural phenomena.

Perhaps Thales was influenced by the creation myths of older civilizations and reshaped these ideas into a more or less self-consistent and wholly natural view of the universe, but Aristotle suggested that Thales's reasoning was primarily physiological. That is, because nutriment and semen were always moist and the warmth of life was a moist warmth, all things must come from and be composed of water. Although sixth-century philosophers seem to have been more interested in meteorology than physiology, human beings were probably asking questions about reproduction and birth long before they began to ponder other aspects of natural philosophy. In addition to a primary material substance, Thales's cosmology required two processes or forces: *consolidation* and *expansion*. Water could be expanded until it turned into air and consolidated until it turned into earth. Anyone who has observed the appearance of a kettle from which hard water has boiled away will know that everyday experience appears to validate the idea that water has been transformed into steam (air) and mineral deposits (earth).

Other philosophers agreed with Thales that the cosmos could be explained in purely naturalistic terms, but disagreed with him about the nature of the primary element and the forces that shaped the cosmos. Anaximander (ca. 611–547 B.C.), who may have been a pupil and perhaps a nephew of Thales, argued that the primary material basis of the universe could not have been any of the individual characteristics of the things that it becomes. To avoid the trap of the primary element being the same as any one of the four elements and a part of the other three, Anaximander's cosmology is based on the existence of *aperion*, a rather ill-defined universal stuff. When acted on by heat and cold this primary material was transformed into the four elements which made up the world: earth, air, fire, and water. Anaximander also provided a dynamic but naturalistic explanation for the formation of the universe out of the original chaos by means of an indeterminant primary force, which caused the formation of a vortex. The motion of the vortex led to the separation of our world from the infinite and to the separation and stratification of the elements according to their density. Because earth is the heaviest element, it tends to remain at the center of the universe. Water covers the earth and is in turn enveloped in a layer of mist. Fire, which is the lightest element, escaped to its natural place at the outside of the universe where it became the heavenly bodies. Eventually the whirling motion disrupted the orderly arrangement of the elements and created wheels of fire enclosed by tubes of mist. The sun and moon were ring-shaped bodies made of fire surrounded by air. Tubes in this mass of air allow light caused by the fires to reach the earth. Because the tubes are of various sizes, we see the sun as larger than the moon, and the moon appears to be larger than the stars.

Eventually the action of the vortex caused the earth and its creatures to become formed and organized. The earth took on the shape of a rather flat cylinder. While the separation of the elements was occurring, fire acted on mud to produce dry land and mist. As the sun warmed the mud, which was a mixture of earth and water, it bubbled and heaved and brought forth various animals. The first creatures were fish formed during the period when water predominated. As separation proceeded and dry land appeared, some of these fishy creatures came onto the land and changed as conditions on earth changed. Man developed from a fish-like creature which had adapted to life on land and discarded its fishy skin. Because man is originally helpless, he could not emerge as an infant, but had to be nurtured inside the fish skin until the land was ready to receive him. Anaximander did not believe that such changes were directed along a unidirectional path. On the contrary, change was a cyclical process in which the forces that led to the separation of the elements and the development of life could be reversed so that chaos would once again predominate. Worlds come into being, perish, and are ultimately reabsorbed into the eternal infinite.

Anaximenes (fl. 546/545 B.C.), the last of the Milesian philosophers, accepted Anaximander's idea that all that exists has come from the infinite, but Anaximenes chose *air*, or *mist*, as the principle from which all other things are produced. Other substances were formed from air by rarefaction and condensation, that is, the amount of air packed into a particular place determined its form without altering its original nature. The process of felting, in which wool fibers are packed down to form a fabric, may have served as a simple model for this idea. Air is always in motion because there would be no change without movement. When air is in its most uniform or evenly distributed state, it is invisible to our senses. When air is rarefied and made progressively finer, it becomes fire. Air can be revealed to us as it becomes condensed by means of movement, cold, warmth, and moisture in the form of wind, mist, and clouds. If air is progressively condensed, it becomes water and earth. When it is condensed as far as possible, it becomes stone.

The concept of the earth as a broad, flat, shallow disk floating on air and situated at the center of the world seems to have been formalized by Anaximenes. The heavenly bodies, which were created by the rarefaction of mist into fire, were also supported by air. An infinity of different worlds came into existence and eventually passed away to be resorbed into the infinite air, which is all-encompassing and in perpetual motion.

The speculations of the earliest natural philosophers were subjected to further analysis by Xenophanes and Heraclitus, who were also Ionians, and by Anaxagoras, Leucippus, and Democritus.

Xenophanes of Colophon (576–490 B.C.) apparently taught that everything, including man, originated from water and earth. Earth and sea, he suggested, had changed places in the past and would do so again. Fossil evidence supported this idea because shells are found inland, in the mountains, and in the quarries of Syracuse and other places. Fossils proved that the earth had once been covered with mud, so that impressions of fish and other marine objects remained when the mud dried out. Previous philosophers had also suggested that living creatures had arisen from the mud, but Xenophanes appears to be the first to call attention to the physical evidence provided by fossils. Primarily interested in religion, Xenophanes recognized the tendency of man to make gods in his own image and the cultural relativity of human judgments on supposedly universal problems. "The Ethiopians say that their gods are snub-nosed and black, the Thracians that theirs have light blue eyes and red hair," he wrote, "but if cattle and horses or lions had hands, horses would draw the forms of the gods like horses, cattle like cattle." Although Xenophanes emphasized the limitations of human knowledge, he believed that through actively seeking the truth one could reach a higher state of wisdom. Not all of his contemporaries believed that he had reached such an exalted state. Heraclitus of Ephesus (556–469 B.C.) dismissed the work of Xenophanes with the statement that "a great deal of learning does not bring wisdom."

Bright fire was the primary element of Heraclitus, but his contemporaries called him the "dark one" because they found his work difficult and obscure. The doctrines for which he is best known are those attributed to him by Plato and Aristotle, who considered Heraclitus very important in terms of the development of cosmology and the separation of philosophers into those who thought of the universe in terms of change and those who thought about it in terms of stability. Others called him an arrogant misanthrope and charged him with adopting the style of the Delphic oracle in his collection of aphorisms. While Heraclitus may have impressed his contemporaries as a pretentious, supercilious aristocrat, he appears to have modestly referred to himself as the vehicle, rather than the author of revelations concerning the relationships that governed the natural world just as civil laws regulated relationships among citizens.

If Heraclitus's system seems more unified than those of his predecessors, perhaps this is merely because it is better preserved. His major premise was that there is a regularity and balance that governs all changes in nature. Balance in the universe was based on pairs of opposites in a state of tension with each other, much as harmonious music is created by the tension and opposition of the bow on the strings of the lyre. This balance existed in the mundane world of business transactions as it did in cosmic transactions. "All things are an equal exchange for fire and fire for all things," wrote Heraclitus, "as goods are for gold and gold for goods." The *opposites* of the Heraclitean system seem to be different from the *contraries,* such as heat and cold, or wet and dry, that are found in most pre-Socratic philosophies. The Heraclitean opposites are pairs generally related to the properties of living things, such as life and death, or health and disease, or verbal expressions, such as present and absent. Like a Zen master, Heraclitus confounded his audience by telling them that binary opposites such as youth and age, day and night, "the way up and the way down" are the same. Similarly, the ever-living fire that is the common element of all things both kindles and quenches.

It is not clear whether Heraclitus selected fire to represent the substance from which the whole world formed, just as Thales chose water, or whether he saw fire as a paradigm for explaining natural processes. Fire is both an element that changes and the process that causes change, whether in cooking, pottery making, or metallurgy. When things are consumed by fire they are transformed into fire, smoke, and vapors, which when condensed can form liquids and solids. Images of fire and rivers appear frequently in the Heraclitean fragments, reflecting the principle that all things are in flux and nothing is stable. The world we think we are seeing is actually like the flow of a river and, according to Heraclitus, we can never step twice into the same river. The endless flux could be seen in the changing seasons and the human life cycle.

Heraclitean philosophy could explain everything from the changes in the external world to the complexities of human thought. Indeed, understanding nature was not separate from understanding man, because the same materials and laws covered both. Life and health were associated with fire, while death and decline were associated with water. If men did not struggle towards an understanding of the true constitution of things, their souls, which were made of fire, would become excessively moistened. A dry soul was the wisest and best, and unfortunately for a civilization that treasured the fruit of the vine, wine in particular moistened the soul.

Weary of attempts at human discourse, the weeping philosopher withdrew from the world to live in the mountains, where he ate grasses and plants. This caused him to develop dropsy. Seeking aid from doctors, he tested them with a riddle. Because they could not answer him, Heraclitus rejected their medicines and buried himself in a cow stall. Theoretically, the heat of the decaying manure should have evaporated the pathological accumulation of water, but it did not. And so the philosopher died, buried up to the neck in cow manure. This bore out his statement that "it is death for souls to become water" as well as his comment that corpses are more worthless than dung.

All the speculative thinkers of the sixth and fifth centuries can be referred to as the pre-Socratic philosophers, but there are significant differences between the group known as the Milesians, those referred to as the Pythagoreans, and those known as the Eleatic philosophers, who accepted the teachings of Parmenides of Elea (fl. 500 B.C.). Very little is known for certain about Pythagoras of Samos (ca. 582–500 B.C.), whose name has been immortalized through the Pythagorean theorem. He is said to have escaped from Samos during the reign of the tyrant Polycrates and to have settled in southern Italy at Croton where he founded a brotherhood concerned with religious beliefs and practices, as well as mathematical inquiries. The Pythagorean community was remarkable in that both men and women were admitted on equal terms. Eventually the Pythagoreans seem to have split up into scientific and religious factions. By the time of Plato, Pythagoras and the brotherhood had become enmeshed in mystery and legend. Pythagoreans apparently believed in the immortality of souls and thought that human souls could be reincarnated in the form of other animals. This has been taken to mean that the Pythagoreans believed in the kinship of all living things.

Apparently inspired by mathematical principles, the Pythagoreans believed that all things in the universe could be thought of in terms of numbers.

Numbers could be divided into the categories odd and even; harmony was found in the ratios of the numbers that made up the musical scale. For the Pythagoreans, numbers seemed more fundamental to explaining the nature of the universe than fire, earth, and water. The appropriate harmony and balance of pairs of opposite qualities, such as hot and cold, moist and dry, could then be invoked to account for human health and disease. Pythagoras seems to have believed that religion and science were inseparable aspects of the complete life. But there is no direct reliable evidence about the scientific teaching of Pythagoras himself and no satisfactory way to separate his ideas from those of his followers. Although few of the members of the Pythagorean community are even known by name, Alcmaeon of Crotona (fl. ca. 500 B.C.), a physician with a special interest in biological questions, is said to have been a pupil of Pythagoras. In addition to medical theory, physiology and anatomy, Alcmaeon may have written on meteorology, astronomy, philosophy, and the nature of the soul. As would be expected of a Pythagorean, Alcmaeon apparently believed that the soul was immortal and always in motion.

Alcmaeon may have been the first natural philosopher to carry out dissections and vivisections of animals for the sake of learning about their nature rather than for purposes of divination. Moreover, he may have introduced the practice of examining the developing chick egg as a way of studying embryology. In order to understand the nature of sense perception Alcmaeon conducted dissections that led to the discovery of the optic nerves and the eustachian tubes (which were rediscovered by Eustachius in the sixteenth century). Despite later criticism of Alcmaeon's theories, he was referred to as one of the first to clearly define the difference between man and other animals. Man is the only creature that has understanding, while other animals could perceive but did not understand. If the senses were connected with the functioning of the brain, then when the brain moved or changed its position, the passages which admitted sensations would be blocked and the corresponding senses would be incapacitated. In explaining the nature of health and internal disease, Alcmaeon contended that health required a harmonious balance of contrary qualities, such as moist and dry, hot and cold. Disease was caused by an excess of one of the qualities. There is suggestive evidence that Alcmaeon was not dogmatic about his ideas and admitted that his theories about health and disease were fundamentally speculations about invisible phenomena.

An intriguing answer to the problems posed by competing speculations about the nature of the primary element, the reliability of sense perceptions, and the nature of change was proposed by Empedocles of Agrigentum in

Sicily (ca. 492 to ca. 432 B.C.), who seems to have been influenced by the Pythagoreans as well as by Parmenides of Elea, the philosopher who argued, contrary to Heraclitus, that change does not occur at all. Parmenides limited philosophical discourse to two general ideas, *Being* and *Not-Being*. Since all change must involve the addition of Not-Being to Being (that which exists), change is logically impossible. Author of major poems dealing with nature, medicine, and religion, Empedocles was said to be the founder of the four-element theory of matter, the inventor of rhetoric, a great orator, fervent democrat, pioneer of city planning and public health medicine. He appeared as a heroic figure in the work of Romantic poets of the eighteenth and early nineteenth centuries as the supposed champion of democracy.

Like Pythagoras and Heraclitus, Empedocles inspired many legends. He allegedly boasted about his supernatural gifts, claiming to have the power to heal the sick, cure the infirmities of old age, raise the dead, change the direction of winds and rivers, and bring the rain and the sun. Numerous references to him in later medical writings suggest that he was truly a famous and successful physician. Empedocles may, indeed, have changed the balance of health and disease in his native city by draining swampy areas, improving water supplies, and making a breach in the surrounding mountains to let in the cool north wind. According to another legend, he was offered the kingship of his city but refused the opportunity because of his commitment to democratic principles.

Only fragments of two poems by Empedocles have survived; "On Nature," which seems to have incorporated and refuted elements of the metaphysics of Parmenides, was a complex physical explanation of the universe, while the religious poem "Purifications" apparently dealt with the question of personal salvation in a manner influenced by the Pythagorean belief in transmigration of souls. Parmenides had taught that the real world must be "unborn and imperishable, one and indivisible, immobile, and a complete actuality." In contrast, that which is found in the world of sense perception was merely a man-made illusion. In accordance with the Parmenidean principles that Being cannot come from Not-Being and that some original unity cannot produce a subsequent plurality, Empedocles taught that there never was an original unity such as the Milesians had assumed. Instead, there were four eternal, unchanging, and distinct elements or roots: earth, air, fire, and water. Mixing these four eternal principles created new things, just as artists mix pigments to produce new colors. Creations and destructions are merely changes that are due to the mixture and separation of the "roots of all things." But what produces the motions of the four elements that lead to

Empedocles estoit un Philosophe, qui à l'imitation de Pytagore, ne vouloit point manger de tout ce qui viuoit et se mouuoit.

Empedocles of Agrigentum

changes in their local combinations? According to Empedocles these changes are due to two forces, Love and Strife (or Hate). Love causes attraction, aggregation, and the creation of new worlds, but when Strife predominates, worlds are torn apart. Thus, in the Cosmic Cycle of Empedocles the universe oscillates between stages of creation and destruction.

Unlike Parmenides, Empedocles believed that the senses were a valid guide to the philosopher speaking the truth, if each sense was carefully and critically employed for its appropriate purpose. Perception was ascribed to the recognition of similar elements and forces, so that we see earth with earth, water with water, and Love with Love. This mechanism of perception was possible because all things, including plants and animals, earth, sea, and stones give off subtle emanations of varying sizes. Effluences of particular sizes could fit into the pores or passages of the various sense organs so that the meeting of elements, which constituted perception, could occur within the body. Consciousness and thought were also products of this sorting of the elements. Thought processes take place mainly in the sea of blood surging back and forth around the heart. Because the four elements were most intimately blended in this blood, it was able to serve "as the chief seat of perception." A kinship among all living things is suggested by the assertion that "all things possess thought," however, Empedocles stipulated that not all things or all men possess the power of thought to the same degree.

Man and all other creatures, made up of the same four elements, were the products of the complex and peculiar evolution that characterized the cosmic cycle. There is considerable ambiguity about the details of Empedoclean thought, but basically the cosmic cycle consisted of two polar and two transitional stages and two distinct stages in the evolution of living things during the transitions. In stage one the four elements are completely mixed in a homogeneous sphere. Stage two is a transition in which the elements become increasingly separated under the influence of Hate. During stage three the elements are completely separated. Stage four involves another transition during which the elements coalesce under the influence of Love.

At an earlier stage of the great cycle the earth had powers she now lacks and was able to create a great variety of living things. The first products were so incomplete and crudely formed that arms wandered about without shoulders, faces without necks, and eyes were not attached to heads. In the second stage, the solitary parts began to unite in random combinations. Sometimes monstrosities were formed: man-faced oxen, ox-headed men, and sterile creatures that were part male and part female. A great variety of living forms appeared, including some "whole-natured forms," but many creatures were unable to compete with others that had a better assortment of parts, and few creatures contained all the parts necessary to reproduce their own kind. As separation continued, the sexes eventually become distinct and separate, leading to the next stage. In the fourth stage, the one in which we are living, new forms no longer arise directly from the earth or water, but only by means of the powers of generation. Moreover, all the species that survived into this stage were those with special courage, or skills or speed that allowed them to protect, preserve, and reproduce themselves.

All living things are composed of a particular mixture of elements. Each kind of being seeks its natural place according to the mixture of elements it contains. Trees are still attached to the earth because they have more earth in their nature, fish obviously have more water, while birds have more air and fire in their composition. Although there are only four elements, they can combine in many different proportions to form different kinds of compound substances. Bone was composed of fire, water, and earth in the ratio of 4:2:2. Unfortunately, the recipes for other entities did not survive. In plants, as in the case of the whole-natured forms of the previous stage, the two sexes are combined in one individual, but human beings and the higher animals reproduce through sexual generation. In contrast to the prevalent doctrine that the child was the product of the father, Empedocles suggested that some parts of the embryo came from the father and others from the mother. Male offspring, however, were conceived in the warmer part of the womb and contained a greater proportion of warmth than females.

Imaginative speculation on a grand, even grandiose, scale is part of the legacy of Empedocles. Another aspect is the use of careful observations of simple, ordinary events to provide insight into the workings of nature. One of Empedocles's most significant insights, in terms of the history of science, was his attempt to explain the process of respiration in terms of an experimental demonstration concerning the physical nature of air. Empedocles seemed to think that the blood oscillated in the body like the tides at sea; the surging of the blood drove air into and out of the body. Empedocles compared the movement of the air into and out of the body with the movement of liquid in a *clepsydra*, or "water-stealer," a device rather like a pipettc. Essentially a hollow cylinder with a strainer at the bottom, the clepsydra was used to transfer small quantities of liquid from one vessel to another. In "On Nature" Empedocles described a young girl playing with a gleaming brass clepsydra. When she covered the mouth of the cylinder and dipped the device into a fluid, no liquid entered the vessel because air inside prevented liquid from entering. Uncovering the device allowed air to escape as liquid entered. These observations, Empedocles argued, proved that air is not simply empty space, but that air has the nature of a *substantial* although *invisible* body. Once aware that nature works by unseen bodies, we can overcome the limitations of the senses by combining reason and observation in order to learn about things that are not directly perceptible.

Exactly what this insight meant in terms of the history of science is controversial. Some historians of science see Empedocles as the originator of the experimental method, while others contend that he merely applied a common observation to a preconceived theory. If Empedocles had really thought that the experimental method was the key to understanding nature, he could have invented many other experimental tests of his theories. Perhaps, given the literary device employed in making his case, Empedocles actually deserves credit for observing that children of both sexes are natural experimentalists. Certainly many of Empedocles's theories can be thought of as fruitful, in terms of inspiring large new ideas about natural phenomena, which is often more important than being correct about small matters.

In any case, Empedocles was not the only philosopher to offer a defense of the use of the senses, in conjunction with reason and logic, as a guide to understanding the world of natural phenomena. Anaxagoras of Clazomenae (ca. 500 to ca. 428 B.C.), represents the new Athenian phase in the development of Greek philosophy and the dangers that would haunt those who pursued unconventional ideas. Despite his friendship with Pericles, Anaxagoras was persecuted for impiety, allegedly for saying that the sun was only a red-hot rock and not a god, and was forced to leave Athens for Lampsacus. When asked what privilege he wished to be granted by his adopted city, the philosopher is said to have requested a yearly holiday for children in the month in which he died.

In his defense of the value of evidence obtained by the senses through observation and experience, Anaxagoras resembles Empedocles. Like the early Ionian natural philosophers, Anaxagoras combined an interest in practical matters with a passion for understanding how the universe worked. Wrestling with the logical and abstract arguments of Parmenides concerning the impossibility of change, Anaxagoras insisted that there never had been a unity with respect to cosmology, but had always been a plurality of eternal, different substances that totally filled all of space. The eternal substances could aggregate and separate under the influence of Mind and motion.

For Anaxagoras, the first principles were homogeneous entities or *seeds*. The seeds that were the first principles in his system were infinite in both number and variety, and yet every one contained a little of all the qualities to which our senses responded. While all things contain a portion of everything, according to Anaxagoras *Mind* alone was pure and unmixed, as well as infinite and self-ruled. Mind contained all knowledge, controlled all living things, and Mind alone possessed the power of initiating motion. In

Anaxagoras's scheme Mind plays the role Empedocles ascribed to Love and Strife. In the beginning all things were essentially in a uniform, motionless mixture. The universe as we know it began when Mind initiated a localized rotational motion, or vortex, which continuously increased so that all things were mixed, separated, and divided. The dense was separated from the fine, hot from cold, light from dark, and moist from dry. Dense, cold, wet, dark materials were driven to the center by the vortex, while the hot, dry, light, fiery materials were thrown to the periphery. A disk-like earth formed from the dense matter in the lower region, but the vortex separated out the sun, moon, and stars and sent them into the fiery upper regions.

One fragment associated with Anaxagoras poses the question: "How could hair come to be from what is not hair and flesh from what is not flesh?" His answer was that everything that is assimilated by human beings must preexist in their food because there are no simpler elements which form common natural substances by aggregation. The foods we consume must contain the seeds or entities that are capable of producing blood, muscle, bones, and so forth. These seeds must exist in foods in a form that is comprehended by means of reason, but invariably hidden from the senses. During the process of digestion, the seeds or elements are sorted out. Growth occurs by assimilating the seeds into their proper places by the natural attraction of like to like. Any given item must contain within itself a mixture of substances; the predominant constituent would determine its special characteristics. Empedocles, in contrast, held that flesh is a mixture of the four elements; if we divide a specimen down until a minimum bit of flesh remains, any further division will yield particles of the four separate elements.

Like Empedocles, Anaximander called attention to the use of observation to understand the nature of air, but he seems to have engaged in more deliberate experimentation, such as blowing up bladders and demonstrating their resistance to compression. From such observations, he returned to the realm of speculation and argued that the substantiality of the air allowed it to support the disk of the earth. His explanation for perception, however, was opposite that of the theory of like recognizing like put forth by Empedocles. Anaximander believed that perception occurs through opposites; experience teaches us that we do not notice things unless there are contrasts. For example, something that is warm or cold to the touch would command our attention, while that which was as warm or as cold as we are would not warm or cool us. Nevertheless, Anaxagoras insisted that the evidence presented by our senses might not reveal the truth. His warning was accompanied by what was probably a thought experiment involving imperceptible gradations of color. Imagine two vessels: one contains a white liquid and the other a black one. If either liquid is added drop by drop to the other, the eye cannot perceive any change in color until a significant quantity has been added. In this situation our senses tell us no change has occurred, reason tells us our senses must be deceiving us. Nevertheless, in a fragment dealing with the difference between humans and animals, Anaxagoras said that it was "possession of hands that makes man the wisest of living things."

ATOMS AND THE VOID

The speculations of the early natural philosophers may be thought of as ingenious approaches to understanding the nature of the universe, the earth, and its living beings, but their theories seem to have no direct counterparts in modern science. Atomic theory, on the other hand, immediately appeals to modern scientists as a brilliant, if somewhat crude anticipation of one of the most fundamental concepts of modern chemistry and physics. When more closely examined, the atomic theory of the ancients emerges most clearly as an elegant solution to problems that had puzzled philosophers since the time of Thales, rather than a primitive precursor of John Dalton's (1766–1844) atomic theory.

Much is uncertain about the lives of the fifth-century atomists Leucippus (fl. ca. 440 B.C.) and Democritus (ca. 460–370 B.C.). Leucippus may have been the first philosopher to develop a cosmology in which atoms served as the first principles and was the teacher of Democritus, who went on to refine the theory and explore its implications. Both Leucippus and Democritus postulated a world composed of innumerable atoms and the infinite void of empty space. This theory requires the bold assertion that the void, which is nonbeing, does exist and provides the place through which the atoms are perpetually moving. The atoms were uncreated and eternal and did not undergo actual change or destruction, but things appear to come into being through their combination, whereas their separation is interpreted as destruction. Atoms and the void constituted the material causes of all things. The differences in shape, arrangement, and position of the atoms accounted for all apparent modifications in the universe. In essence, varying the arrangement of atoms produced different materials, just as different combinations of letters produced different words. The idea that variety and change depend on the combination and separation of some primary material that was itself unchanging is found in Empedocles and Anaxagoras; these philosophers, however, had not postulated the existence of the void.

Atoms were compact, indivisible, and impenetrable, because they had no void within them, but were too small to impinge directly on the senses.

Whether or not the first atomic theorists thought that weight was a primary characteristic of atoms is unclear. Compound bodies, on the other hand, were divisible because of the void between their constituent atoms. The number of shapes atoms could assume was probably infinite because there was no reason that any atom should be of one shape rather than another.

From the void and the atoms, Democritus believed that innumerable worlds came into existence through mechanical means. Some of these worlds had no sun and moon; in others there were suns and moons more numerous or larger than in our world. Sometimes worlds were destroyed by colliding with each other. Some worlds had living creatures, whereas others were devoid of plants, animals, and moisture. Exactly how things happened is unclear, but the atomists argued that nothing occurred at random, but only "for a reason and by necessity." Events that seemed to occur at random to us were actually the result of a chain of collisions between atoms. The shape and motion of the atoms involved determine the outcome of the unseen interactions among atoms.

Later critics ridiculed Democritus's explanation for sensation because it attempted to reduce all aspects of perception and sensation to touch, contact, or interaction with the emanations allegedly given off by all objects. According to Democritus, all perceptible things are arrangements of atoms which differ only in size and shape. We assign certain qualities to these arrangements and call them color, taste, odor, texture, and so forth. But, he said, these qualities are not in the bodies themselves, they are the effects of the bodies on our sense organs. That is, qualities such as sweet and bitter, hot and cold, and even colors exist only by convention. Sight, for example, was the product of emanations given off by various objects which caused the air between the eye and the object of sight to become contracted and marked. The visual image, therefore, does not originate in the pupil, but in the air which is admitted to the eye; that is, images from outside were the cause of perception and thought. The other senses were explained in terms of the effects of different sizes and shapes of atoms on the appropriate sense organs. Thus, a bitter taste was the result of contact between the tongue and small, smooth, rounded atoms, while larger atoms with jagged edges caused a salty taste.

Democritus taught that the human being was a world in miniature, a *microcosm*, reflecting the whole universe, the *macrocosm*. As a reflection of the macrocosm, the microcosm contained atoms of every kind. Thus, even thought, consciousness, sleep, sickness, and death could be explained in terms of the quality of atoms or loss of atoms. He was especially interested in the nature of respiration because *pneuma*, or "vital air," composed of light, highly mobile soul-atoms, served as the vehicle of life. The soul-atoms tended to escape the body, especially during sleep, but they were replaced by other atoms until respiration ceased and the soul flowed out of the body. Although the very mobile spherical atoms that comprised the soul were diffused throughout the body, the soul-atoms were concentrated in the mind. Thought was the product of the motion that resulted from collisions between appropriate atoms. Tradition has it that Democritus tested his own sense impressions by withdrawing into solitary places such as caves and tombs.

In order to understand the nature of living things and sense perception, Democritus carried out systematic dissections of various animals. He seems to have established the practice later adopted by Aristotle of dividing animals into two major categories: *sanguiferous* (red blooded, essentially the vertebrates) and *bloodless* (invertebrates). Democritus pursued a great range of physiological studies and may have written works on human anatomy, regimen, prognosis, reproduction, embryology, fever, and respiratory diseases, and may have developed a theory for the generation of epidemics. Unfortunately, most of his biological writings were lost and his views are known primarily through Aristotle's polemics against them. Although this surely must distort much of his actual achievement, glimpses of his original insights remain. In contrast to Aristotle, Democritus called the brain the organ of thought, whereas Aristotle believed that the role of the brain was to cool the blood. Democritus apparently described the heart as the organ of courage and ascribed sensuality to the liver. He also called attention to the mule and its peculiar sterility, a topic of considerable interest in the history of genetics. We are perhaps fortunate to have even a few fragments of the work of Democritus, for one tradition holds that Plato tried to burn all of his writings.

Democritus was a contemporary of Hippocrates (ca. 460 to ca. 361 B.C.) and, according to ancient tradition, the two did meet in a professional capacity. Convinced that excessive interest in anatomical dissections and experiments on sense deprivation were indicative of madness rather than signs of scholarly eccentricity, neighbors of Democritus are said to have sent for Hippocrates. But after speaking to Democritus, Hippocrates told the people of Abdera that they were more likely to be insane than Democritus; the philosopher was not only in complete control of his senses, he was the wisest man living. Some historians argue that Hippocrates was neither the author of the Hippocratic texts, nor even a real person. Nevertheless, it is pleasant to imagine such a meeting, for atomism along with the rational medical theories of the Hippocratic school were the most exciting achievements of Greek thought until the time of Plato and Aristotle.

Hippocratic medicine emphasized the patient rather than the disease, observation rather than theory, respect for facts and experience rather than

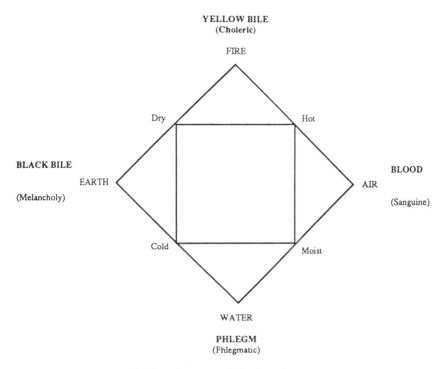

The four humors and the four elements

philosophical systems, and the admirable Hippocratic motto "at least do no harm." Above all, Hippocratic medicine was a craft in which success in helping nature cure the patient was the ultimate measure of the physician. Philosophy was less important to medical practice than *technique*, but the Hippocratic school did develop a theory of health and disease based on the four elements and the four humors. The microcosm that is the human being, like the macrocosm, was composed of the four elements—earth, air, fire, and water. As shown in the figure above, the four elements are related to the four bodily humors—black bile, blood, yellow bile, phlegm—and the four associated qualities—hot, cold, moist, dry. Individual variations in composition determined temperament—melancholy, sanguine, choleric, phlegmatic—and vulnerability to disease. When the four humors were well balanced, a state of health existed. Deviations from perfect balance produced disease.

A modern physician might think it naive to reduce the complex phenomena of physiology and pathology to the balance of four humors, but humoral

Hippocrates

theory was enormously successful, eminently subtle, infinitely flexible, and capable of incorporating and resolving an otherwise bewildering variety of problems. For those who were skeptical about its theoretical rationale, physicians could provide justifications based on common observations. Phlegm could be related to mucus, yellow bile to the bitter fluid stored in the gallbladder, and black bile to the dark material sometimes found in vomit, urine, and feces. Because the only analytic laboratory available to the physician until very recent times was the sum total of the five senses, the odor, and perhaps taste, of all the patient's excretions, secretions, and effluvia had to be studied directly by the earliest analytic chemist, that is, the human nose, which might have been assisted by the tongue. The linkage between theory and observation is also revealed by thinking about the appearance of clotted blood: the darkest part of the clot at the bottom of the collection vessel corresponds to black bile, the serum above the clot is apparently yellow bile, and the light material at the top is phlegm. The Hippocratic conviction that humoral pathology could provide a natural explanation for mental as well as physical illnesses is expressed most forcefully in *On the Sacred Disease*, an account of the illness known as epilepsy, or the falling sickness. Rejecting all supernatural and magical accounts of the cause and treatment of the disease, Hippocrates suggested that epilepsy might be due to an abnormal accumulation of phlegm during gestation that injured the developing brain. Many philosophers thought that the heart was the seat of consciousness, but Hippocrates assigned that role to the brain. Afflictions of the brain, therefore, presented the most formidable challenge to the physician. For hundreds of years, humoral theory rationalized a therapeutic regimen designed to restore humoral equilibrium by bleeding, purging, and dietary management designed to remove putrid humors and prevent the formation of additional bad humors.

THE AGE OF SOCRATES, PLATO, AND ARISTOTLE

Greek philosophy entered into a new phase with the end of the pre-Socratic period. The changing climate of Greek thought is best appreciated from the vantage point of Athens, the turbulent scene of the next phase in the development of Greek philosophy. While the work of Socrates (470–399 B.C.), Plato (429–347 B.C.), and Aristotle (384–322 B.C.) created the very core of Western philosophy, their influence on the history of science was almost equally profound.

According to Plato, philosophy treated only physics before Socrates added a second subject: *ethics*. Socrates was said to have rescued Greek thought from the abstract and materialistic concerns of his Ionian predecessors by bringing philosophy down from heaven to earth. His teacher, Archelaus of Athens, had been a pupil of Anaxagoras, and as a youth Socrates had studied the competing claims of the various natural philosophers. But the mature Socrates is said to have warned his disciples that the study of science could drive men mad, as in the case of Anaxagoras. Presumably Socrates realized how dangerous the study of nature could be when he saw Anaxagoras banished from Athens for impiety. Unorthodox enquiries into ethics and politics were to prove no safer than speculations about nature. Accused of corrupting the youth of Athens, neglecting the official gods of the city, and introducing new and different gods, the 70-year-old philosopher was condemned to death for impiety. In responding to his prosecutors, Socrates argued that he was actually a public benefactor, impoverished because of his dedication to his mission, who should be honored and supported by the state. Angered by the philosopher's apparent arrogance and insolence, the court of some 500 citizens reconfirmed the sentence by an even larger majority. Although his friends urged him to escape, Socrates accepted the judgment of the court and drank the hemlock, the traditional potion of the condemned.

The work of Socrates pertains primarily to moral philosophy rather than to science, but the core of his teaching is essential to both areas; Socrates taught that there was only one good, which is knowledge, and one evil, which is ignorance. He claimed that he had been given a divine mission to question and test the assumptions and beliefs of his fellow men. In carrying out his mission he professed to act the part of a midwife rather than a teacher, because by making others think, he helped them give birth to their thoughts. This image seems strikingly imaginative, but it might have been grounded on the fact that the philosopher's mother, Phaenarete, was a midwife. Socrates left no writings, but his doctrines survived in the work of Plato, who used Socrates as the protagonist in his own dialogues. A few additional insights into his ideas and personality are preserved in the stolid memoirs of Xenophon and the ribald comedies of Aristophanes.

Like Socrates, Plato's main concern was moral philosophy, encompassing ethics, politics, and theology, but Plato also exerted a profound influence on the development of Western science and the characteristics and goals of scientific investigation. Plato is the first of all the Greek philosophers whose writings survived essentially intact. After the death of Socrates, Plato is said to have traveled widely to learn from other philosophers, mathematicians, and even the priests of Egypt. Returning to Athens, he founded the Academy in an olive grove on the outskirts of Athens. Here, surrounded by his pupils, he lectured, developed his own philosophical system, and composed accounts of the doctrines he attributed to Socrates. To physics and ethics as the components of philosophy, Plato added dialectics, and taught his disciples that philosophy was a divine hunger for wisdom. Apparently not all of his students were equally famished, for when Plato performed a reading of his dialogue *On the Soul,* only the faithful Aristotle is said to have stayed to the end.

The Academy provided the means by which Plato could begin his primary mission, the training of political philosophers who would further his vision of the ideal society. The task of the philosopher, according to Plato, was to elucidate the eternal values. Knowledge of these was reached not through direct observation of nature; only by exercise of pure thought could the mind arrive at pure truth. Fundamentally, Plato instilled into philosophy a profound distrust of sensory experience; reason alone could reach beyond the imperfect shadow world of sensory perceptions into the world of transcendent eternal reality. Despite the considerable triumphs of empiricism, Plato's ideas continue to shape, direct, and challenge modern-day Platonists, scientists, and philosophers in search of the "grand design" and transcendent laws of the universe.

As revealed in the *Timaeus,* Plato's chief cosmological treatise, Plato's inquiries into nature were undertaken in support of his ethical precepts. In contrast to the cosmologies of the Ionian philosophers, his account of the creation of the world is highly religious and teleological; it is an extended argument based on purposeful design in nature. Plato's world was created out of an original chaos by an eternal and perfect god who imposes order on disorderly matter in accordance with a rational, intelligent design.

In Plato's version of the creation, man was not derived from simpler creatures, but was the first being to appear. In this case, as in so many others, *man* definitely means *man,* not male and female human beings. Borrowing freely from the Pythagoreans, Alcmaeon, Empedocles, Hippocrates, and so forth, generally without mentioning their names, Plato presents a wealth of biological ideas and observations, but focuses almost exclusively on human beings. Accounts of the causes of disease, respiration, and human anatomy are used to explicate his all-pervasive teleological principle. Thus, the first part of man to be formed was the head because it was most nearly spherical and served as the organ of the immortal soul. Because of the problem caused by the passive resistance of disorderly matter, the Creator achieves the best possible results despite what might appear to be some compromises with the most ideal design principles. The covering of the head is a rather thin layer of bone, even though a thicker covering would offer more safety, because a thicker covering would also reduce sensitivity and intelligence. The soul was tripartite, with the rational part located in the head, the heart the locus of the passionate part, and the liver the site of the appetitive part. To protect these vital organs, the lungs were designed to serve as a buffer against the pulsation of the heart and the spleen was designed to absorb impurities from the liver. In a remarkable reversal of modern evolutionary theory, Plato suggests that other animal species were created by degeneration from man. Even woman was a secondary production, being created from the souls of habitually cowardly and criminal men. With more wit than compassion for the predecessors from whom he had borrowed so much, Plato wrote that "harmless but light-witted men, who studied the heavens but in their simplicity supposed that the surest evidence in these matters is that of the eye" were turned into birds. Men who had no use for philosophy were transformed into four-legged beasts. Poor Anaximander was given an even worse fate for his speculation about men evolving from fish-like creatures. According to Plato, creatures that lived in the water were derived from the most foolish and stupid of men. These souls were so polluted by their transgressions that the gods found them unfit to breathe pure air. Thus Plato dealt with men who had tried to develop a thoroughly naturalistic explanation of the universe and those who believed that the data obtained by the senses were at least of equal value to abstract thought. In the world of Plato, true reality is found only in the sphere of abstract thought. What is perceived by the senses is but an imperfect image of the eternal ideal, like shadows cast on a wall.

Perhaps the greatest gift Plato bestowed on science was his demand that philosophers take the study of nature very seriously because such studies were a means of revealing the intelligent design of the universe. Such a directive led to the search for the elegant mathematical relationships and abstract laws that lie behind the apparently disorderly empirical data. On the other hand, the great successes of this approach in astronomy and physics seemed to reflect badly on the study of biological phenomena and created what has been called the problem of "physics envy" among biologists. In terms of the immediate history of biology, the concept of species and genus and the beginnings of systematic taxonomy, which were established in the work of Aristotle, were inspired by Plato's search for the eternal ideal forms.

Aristotle, disciple of Plato and disappointed heir-apparent, seems to have diverged more and more from the views of his mentor during a long and rather turbulent life. Just as Plato is the first of the philosophers whose writings have survived in bulk, Aristotle is the first great scientist to have this distinction, and his writings are voluminous. Many aspects of Aristotle's life provide insight into the complexities of the politics, customs, philosophy, and science of the Greek world in the fourth century B.C. Aristotle was born in Stagira, a town in Macedonia regarded as semibarbaric by the more sophisticated, older Greek states. Macedonia, largest of the Greek states at that time, was governed by Philip, father of Alexander the Great. Nicomachus, Aristotle's father, was a successful and wealthy physician at the royal court. As a youth, Aristotle must have learned the secrets of the Asclepiads (priest-physicians serving Asclepius the god of medicine), the theory and practice of Hippocratic medicine, and the advantages of royal patronage. Although a youth could be instructed in the rudiments of medicine in Macedonia, a properly educated physician also needed training in philosophy. Therefore, at 17 years of age Aristotle was sent to Plato's Academy where he was immersed in a system of thought very different from that of the practicing physician, a system that valued the abstract over the practical. Apparently this new world suited Aristotle, for he remained at the Academy until the death of Plato 20 years later.

During this time, Aristotle established himself as an independent philosopher and author, even expressing some reservations about some aspects of his mentor's doctrines. Probably because of his reaction to such criticism, Plato appointed his nephew as his successor at the Academy instead of Aristotle. Embittered and disappointed, Aristotle abandoned the Academy, Athens, and mathematics. Biological researches, carried out mainly on the island of Lesbos, were his main interest during this period of his life. When a revolution deposed his patron, he fled to Macedonia where he served as tutor to Philip's son Alexander (the future Alexander the Great) for 3 years. This was not the most successful student-teacher relationship in history, although it is one of the most famous. Certainly, Alexander was more interested in war and conquest than in philosophy, but he did remember his old tutor well enough to send back reports and specimens of peculiar animals found in exotic places during his triumphant military career. In 334 B.C., at the age of 50, Aristotle returned to Athens and established his own school, called the Lyceum. Unfortunately this productive and relatively serene period soon came to an end, when political turmoil again made him a refugee. After the death of Alexander, the Athenians rebelled against Macedonian rule. Faced with an indictment for impiety, Aristotle fled the city to "prevent a fresh crime against philosophy." After a year spent in exile at Chalcis, the philosopher died. Although the bulk of Aristotle's works survived, there are gaps and uncertainties as to the order in which they were written. Some of the texts were apparently edited by his collaborators, and some may be student notes, which might present a garbled version of what

Aristotle

he really said and thought. A gift for organizing, arranging, and even appropriating other people's material and recognizing affinities enabled Aristotle to discuss almost every field of human knowledge. Very often all that is known of his predecessors comes from Aristotle's comments on them. Where he disagreed he could be quite sarcastic and perhaps misleading.

Aristotle's work falls into four major divisions: physics, ethics and politics, logic and metaphysics, and biology. It was Aristotle who formalized the earth-centered, closed, finite universe that dominated Western thought until the time of Nicolaus Copernicus (1473-1543). In clarifying difficult aspects of his views on ethics and politics, nature and society, Aristotle often used biological examples and analogies. While explicating his own principles of logic and metaphysics, Aristotle often called attention to deficiencies in the work of his predecessors, particularly Plato's theory of the ideal forms. Aristotle's biological writings reveal a remarkably acute observer of nature who, despite contemporary prejudice against manual labor, was willing to dirty his hands in dissections of even the lowliest creatures. Almost one quarter of his surviving writings are essentially biological; his major biological works include *Natural History of Animals, On the Parts of Animals*, and *On the Generation of Animals*.

Given the competing demands of abstract thought and the complexity of biological phenomena, it is not surprising that two Aristotles seem to emerge from his collected writings. One was the theorist and logician, who believed that the logical and the real were likely to be identical. The second Aristotle was the observer, who understood that he was working in the still unchartered territory of biological phenomena and advised his followers to rely on their own observations rather than his words. The extent to which Aristotle was influenced by the differing claims of abstract philosophy and scholarly medical tradition is unclear, but it should be remembered that, long before Plato established a preference for pure thought over the evidence of the senses, the author of *Ancient Medicine* had argued that the only way to obtain clear knowledge about natural science and the nature of human beings is through medicine, not philosophy. Although not an unthinking empiricist, the Hippocratic physician could not rely on abstract theory; the successful practicing physician had to confront the evidence of his senses in order to evaluate the symptoms presented by the each patient. In contrast, the physicist could search for the laws that govern the motion of all falling objects without considering their individuality. But in biology, as in medicine, close attention to details is essential.

Aristotle's biological investigations must have been very enjoyable and satisfying to him, but in terms of generating theories they were subordinate to his general philosophical system. Thus, although his descriptions of animals, even those as small and delicate as the eggs and embryos of fish and mollusks, are remarkably accurate, sometimes his powers of observation were distorted by his preconceived notions about the way things should be. As Einstein said, "It is the theory that decides what we can observe." Dissection is not invariably an instrument of understanding because function is not obvious in structure. For example, Aristotle believed that the function of the brain was to secrete mucus and cool the blood by balancing the heat of the heart, which served as the seat of the soul and the intellect. When the heat of the heart overpowers the cooling power of the brain, the blood is hot and man becomes passionate and excitable; when the brain predominates, the cooling of the blood results in sleep. A balanced state allows for rational life. The brain must be composed of earth and water to perform its cooling function because the element earth is cold and dry, while the element water is cold and wet. The vague metaphor of cooking or coction was used to explain nutrition as well as embryonic development. Some foods were cooked in the stomach and produced food vapors which ascended to the heart. This organ transformed them into blood and, through the movement of the blood, nutriments were transferred to the body and directly assimilated.

Both Plato and Aristotle could agree that the highest faculty of man was reason and the most sublime human activity was contemplation, but Aristotle also attached great significance to the collection of facts and data and the observation of general phenomena and particular examples. Aristotle took Plato's highly abstract theory of the Ideas or Forms that existed only in the world of pure thought and transformed it into a guide to the study of nature. Deliberate and systematic inquiries, Aristotle insisted, would reveal the causes of things because for Aristotle the Ideas or Forms exist in nature and not apart from it. Thus, although Aristotle's world is certainly not one without a god, his inquiries into biological phenomena do not require the mythic apparatus that is at the core of Plato's *Timaeus*. Aristotle considered Plato's development of teleology (argument from design) an important innovation having its roots in the work of Anaxagoras and Empedocles. Aristotle's use of teleology, unlike that of Plato, does not rely on the intentional actions of a creator God. Still, Aristotle said in various passages that "Nature does nothing in vain" and that "in all its workings Nature acts rationally."

Thinking about the why of things, that is, thinking about the primary cause of things and the cause of all kinds of physical changes, was, Aristotle asserted, the beginning of philosophy. His predecessors, from Thales to Plato, had speculated about the causes of events in nature in vague and often erroneous ways. Aristotelian philosophy incorporated four causes of phenomenal changes in nature: the *material, formal, efficient* (moving, or dynamic), and *final* causes. In other words, in order to explain an object or event, natural or artificial, we must describe four essential factors: the matter of which it is made; the form or shape in which it exists; its moving or dynamic cause, which is the active agent; and its final cause, which is its purpose or goal. In common modern usage only the efficient cause, and sometimes the final cause, are generally thought of as causes. Actually, Aristotle's "causes" are conditions necessary to account for the existence of a thing or event; he does not use the term cause in the sense of that which is both necessary and sufficient to produce a specific effect. His causes, therefore, include form and matter and can be applied to a description of the fabrication of a table by a carpenter as well as the generation of a new human being.

By recognizing the value of studying organisms in and of themselves, Aristotle initiated an approach to philosophy and biology of great influence and fruitfulness. Moreover, the scope of his own biological research and the extent of his knowledge were so extensive that an account of the history of almost any branch of the life sciences must begin with Aristotle. The diverse sources of Aristotle's biological information included the written works of his predecessors, inquiries put to farmers, hunters, and fishermen, and his own observations and dissections. The true natural philosopher, according to Aristotle, would find every realm of nature to be marvelous and would not recoil with "childish aversion from the examination of the humbler animals." The method adopted by such a philosopher enabled him to pursue the study of every kind of creature "without distaste," because each and every animal would reveal "something natural and something beautiful."

Many biological facts were assembled in *History of Animals*, *Parts of Animals,* and *Generation of Animals* raised fundamental questions about reproduction, the nature of species, and the relationships among organisms. Aristotle assumed that fairly definite species of animals existed and that each species had its own nature, or essence. His species were those which had common names in Greek and whose nature could be discovered by investigating members of the species. Some scholars have attempted to recreate strict zoological hierarchies or construct a scale of nature from passages in Aristotle's work. But Aristotle recognized the difficulties entailed in establishing a complete classification system, especially when the information available to him was so limited; only about 500 species of animals are cited in Aristotle's works, and many of them seem to be subdivisions of common types. His own investigations probably included dissections of about 50 kinds of animals.

Broadly equivalent to genus and species, the terms *genos* and *eidos*, the only taxonomic categories Aristotle used, are the subject of much scholarly debate; *genus* refers to family or kinship, and *species*, which comes from the Greek for "form," was used to deal with the common names of things. Many so-called primitive peoples could distinguish many more plant and animal species than Aristotle, but this does not mean that Aristotle was an incompetent classifier. Rather, it reflects that fact that he was a philosopher outlining methods and systems, not a taxonomist actually constructing a detailed system. Aristotle's goal was to find general principles that would allow scientists to organize many diverse organisms. He suggests the use of "essential traits" to form natural groups. It is the species rather than the individual that exhibits the *essence* that is eternal; therefore, the scientist must study the nature of the species and its specific characteristics. For example, the biologist must study the nature of man, rather than the accidents that make up a particular man such as Socrates or Aristotle.

A natural scheme of classification is one that gives attention to many characteristics and weighs them to determine the natural affinities beneath the variety of structures and functions. For example, it would be absurd to use habitat or means of locomotion alone to order animals; instead natural affinities must be determined by investigating many traits, including structure, behavior, habitat, means of locomotion, and reproduction. In searching through all the characteristics of animals, Aristotle found one major factor that suited him: means of reproduction. Like Democritus, Aristotle began by dividing animals into groups with and without red blood; this essentially corresponds to a division between vertebrates and invertebrates. A more refined hierarchy could be constructed by arranging animals according to their level of development of birth. The group that we call mammals was at the top because the members of this group are warm and moist, not earthly. Their young were born alive and complete and were nursed by the females of the species. Next were creatures that were less warm but were still moist. These animals were born alive, but developed from eggs that grew inside the body of the mother. Lower down the scale were creatures that were warm and dry and laid complete eggs such as the birds and reptiles. Lower still were the cold and earthly animals that produced incomplete eggs, such as the frog and the ink fish. The lowest of all sexually reproducing forms were worms that produced eggs. Below these were the vermin that came into being by spontaneous generation. The quality of the soul or souls determined the habits and functions, as well as the structure and perfection, of each species. Plants have only a vegetative soul for growth and reproduction; animals have also a sensitive soul for movement and sensation; and man has, in addition, a rational soul, whose seat is in the heart.

Three major means of generation were known to Aristotle. Not surprisingly, sexual reproduction is due to the occurrence of the two sexes. In Aristotle's scheme the male is the more complete and warmer entity, while the female is less complete and colder. The higher animals and even some of the insects, such as the grasshopper, were produced by sexual reproduction. Animals such as the starfish, worms, and shellfish arose through asexual reproduction. Regeneration, common in marine organisms, was regarded as part of generation. Spontaneous generation routinely produced fleas, mosquitoes, and various kinds of vermin. Even spontaneous generation obeys rules; certain kinds of mud or slime give rise to specific kinds of insects and vermin.

Plato found a solution to the challenge concerning being, nonbeing, and change issued by Parmenides in his theory of the eternal and unchanging Forms, which were the only truly existing things. The phenomenal world was quite a different set of attributes, only temporarily found in associations. Aristotle objected to this solution, primarily on the grounds that it was obvious that the phenomenal world was composed of *things*, men, animals, and other objects. The *eidos* was for Aristotle the true unchanging object, but it existed in and through individuals of the phenomenal world. Aristotle's solution to the problem of Parmenides and nonbeing was based on his persistent use of the distinction between *potentiality* and *actuality* for each variety of change. Perhaps the most intriguing aspects of the challenge of change, potentiality, and actuality are those raised by reproduction and the development of new organisms.

How could philosophy and science explain the generation of like by like, that is, the production of an animal by an animal, or a plant by a plant? Aristotle's *Generation of Animals* applies his concepts of actuality and potentiality, matter and form, to one of the great problems of biology and philosophy: understanding how a new life begins and develops. "The formation of that which is *potentially*," according to Aristotle, "is brought about by that which is *in actuality*." That is, the embryo is not a preformed entity that expands during development, but a *potentiality* that is imposed on matter and expressed part by part over time.

Aristotle believed that man is the most rational animal and therefore all other animals are inferior. When Aristotle describes characteristics which are natural to particular species, but different from those of human beings, such as lack of external ears in seals, he refers to them as deformities. Similarly, Aristotle asserts that the human female is a deformed male, even though the female is apparently necessary for the continued existence of the species. In the process of the generation of a new individual, the male provides the form principle (the moving cause of generation and development) via the semen; the female contribution is passive matter in the form of the menstrual blood. To explain the essential but subtle contribution of the male to generation, Aristotle offered the famous carpenter analogy: wood is passive and cannot transform itself into some useful entity, but the action of the carpenter can transform it without becoming part of it; thus, the form is present in the soul of the carpenter. This appears to be a compelling analogy, but analogies are often used as illustrations that divert attention from the fact that a real explanation is lacking. During pregnancy the blood normally lost during menstruation served to build up and increase the size of the embryo. Aristotle's theory of generation had the major virtues of accounting for the known facts, accommodating the prejudices of his time, and the minor defect of being completely wrong. Although Aristotle's form principle has been compared to DNA as a blueprint for development, his theory neglected the contribution of the female parent, a contribution that predecessors of Aristotle, like Hippocrates, had already acknowledged. Aristotle did note that men lived longer on average than women, and that women who had many pregnancies had shorter life spans.

Although Aristotle had many disciples, he completely overshadows them. His ideas dominated biology as well as other aspects of human thought for generations. Indeed the biology of antiquity and the Middle Ages never really got beyond Aristotle's theories of the phenomena of life. The great tragedy is that while Aristotle's own approach to the study of nature called for close and accurate observations, respect for facts, systematic collection of evidence, and deductions from facts, this part of his teachings was almost completely neglected. Later scholars tended to become slavish followers of the words, rather than the spirit, of Aristotle. The disciples of Aristotle tended to take more modest portions of all possible knowledge for their investigations. The old tradition of formulating a complete cosmology essentially died with Aristotle. Theophrastus (ca. 373 to ca. 285 B.C.), Aristotle's favorite pupil, friend, and chief assistant, and Strato (fl. 286–268 B.C.) are the only followers of Aristotle at the Lyceum with a real claim to a substantive place in the history of science.

Inevitably overshadowed by his teacher, Theophrastus was a prodigious worker, voluminous writer, as well as an independent and original thinker; only a fraction of his writings survived. In addition to his major studies of botany, Theophrastus also wrote about broader problems in metaphysics, biology, medicine, and the doctrine of the four elements. Theophrastus seems to have objected to the preoccupation of natural philosophers with the study of metaphysics (the study of first principles), the reckless use of teleological explanations, and their tendency to denigrate the direct study of nature. If natural phenomena were observed more closely, he argued, it would be easy to demonstrate that nature does not always act with universality of purpose and intelligent design. Theophrastus realized that the teleological explanation was often an oversimplification and at times a disguise for sloppy thinking. This approach provides instant answers to all questions but militates against asking hard questions and answering them in a meaningful way. Scientific inquiry into the universe, Theophrastus argued, required setting limits on the search for final causes and focusing on "the conditions of existence of real things and their relations with one another."

Theophrastus described scholars as true citizens of the world, able to survive changes of fortune and obtain employment in any realm. Strato, who abandoned the study of ethics for researches in pneumatics, apparently absorbed valuable lessons in survival and patronage as well as natural philosophy from his mentor and became the well paid tutor of Ptolemy II of Alexandria. Strato was primarily interested in pneumatics and carried out many ingenious experiments to test the nature of air, motion, and the void. Unfortunately, with the exception of sections apparently appropriated by later scientists, all of his writings were lost. Some of Strato's views on perception and the nature of the senses have been reconstructed. Unlike Plato, Strato believed that true knowledge was the result of experience. He rejected Plato's view of true knowledge as something possessed only by the soul and independent of experience. Through investigations of the relationship between the senses and the mind, Strato concluded that it is in the mind that a stimulus generated from outside the body is transformed into a sensation. According to Strato, perception and thought were the province of the same soul. Since animals have sense organs like those of human beings, they must also have minds. Unlike the Aristotelian view that animals are degenerate forms of man, Strato regarded man as merely a better kind of animal.

Strato succeeded Theophrastus as the head of the Lyceum in 287 B.C., and, even though he realized that the institution was declining along with the numbers of students and the quality of their work, he retained that position until 269 B.C. By that time it was clear that the city of Athens was being eclipsed by Alexandria in Egypt as a vital center of intellectual activity.

While Aristotelian philosophy has always been an important component of Western thought, the grip of Aristotelianism has waxed and waned over the centuries since his death. After the Renaissance, Reformation, and Scientific Revolution, alternative ancient philosophies, such as those linked to the Stoics and Epicureans, and others known only through fragments preserved by Aristotle and his followers, attracted the attention of European intellectuals, philosophers, theologians, and scientists. In other words, the history of Western intellectual life can be seen in terms of the interaction of the whole heritage of ancient Greek thought with the body of thought characteristic of each succeeding era. The ideas associated with the ancient Greeks were so creative, intriguing, and fertile that they engaged the attention and confounded the prejudices of early modern scientists; they remain challenging and fruitful today.

Stoicism and Epicureanism arose in late fourth-century B.C. Athens. The founders of these schools, Epicurus (341–271 B.C.) and Zeno of Citium (334–262 B.C.), developed their ideas in reaction to the prevailing schools of Plato and Aristotle. Both Epicurus and Zeno were primarily concerned with ethical issues and insisted that understanding ethics required an understanding of the nature of the universe and human beings. Only a few fragments from these philosophers have survived, but the writings of their later admirers and followers preserved and transmitted their basic concepts. Both the Stoics and the Epicureans believed that the good life led to a state of mental tranquility. Like the atomists, the Stoics and Epicureans expounded a philosophy of nature in which all natural phenomena, and even the human soul, were explained in terms of chance collisions among atoms moving through the void. The gods were not totally banished from their world picture, but were made far enough removed from human affairs that fear of the gods and punishment after death would no longer disturb humans beings who accepted philosophy.

SUGGESTED READINGS

Africa, T. W. (1968). *Science and the State in Greece and Rome*. New York: Wiley.

Anton, J. P., ed. (1980). *Science and the Sciences in Plato*. New York: EIDOS.

Aristotle (1908–52). *The Works of Aristotle*. Edited by J. A. Smith and W. D. Ross. 12 vols. Oxford: Clarendon Press.

Asthana, S. (1985). *Pre-Harappan Cultures of India and the Borderlands*. New Delhi: Books and Books.

Bag, A. K. (1985). *Science and Civilization in India*. Vol. 1: *Harappan Period*. New Delhi: Navrang.

Barnes, J. (1986). *The Presocratic Philosophers*. New York: Methuen.

Boorstin, D. (1983). *The Discoverers: A History of Man's Search to Know His World and Himself*. New York: Random House.

Boylan, M. (1983). *Method and Practice in Aristotle's Biology*. Lanham, MD: University Press of America.

Brain, P. (1986). *Galen on Bloodletting*. New York: Cambridge University Press.

Chattopadhyaya, D. (1986). *History of Science and Technology in Ancient India: The Beginnings*. Calcutta, India: Firma KLM Private Ltd.

Clagett, M. (1955). *Greek Science in Antiquity*. New York: Abelard-Schuman.

Cockburn, A., and Cockburn, E., eds. (1980). *Mummies, Disease, and Ancient Cultures*. Cambridge, England: Cambridge University Press.

Cohen, M. N., and Armelagos, G. J., eds. (1983). *Paleopathology and the Origins of Agriculture*. New York: Academic Press.

Cohen, M. R., and Drabkin, I. E., eds. (1958). *A Sourcebook in Greek Science*. 2nd ed. Cambridge, MA: Harvard University Press.

Cornford, F. M., ed. (1937). *Plato's Cosmology: The Timaeus of Plato*. London: Paul.

Diogenes Laertius (1969). *Lives of the Philosophers*. Trans. and ed. by A. R. Caponigri. Chicago, IL: Regnery.

Dodds, E. R. (1973). *The Ancient Concept of Progress and Other Essays on Greek Literature and Belief*. Oxford: Clarendon.

Dorn, H. (1991). *The Geography of Science*. Baltimore, MD: Johns Hopkins University Press.

Edelstein, L. (1967). *The Idea of Progress in Classical Antiquity*. Baltimore, MD: Johns Hopkins University Press.

Farrington, B. (1961). *Greek Science, Its Meaning for Us*. New York: Penguin.

Ferejohn, Michael T. (1991). *The Origins of Aristotelian Science*. New Haven, CT: Yale University Press.

Finley, M. I. (1982). *Economy and Society in Ancient Greece*. New York: Viking.

Fortenbaugh, W. W., Huby, P. M., and Long, A. A., eds. (1985). *Theophrastus of Erusus: On His Life and Work*. New Brunswick, NJ: Rutgers University Press.

Furley, D. J. and Allen, R.E. (1975). *Studies in Presocratic Philosophy*. New York: Humanities.

Ghalioungui, P. (1973). *The House of Life: Magic and Medical Science in Ancient Egypt*. Amsterdam: B. M. Israel.

Gilson, Etienne (1984). *From Aristotle to Darwin and Back Again: A Journey in Final Causality, Species, and Evolution*. Notre Dame, IN: University of Notre Dame Press.

Guthrie, William K. C. (1962–69). *A History of Greek Philosophy*. 6 vols. Cambridge: Cambridge University Press.

Harris, C. R. S. (1973). *The Heart and the Vascular System in Ancient Greek Medicine from Alcmaeon to Galen*. Oxford: Clarendon.

Hippocrates (1939). *The Genuine Works of Hippocrates*. Trans. by Francis Adams. Baltimore, MD: Williams and Wilkins.

Hodder, I. (1986). *Reading the Past. Current Approaches to Interpretation in Archaeology*. New York: Cambridge University Press.

Hughes, J. D. (1975). *Ecology in Ancient Civilizations*. Alberquerque, NM: University of New Mexico Press.

Kirk, G. S., Raven, J. E., and Schofield, M. (1983). *The Presocratic Philosophers: A Critical History with a Selection of Texts*. 2nd ed. Cambridge: Cambridge University Press.

Leslie, C., ed. (1976). *Asian Medical Systems. A Comparative Study*. Berkeley, CA: University of California Press.

Lloyd, G. E. R. (1970). *Early Greek Science: Thales to Aristotle*. New York: Norton.

Lloyd, G. E. R. (1973). *Greek Science After Aristotle*. New York: Norton.

Lloyd, G. E. R. (1983). *Science, Folklore, and Ideology: Studies in the Life Sciences in Ancient Greece*. New York: Cambridge University Press.

Lovejoy, A. O. (1936). *The Great Chain of Being: A Study of the History of an Idea*. Cambridge, MA: Harvard University Press.

Luck, G. (1985). *Arcana mundi: Magic and the Occult in the Greek and Roman Worlds*. Baltimore, MD: Johns Hopkins University Press.

Majno, G. (1975). *The Healing Hand. Man and Wound in the Ancient World*. Cambridge, MA: Harvard University Press.

Mayr, E. (1982). *The Growth of Biological Thought*. Cambridge, MA: Harvard University Press.

Mercer, E. H. (1981). *Foundations of Biological Theory*. New York: Wiley.

Morsink, Johannes (1979). Was Aristotle's biology sexist? *Journal of the History of Biology 12* (1): 83–112.

Nakayama, S., and Sivin, N., eds. (1973). *Chinese Science: Explorations of an Ancient Tradition*. Cambridge, MA: MIT Press.

Needham, J. (1954). *Science and Civilization in China*. Cambridge: Cambridge University Press.

O'Brien, D. (1969). *Empedocles' Cosmic Cycle: A Reconstruction from the Fragments and Secondary Sources*. London: Cambridge University Press.

Olson, R. (1982). *Science Deified and Science Defied: The Historical Significance of Science in Western Culture*. Berkeley, CA: University of California Press.

Onians, Richard Broxton (1951). *The Origins of European Thought About the Body, the Mind, the Soul, the World, Time, and Fate*. New York: Cambridge University Press.

Osler, Margaret J., ed. (1991). *Atoms, Pneuma, and Tranquillity. Epicurean and Stoic Themes in European Thought*. Cambridge: Cambridge University Press.

Pellegrin, P. (1986). *Aristotle's Classification of Animals*. Trans. by A. Preus. Berkeley, CA: University of California Press.

Pinault, Jody Rubin (1990). *Hippocratic Lives and Legends*. New York: E. J. Brill.

Porkert, M. (1974). *The Theoretical Foundations of Chinese Medicine; System of Correspondence*. Cambridge, MA: MIT Press.

Preus, Anthony (1975). *Science and Philosophy in Aristotle's Biological Works*. New York: Georg Olms Verlag.

Reingold, N., and Rothenberg, M., eds. (1987). *Scientific Colonialism. A Cross-Cultural Comparison*. Washington, DC: Smithsonian Institution Press.

Saggs, H. W. F. (1989). *Civilization Before Greece and Rome*. New Haven, CT: Yale University Press.

Sallares, Robert (1991). *The Ecology of the Ancient Greek World*. Ithaca, NY: Cornell University Press.

Sambursky, Samuel (1974). *Physical Thought from the Presocratics to the Quantum Physicists*. New York: Pica Press.

Schmitt, Charles B. (1983). *Aristotle and the Renaissance*. Cambridge, MA: Harvard University Press.

Sivin, N., ed. (1987). *Traditional Medicine in Contemporary China*. Ann Arbor, MI: University of Michigan Center for Chinese Studies.

Taylor, Henry Osborn (1963). *Greek Biology and Medicine*. New York: Cooper Square.

Theophrastus (1916). *Enquiry into Plants*. 2 vols. Trans. by Sir Arthur Holt. Loeb Classical Library. Cambridge, MA: Harvard University Press.

Toulmin, S. (1982). *The Return to Cosmology: Postmodern Science and the Theology of Nature*. Berkeley, CA: University of California Press.

Unschuld, Paul U. (1985). *Medicine in China. A History of Ideas*. Berkeley, CA: University of California Press.

Vlastos, Gregory (1975). *Plato's Universe*. Seattle, WA: University of Washington Press.

Waterford, Robin (1989). *Before "Eureka": The Presocratics and Their Science*. New York: St. Martin's Press.

Waterlow, Sarah (1982). *Nature, Change, and Agency in Aristotle's Physics: A Philosophical Study*. New York: Oxford University Press.

Woodbridge, Frederick J. E. (1965). *Aristotle's Vision of Nature*. New York: Columbia University Press.

Zysk, Kenneth G. (1985). *Religious Healing in the Vedas*. Philadelphia, PA: The American Philosophical Society.

2

THE GREEK LEGACY

As the vitality of the Lyceum declined, a new center of scholarship was called into being by the Ptolemies of Alexandria, heirs to the conquests of Alexander the Great (356–323 B.C.). The rulers of Alexandria recognized that science could be made to serve the state and exerted their influence to transfer all that was valuable from Athens to Alexandria. The ascent of Macedonia, once a poor and backward province, began in earnest during the reign of Alexander's father Philip (393–336 B.C.) who became ruler of Macedonia in 359 B.C. Having conquered his barbarian neighbors and the city-states of Greece, Philip was killed by an assassin. Only 20 when his father died, Alexander efficiently disposed of three possible rival claimants to the throne, defeated the Persians, conquered Egypt and Babylonia, and invaded India. Alexander's military forces were followed by a tiny army of scholars and historians, botanists and geographers, engineers, and surveyors. His campaigns thus produced a wealth of information on the conquered areas, their resources, natural history, and geography. Before reaching his thirty-third birthday Alexander fell ill with a fever and died. Immediately after his death his generals struggled for control of the army and the conquered territories. Ultimately, the Empire was divided among three successors: Antigonus Cyclops, the One-eyed (382–301 B.C.), who took the Euro-

Alexander the Great and his physician

pean part, Macedonia, and Greece; Seleucus Nicator (356–281 B.C.), who took the Asiatic part; and Ptolemy Soter (367–283 B.C.), who ruled over Egypt. The dynasty of the Ptolemies lasted for about 250 years and had the most lasting effect on the history of the sciences.

Ptolemy Soter, like Alexander, had studied with Aristotle. A literate man, he provided the only firsthand record of the life of his former commander. Ptolemy arranged for Aristotle's disciple Strato of Lampsacus to tutor his son Philadelphus, who ruled from 285 to 247 B.C. as Ptolemy II. The city of Alexandria in Egypt, symbolized by its lighthouse, one of the seven wonders of the world, became the capital of the Ptolemies, and Alexandria became the intellectual capital of the world. Scholars, poets, and philosophers came from all parts of the Empire, especially from Athens, thus adding the achievements of Greek culture to the ancient heritage of Egypt. The general intellectual climate of Alexandria encouraged by the early Ptolemies and the opportunities made available to scholars willing to relocate provided new forms of patronage that stimulated scientific research.

The Ptolemies founded and supported development of a magnificent Library and House of Muses, or Museum, a term that originally applied to a cult center dedicated to the cultivation and worship of the Muses, the nine goddesses who presided over literature and the arts and sciences. The Muses had been associated with philosophical studies since the time of Pythagoras. Both Plato's Academy and Aristotle's Lyceum could also be thought of as brotherhoods of scholars gathered together to serve the Muses. The evolution of the Museum in Alexandria into the ancient equivalent of a modern research university was, however, unique. Little is known for certain about the Museum because no contemporary accounts have survived. Modern scholars have argued that there is no reliable documentation that medical research was conducted at the Museum. But it must be noted that the extant evidence is so fragmentary that the absence of evidence is not evidence of its absence. Among the benefits that allegedly accrued to the scholars at the Museum were free meals and a tax-exempt salary generous enough to make the Greek poet Theocritus call Ptolemy "the best paymaster a free man can have." While the Museum was nominally under the direction of a high priest, its real goals were those more closely associated with a research institute. As a place for teaching, it was modeled after the Lyceum, but dwarfed that institution in the size, if not the quality, of its permanent faculty members and student body. According to traditions based on later descriptions, the facilities and surroundings were magnificent, including a promenade, rooms for research, lectures, conferences, private and group study, an observatory, zoo, botanical garden, dissecting rooms, and a large dining room.

The Museum served as an efficient and orderly way of producing engineers, physicians, geographers, astronomers, and mathematicians. Scientists and technicians were needed by the state, which was, after all, a Greek enclave in the ancient Egyptian world. Illustrative of the practical emphasis of Alexandrian research are many engineering advances in hydraulics and pneumatics as well as inventions, such as the valve, pump, and screw. Most Alexandrian inventions, however, seem to have remained at the level of toys, except for weaponry, temple magic, and to a lesser extent, medical instruments. The Ptolemies were, not surprisingly, interested in using religion, scholarship, science, and technology to enhance their power and prestige. Although Strato had boasted that he did not need the help of the gods to create a world, the science of pneumatics that he had helped to create allowed the rulers of Alexandria to make science the servant of the gods. By combining religious mysteries with the very practical side of pneumatic science, as developed by Hero of Alexandria, the temple priests were able to perform miracles on demand. Later writers satirized the Museum as a place where large numbers of bookworms were fed and cloistered like birds in a henhouse as they engaged in endless argumentation. In later Roman times, however, the Alexandrian Museum was essentially a secular teaching establishment whose fame and prestige inspired the development of similar centers.

Libraries were not unknown in Greece before the age of Alexander, but given the problems of producing books, significant collections were rare. Aristotle's library, assembled at the Lyceum, may have been a major stimulus and model for the Alexandrian Library, which was said to house about half a million manuscript rolls. In order to expand the Library's collection, travelers were required to declare any books they had in their possession. If the texts were of interest to the librarians, the books were purchased, copied, or seized by force if necessary. There are more extant references to the Library than the Museum, but many controversies surround the interpretation of these sources. For example, in discussing the authenticity of various works in the Hippocratic collection, Galen of Pergamum (130–200) claimed that the rulers of Alexandria were so jealous of the great reputation of their Library that they ordered that all books found on ships unloading in Alexandria should be seized. According to Galen this passion for book collecting led to the production of many forgeries, that is, new texts masquerading as antiquities and titles falsely ascribed to Aristotle and other great figures.

The Library of Alexandria seems to have exerted a major influence on the direction of research and teaching. The concept of the textbook seems to have developed in Alexandria. Experts in various fields systematized their knowledge in a standardized format, which proceeds from general principles, through well-established concepts, and on to the latest ideas at the frontiers of research. Nevertheless, tension between the oral tradition and the written word appears to have continued to exist within Greek culture well into the Early Hellenistic period when the text had already become well established as a systematic source for medical, philosophical, and rhetorical training.

Clearly, the Alexandrian Era was one in which science and scholarship enjoyed unprecedented support. Yet for the most part, the institutions and individuals who contributed to the intellectual life of this brief but brilliant period have been plunged into obscurity. In terms of the history of science, one of the most significant factors was the destruction of the Library and the loss of many of the Alexandrian texts. The exact sequence of events that led to the destruction of the Library is uncertain; it is clear that little of the work carried out at the Library and the Museum survived. During the reign of Cleopatra, the last of the Ptolemies, the Library is said to have contained 700,000 volumes. When Julius Caesar and his army visited Alexandria in 48 B.C., Cleopatra gave him many precious works from the Library as gifts. During the riots against the Romans, fires destroyed tens of thousands of manuscripts. Later losses have been attributed to mob violence unleashed by early Christian leaders intent on destroying pagan institutions. According to traditional accounts of the lives of martyrs, the last scholar at the Museum was a woman mathematician named Hypatia, who was beaten to death in A.D. 415 by a frenzied Christian mob. Further attacks on the ancient pagan centers of learning were attributed to the Arab forces that conquered the city in A.D. 646.

THE ANATOMICAL RESEARCH OF HEROPHILUS AND ERASISTRATUS

Many sciences flourished at Alexandria, although research was primarily oriented towards specialized sciences and their practical applications. Attempts to encourage progress in medicine led to unprecedented advances in anatomical studies that, for the first time in history, included systematic dissection of human bodies. Perhaps the Egyptian tradition of cutting open the body and removing certain organs as part of the embalming ritual helped overcome traditional Greek antipathy to the mutilation of corpses. Egyptian rites of mummification were performed for religious reasons having nothing to do with medicine or scientific research. Opening the body and removing specific organs was part of the mummification procedure, but the embalmers who performed the process were neither physicians nor scientists. Direct evidence of Egyptian science and medicine during the time of the establishment of the Greek city of Alexandria, and the possibilities for interaction between Greek and Egyptian scholars, are, unfortunately, virtually nonexistent.

Despite the paucity of reliable primary sources, there is good evidence that sophisticated anatomical studies were carried out during the Alexandrian Era by at least two famous, or infamous, scientists—Herophilus (ca. 330/320–260/250 B.C.) and Erasistratus (ca. 310–250 B.C.)—but the exact relationship of these two anatomists to Alexandria and to each other is uncertain. When Christian theologians such as Tertullian and St. Augustine wanted evidence of the heinous acts committed by the pagans, they pointed to the notorious Herophilus and accused him of torturing 600 human beings to death. The Roman encyclopedist Celsus charged Herophilus and Erasistratus with performing vivisection on condemned criminals who were awarded to them by the rulers of Alexandria. While these accusers were obviously not eyewitnesses, some historians have accepted their allegations; others remain skeptical. The extent or existence of the practice of human vivisection during the Alexandrian Era remains controversial because the writings of Herophilus and Erasistratus have not survived and are known only through the diatribes of their enemies.

There seems to be little doubt that the Ptolemies gave anatomists access to the bodies of condemned criminals and that Herophilus and Erasistratus were able to overcome the traditional Greek antipathy for human dissection. Performing public anatomies would have offended prevailing sentiments and perhaps gave rise to charges of even greater cruelties. Physician-scientists might also have used their patients and apprentices as experimental subjects in therapeutic research. Whatever conditions made the work of Herophilus and Erasistratus possible did not last long; human dissection was probably discontinued by the end of the second century B.C. On the other hand, some scholars believe that human dissection might have continued until the first century A.D. In any case, Alexandrian anatomists could presumably have used human skeletons for research and teaching even after they were forced to confine dissection and vivisection to other animal species.

Those who argue that it is inconceivable that Greek physicians would have performed human vivisections should reconsider the not uncommon references to the mutilation of bodies in the Greek literature, the public torture of slaves in law courts to extract evidence, and accounts of the use of convicts in tests of poisons and their putative antidotes. Of course the evidence from well-documented twentieth-century history suggests that there is no limit to possible human cruelty, individual or state organized. Herophilus, Erasistratus, and their contemporaries were engaged in constructing an entirely new science of human beings rather than an abstract philosophical system. The argument that Herophilus would not have made certain errors if he had done vivisections fails to allow for the influence of preexisting concepts on perception, awareness, and interpretation, and the intrinsic difficulty of such studies.

Except for the possibility that Herophilus studied with Praxagoras of Cos, little is known of his life, although he is said to have come from Chalcedon. Among his students, according to a story sometimes treated as legend, was an Athenian woman named Agnodice, a female physician who had to disguise herself as a man in order to practice medicine. If the story of Agnodice really applies to Herophilus, it suggests that he must have spent some time in Athens. There is little doubt that Herophilus practiced medicine in Alexandria, but whether or not he performed his dissections and research at the Museum or received financial support from the state is unclear.

Later writers attributed many texts to Herophilus, including *On Anatomy*, *On the Eyes*, and a handbook for midwives. Some of the books written by Herophilus seem to have been in use for several centuries, but only a few fragments have survived. He seems to have had a good overall picture of the nervous system, including the connection between the brain, spinal cord, and nerves. Rejecting Aristotle's claim that the heart was the seat of intelligence, Herophilus argued that the brain was the center of the nervous system. He also described the digestive system, called attention to the variability in the shape of the liver, and differentiated between tendons and nerves.

In his investigation of the circulatory system, Herophilus noted the difference between the arteries, with their strong pulsating walls, and the veins, with their relatively weak walls. Contrary to the prevailing assumption that the veins carried blood and the arteries carried air, Herophilus stated that both kinds of vessels carried blood. Intrigued by differences in the pulse that correlated with health and disease, Herophilus tried to quantitate the beat of the pulse using a water clock that had been developed in Alexandria.

Apparently known as a chronic skeptic, Herophilus regarded all physiological and pathological theories as hypothetical and provisional, including Hippocratic humoralism. While he probably did not totally reject humoral pathology, he seems to have preferred a theory of four life-guiding faculties that governed the body: a nourishing faculty in the liver and digestive organs, a warming power in the heart, a sensitive or perceptive faculty in the nerves, and a rational force in the brain. In clinical practice Herophilus seems to have favored more active intervention than that recommended by Hippocrates, and he may have used the concept of hot-cold, moist-dry qualities as a guide to therapeutic decisions. Vigorous bloodletting and a system of complex pharmaceuticals became associated with Herophilean medicine, but the anatomist appears to have urged his students to familiarize themselves with dietetics, medicine, surgery, and obstetrics.

When scholars of the Renaissance sought to challenge the stifling authority of the ancients, especially that of Galen, the long neglected and much vilified Herophilus was lauded as the "Vesalius of antiquity." The title could also have applied to Erasistratus, another intriguing and rather shadowy figure, who was attacked by Galen for the unforgivable heresy of rejecting the Hippocratic philosophy of medicine. Galen wrote two books against Erasistratus and criticized his ideas whenever possible. The relationship between Erasistratus and Alexandrian anatomy is obscure. Galen claimed that Erasistratus and Herophilus were contemporaries, but Erasistratus may have been at least 30 years younger than Herophilus. Little is known of his life, except the tradition that having diagnosed his own illness as an incurable cancer, he committed suicide rather than suffer inexorable decline. He may have authored more than 50 books, but none of them have survived. Some sources suggest that Erasistratus was a student or disciple of Strato, or even the nephew of Aristotle. Erasistratus was probably active in Antioch about 293 B.C. as physician to the court of Seleucus I Nicator. If this is true, then human dissection must have been, at least briefly, accepted in both areas. There are striking similarities in the anatomical studies attributed to Herophilus and Erasistratus, but Erasistratus seems to have attempted to apply Alexandrian mechanics to his medical theories and to have been more concerned with physiology and experimentation than Herophilus. Both anatomists were associated with human dissection and vivisection and many anatomical and physiological discoveries. Both were interested in respiration, reproductive physiology, ophthalmology, the brain and nervous system, the heart and vascular system, the pulse, fevers and therapeutics, where both seem to have favored treatment by opposites.

Galen accused Erasistratus of rejecting the Hippocratic philosophy of medicine and following the teaching of Aristotle. Like Herophilus, Erasistratus probably tried to replace humoral theory with a new doctrine. In the case of Erasistratus this seems to have developed into a pathology of solids, which perhaps did more to guide his anatomical research than his approach to therapeutics. Erasistratus is said to have been a gifted practitioner who rejected the idea that a general knowledge of the body and its functioning in health was necessary to the practicing physician. Many problems, he argued, could be prevented or treated with simple remedies and hygienic living. Nevertheless, he believed in studying pathological anatomy as a key to localized causes of disease, in particular the possibility that disease and inflammation were caused by the accumulation of a localized plethora of blood which caused blood to pass from the veins into the arteries. A local excess of blood formed from undigested food could damage surrounding tissues because blood in the overloaded veins, vessels designed to carry blood, could spill over into the arteries. When this occurred, the flow of pneuma, or vital spirit, which was supposed to be distributed by the arteries would be obstructed.

One remarkable aspect of Erasistratus's theory was the assumption that invisible connections existed between arteries and veins; the direction of flow was, however, from veins to arteries, and the phenomenon occurred only in pathological states. If a living animal was injured so that an artery was cut, the air escaped and pulled some blood after it. Nature's horror of vacuums, as taught by the Aristotelians, was thus incorporated into Erasistratus's physiology. The veins, which supposedly arose from the liver, and the arteries, which were thought to arise from the heart, were generally thought of as independent systems of dead-end canals through which blood and pneuma seeped slowly to the periphery of the body so that each part of the body could draw out its proper nutriment. Observations of engorged veins and collapsed arteries in the cadaver would appear to support these ideas. It logically followed that the goal of therapy was to reduce the plethora of blood. One way to accomplish this was to interrupt the production of blood at its point of origin by eliminating the supply of food. In addition to general starvation, a form of localized starvation could be accomplished by trapping blood in other parts of the body by tying ligatures around the roots of the limbs until the sick part had used up its plethora. The use of the ligature to stop the flow of blood from torn vessels was ascribed to Erasistratus.

Perhaps because of his theory that disease was caused by a local excess of blood, Erasistratus paid particular attention to the heart, veins, and arteries. In his lost treatises, he apparently gave a detailed description of the heart, including the semilunar, tricuspid, and bicuspid valves. Mechanical analogies, dissections, and perhaps vivisection experiments apparently suggested to Erasistratus that the heart could be seen as a pump in which certain "membranes" served as the flap valves. Using a combination of intuition, logic, and imagination, Erasistratus traced the veins, arteries, and nerves to the finest subdivisions visible to the naked eye and speculated about further subdivisions beyond the limits of vision. He also gave a detailed description of the liver and gallbladder and initiated a study of the lacteals that was not improved upon until the work of Gasparo Aselli (1581–1626).

Although Erasistratus was sometimes called a materialist, atomist, or rationalist, he did not reject the concept of animating spirits. Apparently he believed that life processes were dependent on blood and pneuma, which was constantly replenished by respiration. Two kinds of pneuma or spirits were found in the body: the vital pneuma was carried in the arteries and regulated vegetative processes. Some of the vital pneuma got to the brain and was changed into animal spirits. Animals spirits were responsible for movement and sensation and were carried by the nerves, a system of hollow tubes. When animal spirits rushed into muscles, they caused distension, which resulted in shortening of the muscle and thus movement. Perhaps inspired by Strato's experimental methods, Erasistratus is said to have attempted to provide quantitative solutions for physiological problems. In one experiment Erasistratus put a bird into a pot and kept a record of the weight of the bird and its excrement. He found a progressive weight loss between feedings, which led him to conclude that some invisible emanation was lost by vital processes.

The Alexandrian age provided unprecedented opportunities for scholarship and scientific research. The Library of Alexandria created the basic tools of scholarship: a storehouse of knowledge, a workplace for the duplication of manuscripts, a tradition of textual criticism, and a reverence for the written word as a gift from the past and a legacy for the future. The Museum and Library flourished for barely two centuries, an oasis in the intellectual wastelands of much of Western history. The later Ptolemies were not scholarly, enlightened, or wise and seem to have become increasingly divorced from Greek intellectual traditions. By the second century B.C. Greek and foreign scholars were unwelcome in Alexandria, and many of them created new centers of learning in other areas. Much of the deterioration can be blamed on political forces, but scientists and scholars themselves have been accused of bringing down the structural supports of their houses of learning. Rivalries between different schools of thought were intense, with members of various factions impugning the work, methods, conclusions, and character of members of other groups. The tension that always exists in medicine between disinterested scientific research and the immediate needs of the sick intensified and disrupted the ancient search for harmony and balance between the scientist and the healer. Critics of anatomical research and studies of the nature of disease claimed that such pursuits distracted physicians from caring for patients as people. Finally, the worst of fates fell upon the Alexandrian scientists; they were persecuted, their grants were cut off, and they had to turn to teaching to eke out a living in new places. After Caesar conquered Egypt, Alexandria was reduced to the status of a provincial town in the great Roman Empire and the Museum and Library were fully eclipsed. The annexation of Egypt in 30 B.C. is generally regarded as the end of the Hellenistic period.

THE ROMAN WORLD

One of the major accomplishments of the Hellenistic era was the extensive diffusion of Greek culture. Long after this period came to a close, literature, art, philosophy, and science were dominated by Greek ideas, but the Romans became the undisputed masters of political and military power. Like the Greeks, the Romans had the advantage of coming to civilization in the Iron Age. Narrowly defeated by Pyrrhus of Epirus in 279 B.C., 4 years later the Romans defeated Pyrrhus, the king who had expected to become the Alexander of the West. During the second century B.C. the Romans overcame the successors of Alexander in Macedonia and Syria. Soon the Greek cities of Asia Minor and the mainland succumbed to Roman rule. Although the Romans were highly successful in mastering practical aspects of warfare, administration, agriculture, engineering, sanitation, and hygiene, Roman contributions to biology and medicine were modest. Despite considerable suspicion, hostility, and resistance, by the first century B.C. the Roman world had generally adopted and assimilated many aspects of Greek culture. Indeed it was not too difficult for an upper-class Roman to acquire a veneer of Greek sophistication; Greek tutors could be purchased in the slave markets for less than the price of a good chef.

Roman writers were especially interested in collecting and recording information on a grand scale. For some Roman authors the motive for writing was to show that Roman work could be superior to that of the Greeks. This was the impulse that drove Cato the Censor (234–149 B.C.) to write about medicine and agriculture. Although his medicine was heavily imbued with the kinds of magical and superstitious formulas that had been rejected by Hippocrates, Cato boasted that Rome had long been healthy without doctors. On the other hand, the writings of the Roman poet Lucretius (99–44 B.C.), a follower of the philosophy of Epicurus (341–271 B.C.), indicate that some Roman authors were willing to explore philosophical and theoretical ideas. In his poem *De rerum natura* (*The Nature of Things*) Lucretius uses Epicurean teachings on ethics, atomic theory, the universe, and the nature of human life to combat the persistence of antiquated religious and superstitious ideas that he thought unworthy of intelligent beings. The superstitious fears that tormented men, Lucretius argued, were deliberately manipulated by politicians and demagogues. In assuring his readers that a true philosophy of nature would allow them to defend themselves against state sponsored mythology, Lucretius was making a deliberate political statement as well as defending Epicurean philosophy and atomist theory.

Unlike Aristotle, Lucretius did not believe that the world was the finite, eternal creation of a perfect God. Rather, he thought that the earth was fated to change and eventually perish. Like the pre-Socratics, Lucretius provided a natural explanation for the history of the universe and the nature of human society. Human beings were products of nature like all other creatures; even dreams and the soul could be explained in terms of atoms. Fundamentally pessimistic, Lucretius advocated a return to a simpler mode of life and saw no promise in technological advances. He blamed religion and superstition for the decline of Roman civilization. Accustomed to consulting the will of the gods for all things, the people of Rome were easily manipulated by politicians who were willing to exploit the messages of the gods for their own purposes. Early Christian theologians disapproved of the materialistic concepts associated with the Epicureans, and few copies of *On the Nature of Things* survived the Middle Ages. In addition to the charge that they neglected the gods, the Epicureans were attacked for the idea that the goal of ethics was pleasure. Epicureanism, also known as hedonism (from the Greek word for pleasure), was misinterpreted as the pursuit of unrestrained sensuality; actually, the Epicureans believed that the good life would result in a state of mental tranquility. During the Renaissance the work of Lucretius was highly regarded as a major source of Epicurean philosophy and a model for lucid scientific writing.

Unlike the systematic philosophical exposition so carefully crafted by Lucretius, the collection of facts and marvels compiled by Pliny the Elder (A.D. 23–79), one of the most famous of the great Roman encyclopedists, was immensely popular and often imitated by medieval writers. Despite their obvious differences, Pliny and Lucretius shared the goal of freeing their contemporaries from the grip of superstition and fear by providing information about the world and natural explanations of its phenomena. Until the sixteenth century Pliny's much admired *Natural History* was regarded as a comprehensive and complete collection of information about the natural world. In 37 books, Pliny assembled facts and observations about all the natural sciences and human arts from some 2000 previous works written by 146 Roman authors and 326 Greeks. In every way a model Roman citizen, Pliny lived during the height of Roman power and was a member of a family of respected officials. Born in Novum Comum in northern Italy, Pliny practiced law before joining the military service. If travel is truly educational, Pliny was indeed well educated. He traveled through Germany, Gaul, Spain, and Africa, performing many official duties for Vespasian and his successor Titus. Curiosity and efficiency seem to have been Pliny's personal deities. To avoid wasting time that could be employed in reading and writing, he worked while traveling by carriage rather than spend time on horseback. Legend has it that his secretary read to him while he took his bath. It was curiosity that caused his death. When Mount Vesuvius erupted in A.D. 79, demolishing Pompeii and Herculaneum, Pliny insisted on studying the awesome phenomenon himself. He was overcome by the poisonous fumes of the volcano and died without sustaining any external wounds.

Enthusiastically collecting materials "as full of variety as nature itself," Pliny emphasized the Roman viewpoint that knowledge and science were useful either directly or as a means of teaching moral lessons. Unlike the abstract and abstruse works of the philosophers, the *Natural History* was a compendium of subjects of immediate importance for human life, presented in a style accessible to farmers and artisans, as well as men of leisure. Subjects such as mathematics, geometry, and the theory of music, so dear to the philosophers, were omitted. Nevertheless, Pliny could not bear to omit even the most dubious and bizarre stories. The *Natural History* is a rather like a random walk through observations about the heavens and the earth, geography and ethnography, natural history, stories of animals, fish, birds, and insects. The botanical sections deal with forestry, agriculture, and horticulture, as well as the manufacture of useful products from the plant world, such as wine, oil, and medicines. Nature is so generous, Pliny assures us, that even the desert can provide the well-informed individual with a veritable drugstore. Following a wide-ranging discussion of the medicinal uses of plants, Pliny describes drugs derived from man, as well as superstitions, magical practices, and physicians. Even his recommendations for pest control involve a mixture of religion, folk magic, and chemistry that encompasses the use of green lizard's gall to protect trees from caterpillars to soaking seeds in wine to prevent fungal diseases.

Beginning his discussion of animal life with humans, as appropriate for "the highest species in the order of creation," Pliny argued that all other creatures seem to have been produced by nature for their sake. Among all the human races, Pliny asserted, the Roman race was the most virtuous in the whole world. From other parts of the world came stories of strange races, marvelous births, and monsters. Among the more fortunate marvels was a race of sun-hardened people over 7 feet tall who never spit and never suffered from headache or toothache. Some "wonder people" were half male and half female, others had dogs' heads and communicated by barking, some had no mouths and derived all their nourishment from airs and odors. Some of the strange fur-covered Satyrs who alternately walked upright and ran about on all fours were probably monkeys or apes. Exotic animal species included the lion and the elephant, as well as the unicorn and the phoenix. The unicorn was a fierce chimera with the head of a stag, the feet of an elephant, the tail of a boar, the torso of a horse, and a 3-foot-long black horn projecting from the middle of its forehead. Pliny noted that the animal could not be captured alive; presumably, the creature has not been captured dead either.

The degeneration of culture was of great concern to Pliny. Rome had established an enormous empire having great wealth and resources, but the sciences seemed to have stagnated. Pliny considered it paradoxical that science had flourished during a period of constant tension and warfare among the Greek states, but went into a state of decline after Rome imposed order and peace on the world. Complaining that his contemporaries were abusing the earth's resources in their pursuit of pleasure, wealth, and security, Pliny warned that the sad state of science was part of a general malaise overtaking the Roman world. Like the Stoic philosophers, and some twentieth-century ecologists, Pliny equated God with Nature, abhorred extravagance, and favored a return to his vision of a simpler, but better life.

Major sections of Pliny's text dealt with drugs derived from plants, a subject of great practical importance. A more systematic treatment of botany and herbal medicine was composed by Pliny's contemporary, the Greek physician Dioscorides (ca. A.D. 40–80). Little is known about Dioscorides, except that he probably studied at Alexandria before he became a *medicus* attached to the Roman forces in Asia under the emperor Nero. Military service gave him the opportunity to travel widely and study many exotic plant species. The text now known as *De materia medica* (*The Materials of Medicine*) refers to some 500 plants; the discussion of each plant includes its appearance, place of origin and habitat, its medical uses, and the proper method for preparing remedies. Drugs derived from animals and minerals were also described.

Some of the ingredients in the herbal remedies recommended by Dioscorides can be found among the vegetables and spices sold today in any grocery store, but the medicinal properties assigned to them would surprise modern cooks. Cinnamon and cassia, for example, were said to be useful in the treatment of internal inflammations, venomous bites, cough, diseases of the kidneys, and menstrual disorders. Drinking a decoction of asparagus and wearing the stalk as an amulet purportedly induced sterility. Judging from the large number of remedies said to bring on menstruation and expel the fetus, menstrual disorders, contraception, and abortion must have been among the major concerns that patients brought to the physician. Even if earaches do not respond to remedies containing millipedes, cockroaches, or the lungs of a fox, recent studies of *De materia medica* suggest that the arrangement adopted by Dioscorides reflects a subtle and sophisticated drug affinity system rather than traditional methods such as plant morphology or habitat. Clearly, Dioscorides was an acute observer and keen naturalist, who had gathered much information about medically useful plants, their place of origin, habitat, growth characteristics, and proper uses. Despite the presence of many bizarre ingredients, Dioscorides recorded recipes for many effective drugs, including strong purgatives, mild emetics and laxatives, analgesics, antiseptics, and so forth. His classification scheme must have required close attention to the effects of drugs on a significant number of patients.

In contrast to the popularity of Dioscorides, the work of the Roman writer Celsus (fl. A.D. 14–37) was essentially forgotten until the fifteenth century. Almost nothing is known about the life of Celsus, and his status as author of *De re medicina* has been questioned. He has been honored as the creator of scientific Latin, but historians are divided as to whether he wrote

Dioscorides

or plagiarized one of the most famous Latin medical texts composed during this period. Some say that Celsus was merely a compiler or translator, but the quality of the text and the display of critical judgment seem to militate against this assessment. Contemporaries, however, considered Celsus a man of quite modest talents. Given Roman traditions, it is unlikely that Celsus would have been a physician. Roman landlords were expected to assume responsibility for the medical care of the sick on their estates. Generally this simply meant knowing enough about medicine to supervise the women or slaves who carried out the unpleasant menial work of the healer.

While Celsus may have written a major encyclopedia, with sections on rhetoric, philosophy, jurisprudence, the art of warfare, agriculture, and medicine, only the section on medicine has survived. Because Celsus wrote in Latin during an era in which Greek was considered the language of medicine and scholarship, his work was ignored by the intellectual elite. His manuscript remained in obscurity for some 13 centuries, but in the 1420s two copies of *De medicina* were discovered. The timing could not have been better, because *On Medicine* became the first medical work to be turned out on the new European printing press. To Renaissance scholars, Celsus represented a source of first-century Latin grammar and medical philosophy, uncorrupted by medieval copyists.

A master of organization, clarity, and style, Celsus established the historical context of Greek and Roman medicine. Celsus agreed with other critics of Greek influences that medicine and physicians had been unnecessary in ancient Rome. The Greek art of medicine became a necessity only after Roman society had become thoroughly infiltrated by indolence, luxury, and other Greek affectations that led to illness. While the Greeks had cultivated medicine more than any other people after the death of Hippocrates, Greek medicine had fragmented into various disputatious sects, such as the Methodists, Dogmatists, Pneumatists, and Empiricists. Without the analysis provided by Celsus, the origins and ideas peculiar to various sects that flourished in his time would be totally obscure. The Dogmatists, for example, emphasized the study of anatomy and claimed Hippocrates, Herophilus, and Erasistratus as their founders. Members of other sects accused the Dogmatists of neglecting the practical aspects of medicine while pursuing highly speculative theories. Emphasizing the role of experience in therapeutics, some members of the sect known as Empiricists argued that the study of anatomy was totally useless to physicians.

Taking a balanced view of the rivalry between the various sects, Celsus rejected rigid approaches to medical science and called for moderation in all therapeutic interventions, as well as careful consideration of the needs of the individual patient. For the most part, medical practice consisted of the pre-scription of drugs derived from plants and animals, but Celsus argued that medical practitioners must also know anatomy and surgery, including opera-tions such as amputation, couching the cataract, and the treatment of hernia and goiter. Critical of human experimentation, Celsus denounced Herophilus and Erasistratus for performing vivisections on criminals. Celsus argued that such cruel experiments were unethical and unnecessary because all too often accidents exposed the internal organs of the body and provided opportunities for physicians to learn while engaging in acts of mercy.

On Medicine provides a valuable survey of the medical knowledge and practices of first-century Rome, but Celsus and all the medical writers of antiquity, with the exception of Hippocrates, have been overshadowed by the great physician, anatomist, and writer, Galen of Pergamum (A.D.130–200). After the death of Galen, his writings were abstracted, rewritten, reorganized and transmitted to medieval and Renaissance scholars in the form known as Galenism. For hundreds of years Galen was second only to Hippocrates as a medical authority; as anatomist and experimental physiologist, he had no rivals. Called the "First of Physicians and Philosophers" by the emperor Marcus Aurelius, Galen was referred to as "windbag" and "mulehead" by his critics and rivals. Having written three books by the time he was 13 years of age, Galen claimed to have written a total of 256 treatises. His subjects included philosophy, religion, ethics, mathematics, grammar, law, anatomy, physiology, pathology, dietetics, hygiene, therapeutics, pharmacy, commen-taries on Hippocrates, and some autobiographical material. Fearing that his writings were being corrupted by careless copyists and that mediocre authors were trying to pass off their work as his, Galen composed a guide to the cautious reader called *On His Own Books,* which provides a list of the titles of his books, a description of his genuine works, and a reading program for aspiring physicians. His advice to medical students and teachers emphasized the need to foster a love of truth in the young in order to inspire them to work incessantly to master all that had been written by the Ancients and to devise ways to test and prove such knowledge. Galen portrayed himself as a scholar who understood that despite his own persistent search for truth, he would never be able to discover all that he so passionately wished to know. The only complete modern edition of his surviving works fills 20 large vol-umes; this German edition is said to have taken 12 years to complete and almost as long to read.

Galen was born in Pergamum, in Asia Minor near the Aegean coast op-posite the island of Lesbos. The city boasted a library and temple of Asclepius, the Greek god of medicine, which made Pergamum the cultural rival of Alexandria. The Ptolemies tried to arrest the intellectual growth of Pergamum by forbidding the export of papyrus. In response to this threat parchment was developed and the book as we now know it replaced the papyrus roll. Pergamum was also known as a site of intense Christian evangelism and Galen did not fail to comment on the conflicts between pagans, Jews, and Christians.

In autobiographical allusions, Galen refers to his father, Nikon, a wealthy architect, as a kind-hearted man, but he described his mother as a quarrelsome woman, so violent she sometimes bit the servants during temper tantrums. Despite his resolution that he would emulate the admirable qualities of his father and reject the disgraceful passions of his mother, Galen's writings reveal a penchant for acrimonious quarrels and bitter controversies that suggest he was unable to fully suppress his maternal inheritance. As a youth Galen mastered mathematics and philosophy and began to study medicine after his father learned in a dream that his son was destined to become a physician. When Nikon died, Galen left Pergamum in order to study with famous teachers in Smyrna, Corinth, and Alexandria, a city still renowned for anatomical instruction, despite the Roman prohibition of human dissection. When he was 28, Galen returned to Pergamum and became physician to the gladiators. Because of his success in treating the sick and the wounded, Galen also established a flourishing private practice. Restless and ambitious, in a few years Galen decided to move on to the Imperial City. In Rome, Galen was successful in treating several prominent individuals, won a reputation as a miracle worker, delivered in public anatomical lectures, and composed some of his major anatomical and physiological texts. But 4 years later, complaining about the hostility of less successful physicians, Galen suddenly returned to Pergamum. His critics noted that his departure had coincided with the appearance of a virulent epidemic disease. At the request of the emperor Marcus Aurelius, Galen eventually returned to Rome where he resumed his busy schedule of imperial service, private practice, research, lecturing, and writing.

Like Plato and Aristotle, Galen believed that a form of divine intelligence had created the universe and living beings. In the organization of the human body he found proof of the power and wisdom of the Creator and the revelation of the divine design in all things. Galen was tireless in his tirades against naturalists who would deny the evidence of the great plan and perfection of the human body as the work of the Creator. Supporters of the atomic theory, which he considered materialistic and atheistic, were singled out for particular outrage. For Galen, the body was the instrument of the soul and proof of the existence and the wisdom of God. Anatomical research was the source of a "perfect theology" through which the anatomists demonstrated his reverence for the Creator by discovering his wisdom, power, and goodness through anatomical investigations of form and function. Dissecting any animal, no matter how humble, revealed a little universe fashioned by the wisdom and skill of the Creator. Assuming that Nature acts with perfect wisdom and does nothing in vain, Galen argued that every part of the body was crafted for its proper function. Dissection might be a religious experience for Galen, but he understood that most practitioners needed anatomical knowledge for guidance in performing surgical operations and treating wounds. Where the surgeon could choose the site of incision, knowledge of anatomy allowed him to do the least damage possible. It also allowed him to predict the kind of damage that would occur when muscles were severed in the course of treatment, so that the physician could escape blame. Larger philosophical questions might also be answered by anatomy. For example, Aristotelians argued that the fact that the voice, which is the instrument of reason, came from the chest supported the contention that the seat of reason was in the heart, rather than the head. By demonstrating that the voice is controlled by the recurrent laryngeal nerves, Galen vindicated those who argued that reason resided in the brain.

Even though he could not carry out systematic dissections of human bodies, Galen was quite sure that his findings in pigs, goats, monkeys, and apes could be applied to human beings. Sometimes Galen was the beneficiary of fortuitous accidents that allowed him to inspect human cadavers and skeletons. In his work *On Bones* Galen reported that he had often examined human bones exposed by the destruction of tombs and monuments. On one occasion a corpse had been washed out of its grave by a flood, swept downstream, and deposited on the bank of the river. The flesh had putrefied, but the bones were still closely attached to each other, leaving the skeleton perfectly prepared for an anatomy lesson. While some historians have suggested that Galen's patrons allowed him to carry out surreptitious human dissections, the enthusiasm with which Galen describes such incidents suggests their rarity. Indeed, while Galen was often critical of his predecessors, especially Herophilus and Erasistratus, he obviously envied their resources and privileges. Nevertheless, systematic dissections of other animals would have allowed Galen to take full advantage of the "windows" into the body provided by the injuries of his patients even as he dressed their gruesome wounds.

The importance of Galen's anatomical work can hardly be exaggerated. Galen's word was the ultimate authority on medical and biological questions for hundreds of years. His anatomical studies were not seriously challenged

Galen of Pergamum contemplating the rare opportunity of studying a human body

until the sixteenth century, and his physiological concepts remained virtually unquestioned until the seventeenth century. Even in the nineteenth century physicians who were fighting to reform medical practice had to battle against what they perceived to be the continuing tyranny of Galenism. His errors, which eventually stimulated a revolution in biological research, tend to be overemphasized, but it is essential to try to balance the merits and defects in Galen's work since both exerted a profound influence on the history of anatomy, physiology, and medicine. Galen's anatomy had the obvious defect that he had based his researches on animal models and not on the human body itself. Nevertheless, Galen was an expert and acute observer whose writings could provide an excellent guide to readers who absorbed both the letter and the spirit of his teachings. Galen often told his readers that the only way to learn about the wonders of the human body was by studying the writings of the great physicians, such as himself and Hippocrates, and by actual experience. But even the writings of the great Hippocrates must be confirmed and tested by going directly to Nature and observing the structures and functions of animals.

Not satisfied with purely anatomical research, Galen sought ways to proceed from structure to function, from anatomical analysis to experimental physiology. Through dissection, vivisection, and speculation Galen struggled to solve almost every possible problem in classical physiology. Given the magnitude of the task and the conceptual and technical difficulties inherent in such a program, it is not surprising that Galen was sometimes misled and that his writings were often misleading. Ultimately his errors stimulated revolutions in anatomy and physiology, but for hundreds of years the Galenic synthesis was considered so complete and powerful that it fully satisfied the needs of scientists, scholars, and physicians. Balancing the merits and defects of the Galenic synthesis is difficult for many reasons, not the least of which is the inherent complexity and obscurity of the system. While Galen may well have been misinterpreted and his writings corrupted by copyists, Galen made different assertions in various texts. Attempts to distinguish between what Galen actually said, or probably meant, and the layers of interpretation that ultimately precipitated around Galenic seed crystals are probably futile and counterproductive because not all of his writings have survived.

In essence, Galenic physiology rests on the doctrine of the threefold division of life into vital processes governed by vegetative, animal, and rational souls or spirits. Life ultimately depended on air, or pneuma, the breath of the cosmos. By integrating his anatomical and physiological knowledge with his philosophical system, Galen turned the human body into a place for the elaboration and distribution of three kinds of adaptations of pneuma. These were associated with the three principal organs of the body—the liver, heart, and brain—and the three types of vessels—the veins, arteries, and nerves. Pneuma was adapted by the liver to become the *natural spirits*, which were distributed by the veins and supported the vegetative functions of nutrition and growth. *Vital spirits,* which governed movement, were formed in the heart; the arteries distributed innate heat and pneuma or vital spirits to warm and vivify the parts of the body. Innate heat was produced as a result of the slow combustion process that occurred in the heart. The third adaptation occurred in the brain and resulted in *animal spirits* which were needed for sensation as well as muscular movements; animal spirits were distributed by the nerves.

To work out the details of the elaboration and distribution of the three spirits required much ingenious research and a wealth of speculation concerning the functions of the parts of the body. Because theories of the motion of the heart and blood have played a central role in the history of biology, Galen's writings on these topics have been subject to particular scrutiny. Modern categories are inappropriate in dealing with Galenic physiology; Galen's approach to the workings of the body made it impossibile to separate questions about the circulatory system from those about the respiratory system and the generation and distribution of pneuma and spirits. Even Galen was not always sure about the roles, functions, and existence of parts of his system. Sometimes, for example, he suggests reservations about the nature and functions of the spirits, but in attempting to clarify and simplify his arguments, later commentators tended to transform his tentative hypotheses into dogmatic certainties. Galen's ideas about the circulatory system illustrate both the strengths and weaknesses of his approach to physiological phenomena. By means of simple, direct experiments, Galen easily disproved the idea that the arteries of living beings contain air rather than blood. The artery of a living animal was exposed and tied at two points. When the vessel was cut between the two ligatures, blood and not air spurted forth, indicating that blood had been present in the artery *before* the incision was made. Similarly, the role of the arteries in nourishing the innate heat could be demonstrated by tying a ligature around a limb so that it was tight enough to cut off the arterial pulse, which was generated by the heart. The limb became cold and pale below the ligature because the arteries could no longer supply the innate heart.

In order to provide a complete explanation for the distribution of blood, air, and spirits, Galen asserted that the septum contained pores through which blood could pass from the right side of the heart to the left side.

Blood was said to be continuously synthesized from ingested foods. The useful part of the food was transported as *chyle* from the intestines via the portal vein to the liver. The liver had the faculty of transforming chyle into dark venous blood. Since the liver was the seat of vegetative life, concerned with nourishment and growth; it imbued the blood with natural spirits. As the rich venous blood made its way through the veins, the parts of the body took up their proper nutriments. The useless part of the food was converted into black bile by the spleen.

Despite his detailed knowledge of the structure of the heart, its chambers and valves, Galen's preconceived concept of its function led to ambiguities and obscurities. Galen thought that the right side of the heart was a main branch of the venous system and that the auricles were merely dilations of the greater veins, rather than important chambers of the heart itself. However, he demonstrated that when the chest of a living animal was opened and the heart examined, blood was found in the left ventricle of the heart. From the right side of the heart the blood could enter different pathways. Part of the blood discharged its impurities or sooty vapors into the lung by means of the artery-like vein (the pulmonary artery). Part of the blood passed through putative pores in the septum and entered the left heart where it encountered the pneuma that had been brought in by way of the trachea and the vein-like artery (the pulmonary vein). The mixture of dark red blood and air resulted in the production of vital spirits and the fine and vaporous bright red blood, which were distributed to the body by means of the arteries. Completing the threefold system, some of the arteries led to a network of vessels at the base of the brain called the *rete mirabile* where the vital spirits were transformed into animal spirits for distribution to the body via the system of hollow tubes called nerves. Clearly, the concept of continuous blood circulation is incompatible with a scheme in which blood is constantly synthesized from food by the liver and is then assimilated into the substance of the body as it ebbs and flows through the blood vessels. Of course the nature of the Galenic system is so complex that the term "clearly" is hardly appropriate in any attempt at a brief and simplified reconstruction of Galenic physiology.

Because Galen used various animals in working out the details of his physiological system, he ascribed various anatomical features found in other animals to human beings. For example, he described the liver as five-lobed, which is true in dogs, but not in humans. Similarly, the *rete mirabile* is present in ruminants, but not in humans. Having denied that the heart is a muscle, he argued that the active phase of heart action was dilation rather than contraction. Ignoring discrepancies between the rate of respiration and the heart beat, Galen's system associated respiration with the movement of the heart; dilation of the heart would draw air into the body and contraction would drive it out. Galen's system seemed to provide a complete synthesis of medical and philosophical traditions. It combined the four humors of Hippocrates with Aristotle's threefold division of the spirits and incorporated the pneuma, or great cosmic spirit of the Stoics. Beyond these desiderata, it is deeply religious in the sense that it is wholly imbued with reverence and admiration for the work of the Creator.

It is difficult to imagine a physiological problem that Galen did not at least touch on. His mastery of experimental methods is demonstrated in his studies of the nervous system. Galen worked out the anatomy of the brain, spinal cord, and nerves and proved that the nerves originate in the brain and spinal cord and not, as Aristotle said, in the heart. In a remarkable series of experiments, Galen demonstrated that when the spinal cord was injured between the first and third vertebrae, death was instantaneous. Injury between the third and fourth vertebrae arrested respiration. Damage below the sixth vertebra caused paralysis of the thoracic muscles, while lesions lower in the cord produced paralysis of the lower limbs, bladder, and intestines. To prove that the kidneys and not the bladder made urine, Galen carried out experiments in which the ureters were ligated to prevent the flow of urine from the kidneys to the bladder. His study of the functions of the kidneys actually involved the quantitative falsification of the theories put forth by those who argued that urine was formed in the bladder and that the kidneys played no role in its formation. To study digestion, he put pigs on different kinds of diets and examined the contents of their stomachs after appropriate intervals.

Scientific knowledge of anatomy, physiology, the nature of disease, and the nature of human beings was valuable in itself, but it also provided a formidable rationale for medical practice. In addition to his excellent reputation as a scientist and philosopher, Galen was regarded as an excellent clinician and brilliant diagnostician. By relying on the humoral theory of disease, Galen could rationalize his approach to the prevention and treatment of disease. Because the humors were formed by the action of the innate heat on nutriments, relatively hot and cold foods and drugs could be prescribed as a means of altering the balance of the four humors. Despite his reverence for Hippocrates, Galen was unwilling to stand by doing no harm while waiting for Nature to provide a cure. Galen clearly favored active intervention, and his sophisticated individualized approach to the prevention and treatment of disease eventually degenerated into a system of active bleeding and purging, accompanied by large doses of complex drug mixtures. Such remedies were later called "Galenicals," and apothecary shops might be identified by the sign of "Galen's Head" above the doorway.

In *On the Usefulness of the Parts* Galen argued that human beings are the most perfect of all animals, but man is more perfect that woman because the male has more of Nature's primary instrument, which is excess heat. Reproductive anatomy is destiny because, according to Galen, the task of heating and perfecting the blood was performed by the testicles. It was a deficiency of heat that caused a fetus to become a female. Galen essentially explained the female reproductive organs as defective or inside out versions of the perfect male parts; that is, because of the deficiency of heat, the reproductive parts formed inside the body of the fetus, but could not emerge and project themselves to the outside. Worse yet, brain development required greater heat and more perfected blood than that available to the female fetus. Evidence for such assertions, Galen argued, could be found by studying the effects of castration in humans and animals. Destruction of the testicles resulted in the loss of heat, virility, and manliness. It was the lack of heat in the female that resulted in the accumulation of fat; similar accumulations occurred in the body of a eunuch. As further proof that women were imperfect men, Galen offered ancient accounts of women who were suddenly transformed into men. The reverse transformation was impossible, Galen argued, because nature always strives towards perfection from the lower to the higher form. On the other hand, in defense of his belief that bleeding was the appropriate treatment for almost every disorder that afflicted human beings, Galen asserted that women were immune to many of the diseases that attacked men because their superfluous blood was eliminated by menstruation or lactation. As long as their menstrual cycles were normal, women were supposedly spared the miseries of gout, arthritis, apoplexy, epilepsy, melancholy, and other diseases. Men fortunate enough to rid themselves of excess blood through hemorrhoids or nosebleeds could also enjoy protection against such diseases.

In terms of humoral doctrine, bleeding seemed to accomplish the therapeutic goals shared by patients and physicians. Bleeding and purging seemed to rid the body of putrid, corrupt, and harmful materials. Indeed it is possible to explain the apparent therapeutic efficacy of bleeding in terms of the suppression of the clinical manifestations of certain diseases, such as malaria. Bleeding might inhibit the growth and multiplication of certain pathogens by reducing the availability of iron. Lowering the viscosity of the blood might also affect the body's response to disease by increasing the flow of blood through the capillary bed. Obviously bleeding to the point of fainting, a tactic enthusiastically championed by Galen, would force a period of rest and recuperation on feverish and delirious patients.

As skillful in the art of medicine as in the science, Galen was seen by his patients and admirers as a true miracle worker. Even his rivals acknowledged Galen's intelligence, his productivity, and his passionate defense of his own ideas. Concerned with the dignity of the medical profession, Galen warned against displays of ambition, contentiousness, and greed and urged physicians to distinguish themselves from the hordes of quacks and charlatans by means of skill, learning, and a disdainful attitude towards money. Clever students of medicine could also learn from Galen how to impress patients and observers by eliciting diagnostic clues through casual questions, making note of medicines already in use, observing the contents of all basins removed from the sickroom on the way to the dung heap, and secretly assessing the state of the pulse while in conversation with the patient. When unsure of diagnosis and prognosis, the physician could take the case with a great show of reluctance while warning the patient's friends and family to expect the worst. If the patient died the physician's reputation was safe; if the patient survived the physician would have earned his laurels.

Surprisingly, despite Galen's reputation and the fact that he gave public lectures and demonstrations in the Temple of Peace, he left no school or disciples. While his personality may have been admirably suited to dealing with Roman emperors, it seems to have intimidated or repulsed colleagues and potential students. Most of Galen's voluminous writings were unknown during the Middle Ages in Europe, but his work was not lost or entirely neglected. His medical texts were translated into Syriac and Arabic. Finally, the Arabic versions of his works were translated into Latin and inspired generations of European scholars. Galenic ideas, if not his experimental methods, became the foundation of Islamic and Christian medical studies. For many centuries the teachings of Galen were invested with great authority by theologians who were attracted by his religiosity, his concept of the three spirits, his denunciation of atheistic materialism, his support for a worldview that has been called anthropomorphic teleology, and his willingness to see proof of religious dogmas in scientific findings. It was unfortunate for science, and for Galen's reputation, that his most flagrant errors were treated as inviolable truths. When Galenic anatomy, pharmacology, and physiology were finally subject to serious challenges in the sixteenth and seventeenth centuries, the scientists who revolutionized and reformed those fields took their instruction and inspiration from the study of the newly restored writings of Galen.

THE MIDDLE AGES

No natural scientist comparable to Galen appeared during the European Middle Ages, the period from about 500 to 1500, an era during which ideas about the nature of the physical universe, the nature of human beings and their proper place in the universe, and above all, their relationship to their Creator underwent profound changes and painful dislocations. During Galen's lifetime, the borders of the Roman Empire had encompassed the civilized Western world. By the time of his death, Greco-Roman culture and ancient science could well be characterized as essentially moribund. The medieval era was the outcome of the well-known story of the decline and fall of the Roman Empire after centuries of anarchy, turmoil, and warfare. The Roman era, a period of relative peace and prosperity, might have been a time of great intellectual development, but long before Rome lost its place as undisputed political center of the Western world many writers had complained about the malaise permeating Roman society. Pliny voiced the belief shared by many of his contemporaries that humanity had grown corrupt and degenerate. The decline of Rome seemed to owe more to internal decay than to barbarian invasions. Rome suffered the cumulative effects of misgovernment, corruption, war, endemic disease, a growing obsession with occultism, and a crisis in agriculture. The productivity of Italian lands diminished as a result of deforestation, erosion, and the neglect of irrigation canals; numerous farms and fields were abandoned. Mining enterprises failed because the richer veins of minerals had been exhausted. Observers in Italy noted significant declines in population and in the live birth rate. Not long after A.D. 330 when Emperor Constantine established his capital in Byzantium (now Istanbul), the division of the empire between the East and the West became permanent. The Latin West entered into the era popularly known as the Dark Ages during which Christians were warned to avoid all pagan books that might turn the young from the faith. The world of the Christian West was to be that described in the Bible, a flat, unmoving earth created as described in Genesis. In A.D. 529 by the order of the Emperor Justinian, the Academy, the living heritage of Plato, was closed and pagans were forbidden to teach. Many pagan scholars were sent into exile in Persia, taking with them rare and precious manuscripts. Of course throughout antiquity the number of individuals engaging in inquiries that could be called scientific was always very small; with the introduction of legal sanctions against pagan learning, the few scholars who continued to engage in such activities faced increasing isolation as well as hostility and persecution.

Certainly, the climate of the European Middle Ages was not conducive to original research in the biological and medical sciences. On the other hand, during the Middle Ages European society experienced significant changes in social organization, technological innovations, and the appearance of new institutions of learning. Many labor-saving devices, agricultural implements, techniques, and crops were either invented or successfully exploited for the first time. Significant improvements were made in metallurgy, construction, the production of textiles, and the use of watermills and windmills. By the twelfth century both food production and population were showing significant gains. As productivity and population grew, so did industry, the crafts, commerce, and urban centers.

A revival of learning that gained momentum between the eleventh and thirteenth centuries was also transforming medieval European culture. The heritage of Greece reached Europe through a tortuous route from the Jews and Arabs to the East who had been in closer contact with the ancient learning. The pace of scholarly activities had definitely quickened by the thirteenth century, as mathematics, astronomy, alchemy, and medicine joined theology as subjects worthy of serious attention. Many factors engendered the change in the intellectual climate that characterizes the twelfth and thirteenth centuries. New religious movements undermined the monolithic structure of faith, creating pockets of doubt, confusion, and curiosity. Scholars turned to the philosophy and science of ancient Greece for answers to questions that faith alone could not answer.

Another factor in the transformation of scholarship was the growth of the medieval university. Very different from the ancient centers of learning, the university of the Middle Ages was also unlike modern institutions of higher learning, especially in terms of the relationships among students, faculty, and administrators. The community of scholars was replaced by the concept that those formally licensed to teach would instruct those seeking their own credentials. Informal discussion was replaced by set lectures to large groups of students. General inquiry into areas of mutual interest to independent scholars was replaced by a standardized curriculum.

The exact origins of some of the major European universities are unknown, although ecclesiastical authorities seem to have dominated these nascent institutions. As the old monastic schools declined in prestige in the eleventh and twelfth centuries, cathedral schools and municipal schools began to supersede them. Some of these schools grew quite large by the twelfth century and eventually developed into centers of higher learning. For example, Charlemagne (768–814) had founded schools associated with metropolitan churches where young priests gave lessons in theology and other subjects that the Church believed necessary for clerics. As the number of students and teachers increased, groups of teachers, called *magistri*, formed an organization called a *universitas magistorum*. The term *university* was originally applied to any corporate status or association of persons. The early universities were collections of colleges that sought the advantages of corporate status, rather like a guild. During the eleventh century the words university and guild were interchangeable for describing craft associations. The term *studium generale* denoted a guild or university of scholars and students engaged in the pursuit of higher learning. Such institutions drew students from many countries to study with teachers known for particular areas of excellence. Eventually the term university was restricted to institutions of higher learning. The establishment of universities throughout Europe was very uneven. During the thirteenth century there were 5 significant universities in France, 2 in England, 11 in Italy, and 3 in Spain. Germany had none until the fourteenth century. By the end of the Middle Ages about 80 universities had been established in Europe; of these, two thirds were in France and Italy.

Even though the educated, or merely literate, were still a tiny minority, some of the universities attracted large numbers of students, who, despite different national origin, had achieved competence in Latin, the language of learning. Students as young as 14 or 15 might enter the universities along with more mature scholars. The curriculum consisted of the seven liberal arts: grammar, rhetoric, logic, arithmetic, geometry, astronomy, and music.

Universities wielded considerable power and won large measures of self-government from civic and ecclesiastical authorities, but the relationship between "town and gown" was often quite tense. Some institutions, notably those in Paris and Oxford, were governed by guilds of instructors in the liberal arts. In contrast, Bologna, which specialized in legal studies, was dominated by a guild of students. Students had the reputation of being rowdy, armed, often drunk, and always dangerous. Given the lack of scholarships and student loans, it is not surprising that many were accused of making their living by a combination of begging and stealing. Despite their wild reputation, universities generally attempted to satisfy the eagerness of students to acquire knowledge, or at least access to careers that promised wealth, power, and prestige. Medieval universities generally lacked the kind of campus grounds and buildings associated with the modern institution. Lectures were often held in the professors' apartments or in rented rooms. In the absence of standardized courses, groups of students might hire and pay teachers on a contractual basis. Students who passed an examination after 6 years of study became qualified to teach. Many rewarding and lucrative professions were open to the university graduate. He could take orders and become a church official, seek out the patronage of a wealthy noble, study medicine or law, or work as a copyist. But even in the thirteenth century there were students who rejected conventional careers and adopted the life of wandering scholars.

Subjects deemed appropriate for teaching were limited and regulated by religious dogma and convention. During the turmoil of the thirteenth century, secular authorities bowed to the ecclesiastical powers, and orders of mendicant friars were established to combat alleged heresy at the universities. On the other hand, the writings of Aristotle, as edited by Thomas Aquinas, became a source of authority and inspiration to philosophers and theologians. Aristotle's ideas about the earth as the center of the universe and the site of all change and imperfection, in contrast to the pure, incorruptible, and eternal heavens where the Creator abided, were very congenial to the Church. His view of human life as part of a rigid hierarchy created a welcome justification for the hierarchical and authoritarian arrangement of medieval society.

The structure and goals of the university encouraged a form of education called *scholasticism*, that is, the study of the ancient authorities as the repository of all that could be known; this attitude essentially rejected the possibility of new discoveries and innovations. The increasingly formal organization of education and the professions also created rigid barriers that kept out *all* women and men lacking the advantages of wealth and status, who of course constituted the vast majority. Editing the works of the ancients and providing interpretations of the works of Aristotle that were compatible with the teachings of the Church were the major tasks undertaken by medieval scholars such as Albertus Magnus (ca. 1200–1280), a member of the Dominican order, Bishop of Ratisbon, and a teacher at the University of Paris, known to his contemporaries as "Doctor Universalis." Although his writings reflect energy, originality, and encyclopedic knowledge, Albertus Magnus presented his prodigious scholarly efforts as commentaries on the work of Aristotle. His reputation as the foremost naturalist of medieval Europe is based on the detailed descriptions of plants, animals, and minerals that appear in his *De mineralibus*, *De vegatabilus et plantis*, and *De animalibus*. Albertus described several new plant species and may have been the first to prepare arsenic in a free form, but rather than attempt to form new theories, he forced his observations into a preexisting Aristotelian mold. Despite such caution, his reputation as a magician came to rival his fame as a scholar; various works on astrology and the magical properties of plants, stones, and animals were spuriously attributed to Albertus Magnus. *The Book of Secrets of Albertus Magnus*, one of the most popular texts of this kind, went through many editions well into the seventeenth century.

Roger Bacon (1214–1292), the medieval scholar known as "Doctor Mirabilis," has been called the first modern scientist because of the spirit in which he approached nature. After studying at Oxford and Paris, he joined the Franciscan order, but his views were so unorthodox that after 1257 he worked in isolation; for many years he was virtually imprisoned in Paris. Despite his contention that all branches of science were subordinate to theology, his studies of mathematics, astronomy, optics, and alchemy, combined with his quarrelsome nature, imaginative predictions, and emphasis on experimentation led to a reputation as a magician, as well as accusations of witchcraft and necromancy.

Other scholars, however, who had no possible access to the medieval universities found the cloistered life congenial to intellectual and scientific pursuits. Hildegard of Bingen (1098–1179) was one of the very few medieval women to transcend the barriers that kept women out of the the universities and the community of scholars. Widely known and respected during her lifetime as a writer, composer, naturalist, and healer, she was all but forgotten until nineteenth-century German biblical scholars prepared new editions of her writings. Further studies have brought Hildegard to the attention of historians of science, feminist scholars, musicians, poets, herbalists, and homeopathic practitioners. St. Hildegard has been called a mystic, a visionary, and a prophet; nevertheless, her writings suggest practical experience and boundless curiosity about the wonders of nature.

As the tenth and last child of a noble family, Hildegard was offered to God as a tithe and entered a life of religious seclusion at the age of 8; she took monastic vows in her teens and was chosen abbess of her Benedictine nunnery in 1136. At about 15 years of age, Hildegard began receiving revelations about the nature of the cosmos and humankind. The visions were explained to her in Latin by a voice from heaven. In 1141 a divine call commanded her to write down and explain her visions. When she began writing at the age of 43, Hildegard thought that she was thc first woman to embark on such a mission; the names of other women writers were unknown to her. After a papal inquiry into the nature of her revelations, Hildegard became a veritable celebrity and was officially encouraged to continue her work.

Hildegard's *Liber simplicis medicinae* (*The Book of Simple Medicine*), generally referred to as the *Physica*, was probably the first book by a female author to discuss the elements and the therapeutic virtues of plants, animals, and metals. As the first book on natural history composed in Germany, it can be thought of as a pioneering effort in scientific botanical studies. The text includes much traditional medical lore concerning the medical uses or toxic properties of many herbs, trees, mammals, reptiles, fishes, birds, minerals, gems, and metals. The work of the sixteenth-century German botanists was deeply indebted to Hildegard, who often provided the local German name for various plants; some of the first recorded local names of plants are those preserved in her writings.

The *Liber compositae medicinae* (*The Book of Compound Medicine*), Hildegard's other major work, included material on the nature, forms, causes and treatments of disease, the elements, winds and stars, human physiology and sexuality, and so forth. Like Hippocrates, Hildegard attempted to find natural causes for mental and physical diseases. Some of Hildegard's ideas were expressed in vivid comparisons of the workings of the microcosm and the macrocosm. For example, the movement of blood in the vessels was compared to the way the stars pulsate and move in the firmament of the heavens. Although Hildegard seemed to think of the heart as the seat of the rational soul, she considered the brain the regulator of the vital qualities and the center of life. In exploring mental as well as physical diseases, Hildegard argued that even the most bizarre mental states could have natural causes. Thus, people might think a man was possessed by a demon when the real problem might be a simultaneous attack of headache, migraine, and vertigo. Hildegard probably had a special interest in these disorders; modern medical detectives have diagnosed her visions as classical examples of migraine.

ISLAMIC SCIENCE

The Middle Ages of European history roughly correspond to the Golden Age of Islam, the religion founded by the prophet Muhammad (570–632). When Muhammad was about 40 years old, he received the call to prophethood and a series of visions in which the *Qur'an*, the sacred book of the Islamic religion, was revealed to him. By the time of his death, practically all of Arabia had accepted Islam, and a century later his followers had conquered half of Byzantine Asia, all of Persia, Egypt, North Africa, and Spain.

During this turbulent period, the Arabs adopted and disseminated some of the techniques requisite to the further expansion of literacy and learning. At the beginning of the eighth century the Arabs captured Samarkand and learned the art of paper making from the Chinese, who had been using paper since about A.D. 100. A paper-making plant was built in Baghdad in

794. When the Arabs conquered Sicily and Spain, they introduced paper to Europe. Mathematical concepts that originated in India, such as the zero and the so-called arabic numeral system, were also introduced to Europe through the Arabs. Many words used in science and mathematics, such as alchemy, algebra, algorithm, alcohol, cipher, and nadir, are derived from the Arabic.

According to Muslim tradition, the Prophet respected and encouraged the pursuit of knowledge although he was himself illiterate. "He who leaves his home in search of knowledge," said Muhammad, "walks in the path of God." The followers of Muhammad established elementary schools, usually held in a mosque, where most boys and some girls received religious education. Secondary schools were eventually brought under government regulation and subsidized to serve as colleges where students could learn grammar, philology, rhetoric, literature, logic, mathematics, and astronomy. Just as Latin served as the common language of learning for students throughout Europe, Arabic was the language of learning throughout the Islamic world. Thus, Arabic texts need not have Arab authors; Persians, Jews, and Christians took part in the development of Arabic scientific literature.

From the ninth to the thirteenth century Arab scholars could justly claim that the world of scholarship had been captured by Islam. Books were regarded as great treasures, libraries were attached to most mosques, and wealthy individuals were passionate about amassing great collections of books. Scholars at ancient centers of learning were encouraged to continue their work and to assist in the immense task of translating ancient manuscripts into Arabic. Just as Christians sought to reconcile Aristotle with the teachings of the Church, so did Muslims try to harmonize Greek philosophy with the *Qur'an*. Because the scriptures were supposed to contain all wisdom of importance, Islamic scholars often disguised their work as commentaries on the *Qur'an* to make them more respectable.

Pharmacology, optics, chemistry, and alchemy were of particular interest to Arab scientists. Islamic medicine and pharmacology drew on many sources. For example, the famous medical formulary of al-Kindi (Yaqub ibn-Ishaq al-Kindi, ca. 800–870) included drugs derived from Mesopotamia, Egypt, Persia, Greece, and India, as well as traditional Arab remedies. Al-Kindi's text served as a model for many Arabic treatises on pharmacology, botany, zoology, and mineralogy. Although the theory of vision may seem a rather esoteric branch of knowledge, al-Kindi argued that it would prove to be the key to the discovery of nature's most fundamental secrets. The Latin version of his work on optics, *De aspectibus*, was very influential among Western scientists and philosophers, but it was only one of many Arabic works dealing with the anatomy of the eye, its role in vision, and the treatment of eye diseases. Islamic chemists invented and named the alembic, distinguished between alkalis and acids, accomplished the chemical analysis of many complex substances, and acknowledged the value of the experimental approach to the preparation and evaluation of various drugs. Jabir ibn Hayyan (ca. 721 to ca. 815), the most famous of the Arab alchemists, was known in medieval Europe as Geber. More than 100 alchemical texts were attributed to Geber, but many seem to have been later forgeries.

Discussions of Arab achievements in science, medicine, and philosophy once focused on a single question: Were the Arabs merely the transmitters of Greek achievements or did they make any original contributions? Such a question is inappropriate when applied to a period in which the very idea of a quest for novel, empirical scientific knowledge was virtually unknown. Physicians, philosophers, and other scholars accepted the writings of the ancients as truth, example, and authority to be analyzed, developed, and preserved. Having no attachment to the doctrine of the primacy of originality and progress, scholars saw tradition as a treasure chest, not as a burden or obstacle. Like their counterparts in the Christian West, scholars in the Islamic world had to find a means of peaceful coexistence with powerful religious leaders who took the position that knowledge could come only through the prophet Muhammad and his immediate followers. Nevertheless, despite the competing claims of traditionalists, theologians, and skeptics, Greek philosophy, science, and medicine eventually captivated Arab physicians and scholars, resulting in the formation of the modified Greco-Arabic medical system still practiced by traditional healers in many parts of the world.

After the triumph of the Abbasid caliphs in 750 and the establishment of Baghdad as the capital of the Islamic Empire, the scholars of the Age of Translations made it possible to assimilate the intellectual treasures of antiquity into Islamic culture. A major task for translators engaged in this unprecedented enterprise was to create an appropriate nomenclature for concepts never before expressed in Arabic. At the school for translation established during the reign of the Caliph Al-Ma'mun (813–833), many of Galen's medical and philosophical works were translated into Arabic. One of the most important translators was Hunayn Ibn Ishaq (809–875), a scholar whose work was a major factor in creating the great Galenic synthesis. Hunayn and his disciples prepared Arabic editions of Galenic and Hippocratic texts, the *Materia medica* of Dioscorides, as well as abstracts, commentaries, and study guides for medical students. Medieval physicians and scholars, whether Muslims, Jews, or Christians, generally shared the assumption that Galenism was a complete and perfect system. Nevertheless, scholars engaged in translating and editing the writings of the ancients inevitably colored the products of their labors with their own perceptions concerning the nature of the world.

The writings of Islamic physicians and philosophers, which were often presented as commentaries on the work of Galen, were eventually translated from Arabic into Latin and served as fundamental texts in European universities. But for many European scholars, medieval Arabic writers were significant only in terms of the role they played in preserving Greek philosophy during the European Dark Ages. Much about the history of science and medicine, however, has been revealed through attempts to study the work of Islamic writers on their own terms. For example, Rhazes (865–923; al-Razi) was one of most scientifically minded physicians of the middle ages. The indefatigable Rhazes was the author of at least 200 medical and philosophical treatises, including the *Continens*, or *Comprehensive Book of Medicine*. In describing his own methods and goals, Rhazes explained that he had always been moderate in everything except acquiring knowledge and writing books. By his own reckoning, he had written more than 20,000 pages in one year and had damaged his eyes and hands by working day and night for more than 15 years. This was the model he offered to those willing to commit themselves to the life of the mind. All biographical accounts agree that Rhazes became blind near the end of his life and that he refused treatment because he was weary of seeing the world and unwilling to undergo the ordeal of surgery. Eventually, biographical accounts adopted the spurious but irresistible story that Rhazes lost his vision when his patron al-Mansur had the physician beaten on the head with one of his own books for failing to provide proof of his alchemical theories. Rhazes' book *On Smallpox and Measles* provides valuable information about diagnosis, therapy, and concepts of diseases. Among the ancients, diseases were generally defined in terms of symptoms, such as fevers, fluxes, and skin lesions. By differentiating between smallpox and measles, Rhazes provided a paradigmatic case for thinking in terms of specific disease entities.

Islam's Prince of Physicians, known to the West as Avicenna (980–1037; Abu Ali Hysayn ibn Abdullah ibn Sina), was the first scholar to create a complete philosophical system in the Arabic language. His erudition was widely acknowledged, but his critics contended that his influence inhibited the development of medicine, because no physician was willing to challenge the reasoning of the master of philosophy, natural science, and medicine. Avicenna's great medical treatise, the *Canon*, still serves as a guide to traditional practitioners.

Western scholars long maintained that the major historical contribution of Islamic medicine was the preservation of ancient Greek wisdom and that nothing original was produced by medieval Arabic writers. Because the Arabic manuscripts thought worthy of translation were those that most closely followed the Greek originals (all others being dismissed as corruptions), the original premise—lack of originality—was confirmed. The strange story of Ibn an-Nafis (al-Qurashi, d. 1288) and his discovery of the pulmonary circulation demonstrates the unsoundness of previous assumptions about the Arabic literature. The writings of Ibn an-Nafis were essentially ignored until they came to the attention of twentieth-century historians of science.

Honored by his contemporaries as a learned physician and ingenious investigator, Ibn an-Nafis was described as a tireless writer and a man of such great piety that when he was advised to take wine as a medicine for a dangerous illness, he refused because he did not wish to meet his Creator with alcohol in his blood. Indeed, in his own writings, Ibn an-Nafis invoked religious law and his own natural charity to explain why he did not perform dissections. Just how Ibn an-Nafis was led to question Galenic physiology and discover the pulmonary circulation is unclear. Reading the works of other scholars and studying the medical illustrations typically found in medieval Arabic and Persian manuscripts were not likely to promote novel and unorthodox ideas.

After presenting a fairly conventional discussion of the structure and function of the heart, Ibn an-Nafis departed from Galenic orthodoxy and insisted that there were no visible or invisible passages between the two ventricles of the heart. Moreover, he argued that the septum between the two ventricles was thicker than other parts of the heart in order to prevent the harmful and inappropriate passage of blood or spirit between them. To account for the presence of blood in both sides of the heart, he reasoned that after the blood had been refined in the right ventricle, it was transmitted to the lung where it was rarefied and mixed with air. The finest part of this blood was then clarified and transmitted from the lung to the left ventricle. In other words, the blood can only get into the left ventricle by way of the lungs.

Perhaps some still obscure Arabic, Persian, or Hebrew manuscript contains a commentary on the curious doctrines of Ibn an-Nafis, but there is as yet no evidence that later Arab authors or European anatomists were interested in these anti-Galenic speculations. Thus, although Ibn an-Nafis did not influence later writers because what is hidden and unknown cannot be considered an influence, the fact that his concept was so boldly stated in the thirteenth century should lead us to question our assumptions about progress and originality in the history of science. Since only a small percentage of the

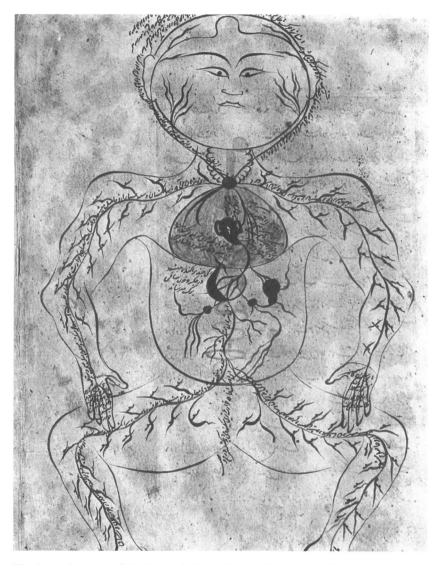

The internal organs of the human body as depicted in a manuscript representative of the golden age of Islam

pertinent manuscripts have been studied, edited, translated, and printed, the question may go unanswered for quite some time.

SUGGESTED READINGS

Africa, T. W. (1968). *Science and the State in Greece and Rome.* New York: Wiley.

Albertus Magnus (1973). *The Book of Secrets of Albertus Magnus.* Ed. by Michael R. Best and Frank H. Brightman. Oxford: Clarendon Press.

Artz, F. B. (1980). *The Mind of the Middle Ages. An Historical Survey, A.D. 200–1500.* Chicago: University of Chicago Press.

Ascher, A., Halasi-Kun, T., and Kiraly, B. K., eds. (1978). *The Mutual Effects of the Islamic and Judeo-Christian Worlds.* Brooklyn, New York: Brooklyn College Press.

Avicenna (1930). *A Treatise on the Canon of Medicine of Avicenna Incorporating a Translation of the First Book.* Trans. by O. C. Gruner. London: Luzac.

Avicenna (1974). *The Life of Ibn Sina.* A critical edition and annotated translation by William E. Gohlman. New York: State University of New York Press.

Bender, T., ed. (1988). *The University and the City. From Medieval Origins to the Present.* New York: Oxford University Press.

Bittar, E. Edward (1955). A study of Ibn Nafis. *Bulletin of the History of Medicine 29:* 352–68, 429–47.

Brain, Peter (1986). *Galen on Bloodletting.* New York: Cambridge University Press.

Brock, Arthur J. (1929). *Greek Medicine. Being Extracts Illustrative of Medical Writers from Hippocrates to Galen.* New York: Dutton.

Cadden, Joan (1984). It Takes All Kinds: Sexuality and Gender Differences in Hildegard of Bingen's "Book of Compound Medicine." *Tradition 40:* 167–69.

Celsus (1960–61). *De medicina.* 3 vols. Trans. by W. G. Spencer. Loeb Classical Library. Cambridge, MA: Harvard University Press.

Crombie, A. C. (1969). The significance of medieval discussions of scientific methods for the scientific revolution. In *Critical Problems in the History of Science.* Ed. by Marshall Clagett. Madison, WI: University of Wisconsin Press, pp. 79–102.

Dioscorides (1968). *The Greek Herbal of Dioscorides.* (Illus. by a Byzantine in 512 A.D., Englished by John Goodyear, 1655 A.D.) Ed. by Robert T. Gunther. New York: Hafner Reprint.

Eastwood, Bruce S. (1982). *The Elements of Vision: The Micro-Cosmology of Visual Theory According to Hunayn Ibn Ishaq.* Philadelphia, PA: American Philosophical Society. Transaction APS, Vol. 72, Pt. 5.

Forster, E. M. (1961). *Alexandria, A History and a Guide.* Garden City, NJ: Anchor.

French, Roger, and Greenaway, Frank, eds. (1986). *Science in the Early Roman Empire: Pliny the Elder, His Sources and His Influence.* London: Croom Helm.

Friedman, John B. (1981). *The Monstrous Races in Medieval Art and Thought*. Cambridge, MA: Harvard Univeristy Press.

Galen (1944). *On Medical Experience*. First edition of the Arabic version, with English translation and notes by R. Walzer. New York: Oxford University Press.

Galen (1962). *On Anatomical Procedures. The Later Books*. Trans. by W. L. H. Duckworth. Cambridge: Cambridge University Press.

Galen (1968). *On the Usefulness of the Parts of the Body*. Trans., intro., and commentary by Margaret Tallmadge May. 2 vols. Ithaca, NY: Cornell University Press.

Galen (1984). *Galen: On Respiration and the Arteries*. Ed. and trans. by D. J. Furley and J. S. Wilkie. Princeton, NJ: Princeton University Press.

Goichon, A. M. (1969). *The Philosophy of Avicenna and Its Influence on Medieval Europe*. Translation, notes, annotations and a preface by M. S. Khan. Delhi: Motilal Banarsidass.

Grunebaum, G. D. (1963). *Lucretius and His Influence*. New York: Cooper Square.

Gruner, O. Cameron (1967). *Commentary on the Canon of Avicenna*. Karachi: Hamdard Foundation.

Hamarneh, Sami Khalaf (1983-4). *Health Sciences in Early Islam: Collected Papers*. 2 vols. San Antonio, TX: Azhra Publications.

Khan, M. S. (1986). *Islamic Medicine*. London: Routledge & Kegan Paul.

Kibre, P. (1984). *Studies in Medieval Science: Alchemy, Astrology, Mathematics and Medicine*. London: Hambledon Press.

Kudlien, F., and Durling, R. J., eds. (1990). *Galen's Method of Healing*. New York: E. J. Brill.

Latham, R. E. (1951). *Lucretius, on the Nature of the Universe*. London: Penguin.

Le Goff, J. (1988). *The Medieval Imagination*. Trans. by A. Goldhammer. Chicago, IL: University of Chicago Press.

Levey, Martin (1973). *Early Arabic Pharmacology. An Introduction Based on Ancient and Medieval Sources*. Leiden: E. J. Brill.

Lindberg, David C., ed. (1978). *Science in the Middle Ages*. Chicago, IL: University of Chicago Press.

Lindberg, David C. (1983). *Roger Bacon's Philosophy of Nature. A Critical Edition*. New York: Clarendon.

Lloyd, G. E. R. (1973). *Greek Science After Aristotle*. New York: Norton.

Luck, G. (1985). *Arcana mundi: Magic and the Occult in the Greek and Roman Worlds*. Baltimore: Johns Hopkins University Press.

Lucretius (1929). *On the Nature of Things*. Trans. by H. A. J. Munro. London: Bell.

Makdisi, G. (1981). *The Rise of Colleges. Institutions of Learning in Islam and the West*. Edinburgh: Edinburgh University Press.

Meyerhof, M. (1984). *Studies in Medieval Arabic Medicine: Theory and Practice*. London: Variorum Reprints.

Momigliano, A. D. (1975). *Alien Wisdom: The Limits of Hellenization*. Cambridge: Cambridge Unviersity Press.

Nasr, S. H. (1968). *Science and Civilization in Islam*. Cambridge, MA: Harvard University Press.

Newman, B. (1987). *Sister of Wisdom: St. Hildegard's Theology of the Feminine*. Berkeley, CA: University of California Press.

Nutton, V., ed. (1981). *Galen: Problems and Prospects*. London: Wellcome Institute for the History of Medicine.

Parsons, E. A. (1952). *The Alexandrian Library: Glory of the Hellenic World*. New York: Elsevier.

Pliny. (1956–66). *Natural History*. 10 vols. Trans. H. Rackham, W.H.S. Jones, and D. E. Eichholz. Loeb Classical Library. Cambridge, MA: Harvard University Press.

Riddle, J. M. (1985). *Dioscorides on Pharmacy and Medicine*. Austin, TX: University of Texas Press.

Roberts, L. D., ed. (1982). *Approaches to Nature in the Middle Ages*. New York: Center for Medieval & Renaissance Studies.

Sarton, G. (1954). *Galen of Pergamon*. Lawrence, KA: Univeristy of Kansas Press.

Shah, M. H. (1966). *The General Principles of Avicenna's Canon of Medicine*. Karachi, Pakistan: Naveed Clinic.

Siegel, R. E. (1968). *Galen's System of Physiology and Medicine. An Analysis of his Doctrines and Observations on Bloodflow, Respiration, Humors and Internal Diseases*. Basel: Karger.

Siraisi, N. G. (1973). *Arts and Sciences at Padua. The Studium of Padua Before 1350*. Toronto, Canada: Pontifical Institute of Mediaeval Studies.

Siraisi, N. G. (1991). *Medieval and Early Renaissance Medicine*. Chicago, IL: University of Chicago Press.

Staden, H. von (1989). *Herophilus. The Art of Medicine in Early Alexandria*. New York: Cambridge University Press.

Temkin, O. (1973). *Galenism: Rise and Decline of a Medical Philosophy*. Ithaca, NY: Cornell University Press.

Temkin, O. (1991). *Hippocrates in a World of Pagans and Christians*. Baltimore, MD: Johns Hopkins University Press.

Ullman, M. (1978). *Islamic Medicine*. Edinburgh: Edinburgh University Press.

Walzer, R. (1962). *Greek into Arabic: Essays on Islamic Philosophy*. Cambridge, MA: Harvard University Press.

Wethered, H. N. (1937). *The Mind of the Ancient World, A Consideration of Pliny's Natural History*. New York: Longmans, Green.

Wickens, G. M., ed. (1952). *Avicenna: Scientist and Philosopher. A Millenary Symposium*. London: Luzac.

3

THE RENAISSANCE AND
THE SCIENTIFIC REVOLUTION

The Renaissance, which literally means rebirth, has been called the age of discovery, but it was also the time of the rediscovery of Greek philosophy, art, and science. Thus, although the Renaissance seems in many ways unique and distinct, it might also be thought of as the natural culmination of medieval thought as well as a transitional period between the Middle Ages and modernity. Broadly speaking, the European Renaissance comprises the period from about 1300 to 1650. The factors that led to and sustained the complex and often painful transformation of European intellectual, social, economic, and political life during this period are complex and manifold. As medieval institutions disintegrated, the unifying principles of medieval society were challenged by new ideas and facts generated by the exploration of the word, the world, the heavens, and the human body. Of course purely intellectual influences were not the only stimuli for change in Europe. The Renaissance was more than a rebirth of art and science; it was also a period of profound social and economic dislocation. Historians now see it as an era where social mobility, both upward and downward, became possible in a manner unthinkable during the Middle Ages. The sense of individualism was also quite unlike the group identification characteristic of the Middle Ages. The new individualism was marked by an obsession with wealth, social status and its symbols, a growing distance between the higher and lower social orders, a new economy whose inflationary tendencies often left peasants and artisans worse off than before, struggles between the traditional feudal aristocracy and the new monarchies. The ravages of the great pandemic of bubonic plague of 1348, which may have killed one quarter to one half of the population in some regions, played a part in the broader transformation of society. Striking down peasants, clergymen, and nobles, the plague led to labor shortages, peasant uprisings, and general confusion for church and state. Among the salient factors that encouraged cultural renewal and scientific advances can be listed the expansion of commerce, mining, the growth of towns and cities, new inventions such as gunpowder weapons and the printing press, the discovery of a sea route to India, and finally the discovery of the New World. Natural philosophers, physicians, and surgeons were confronted with plants, animals, and diseases absolutely unknown to Hippocrates, Galen, and Pliny. Triumphantly exploring and encircling the earth, Europeans left in their wake a series of unprecedented ecological and demographic catastrophes.

The exact relationship of the Renaissance to the Scientific Revolution, the fundamental change in the sciences that took place during the sixteenth and seventeenth centuries, is complex and controversial.

One witness and participant in the great events of this era, philosopher of science Francis Bacon (1561–1639), said that of all the products of human ingenuity the three most significant in terms of their "force, effect, and consequences" were the compass, gunpowder, and printing. Through their combined effects, Bacon said, these inventions had "changed the appearance and state of the whole world." At least in the West the availability of printing and firearms at the end of the Middle Ages had an effect similar to that of the introduction of iron and the alphabet at the end of the Bronze Age. Gunpowder, a mixture of saltpeter, carbon, and sulfur, which was first reported in Europe in the thirteenth century, was probably invented much earlier in China. Documents from the 1320s depict cannons that were primitive but presumably deadly. Further refinements of gunpowder weapons meant that the mounted knight and the fortified castle were no longer invincible; gunpowder weapons created injuries that could not be found in the writings of Galen and Hippocrates. Military power became concentrated in the hands of those who could control the production of gunpowder and cannons. The rise of absolute monarchies and the modern state was, therefore, given some impetus by the development of firearms.

While the printing press did not create the Renaissance, which had begun in Italy more than a century before the appearance of European printing presses in the 1460s, it probably did sustain and accelerate the advance of the movement into northern Europe. Printing and papermaking, like gunpowder, appeared first in China. According to Chinese chronicles, the eunuch Ts'ai Lun presented the art of papermaking to the Emperor of China in A.D. 105. For bringing this great invention to the attention of the Emperor, Ts'ai Lun was honored as the god of papermakers. Paper of many different kinds, colors, weights, and quality had been developed in China by the time the Arabs learned the art at Samarkand and brought it to Spain and Italy in the thirteenth century. The Chinese also invented various kinds of ink, including a special oily ink that was suitable for printing.

At least in part because of the way in which Chinese and European languages are written, the development of printing techniques had a very different impact in China and Europe. For writing systems based on an alphabet, the invention of movable print was the key to the print revolution. The Chinese system, with many thousands of separate symbols or ideographs, was better served by block printing, a very ancient technique that was used to create seals, charms, and decorations, as well as books. Movable metal type was first invented in Korea, probably in the thirteenth century. The official Korean chronicles, however, set the date as 1392 when the Department of Books was established to take responsibility for the casting of type and printing of books. In China and the nations that used the Chinese writing system, printing was valued primarily as a means of preserving and authenticating the text rather than as a way of producing large numbers of relatively inexpensive books. The earliest known document to be printed with movable type in Europe was a letter of indulgence printed by Johann Gutenberg (ca. 1398–1468) in 1454. Although Gutenberg's name is today almost synonymous with the early modern method of printing, it is likely that what he really did was improve techniques for cutting and casting metal type that others had already developed.

By 1500 printing presses had been established throughout Europe, launching a communications revolution that accelerated the trend towards literacy, the diffusion of ideas and information, and the establishment of a popular vernacular literature. Despite some grumbling about the vulgarity of printing as compared to hand copying, and the elitist fear than an excess of literacy might be subversive, books were soon being delivered to eager buyers by the wagonload. While most people remained illiterate, the print revolution initiated a transformation from a "scribal" and "image culture" to a "print culture." Reforms swept through the arena of elementary education as well as the university. Mass-produced books changed the relationship between teachers and students and made it possible for the young to study, and perhaps even learn, by reading on their own. It has been suggested that printing paved the way to the Enlightenment, the American and French revolutions, and democratic movements. On the other hand, both sense and nonsense were published and distributed with unprecedented rapidity.

Arguments continue as to the distinctness of the Renaissance, its innovative character, and its links to the past. While historians have argued that setting definitive borders around the Renaissance as a separate historical period may cause us to misread the flow and meaning of ideas and events, the idea of the Renaissance can be used to deal with very important aspects of the emergence of modern science. While pursuing novel ends and creating works of great originality, Renaissance artists, scholars, and scientists often appealed to classical authorities. The interplay between appeals to the past and aspirations to originality can be illuminated by considering the work of several key figures among the many remarkable artists, scientists, and philosophers of this era.

ART AND ANATOMY

Artists as well as anatomists are inextricably linked to the reform of anatomy during the Renaissance. Both medicine and art required accurate anatomical knowledge. Much of the distinctive character of Renaissance art was generated by its special relationship with anatomy, optics, and mathematics, as well a return to Greek ideals that encouraged the glorification and study of the human body. Dedicated to a more accurate portrayal of nature, Renaissance artists found that knowledge of the exterior of the body was insufficient. Convinced that the study of the dead would make art more true to life, they wanted to study the working of muscles and bones and the organs of the interior of the body. Unlike their Greek predecessors, Renaissance artists were able to satisfy their demands for realism by observing and even performing human dissection. Many artists attended public anatomies and executions, studied intact and flayed bodies, and carried out dissections with their own hands, but probably none equaled Leonardo da Vinci (1452–1519) in terms of his insatiable curiosity and artistic and scientific imagination.

Leonardo was the illegitimate son of a Florentine lawyer and a peasant woman. Although his father married four times, Leonardo, who was brought up in his father's house, remained an only child until he was about 20. At 14 years of age, Leonardo was apprenticed to Andrea del Verrochio (1435–1488), the foremost teacher of art in Florence. Verrochio insisted that all his pupils learn anatomy. This included the study of surface anatomy and flayed bodies, attending public anatomies and executions, and carrying out animal dissections. Within 10 years Leonardo was recognized as a distinguished artist, but instead of settling down and turning out work pleasing to the wealthy and powerful Florentine patrons of the arts, he embarked on a restless and adventurous life, perhaps first triggered by accusations of homosexuality. Retrospective studies have applauded his genius and imagination while attempts to assess his influence on the many fields he investigated generally emphasize his penchant for secrecy and his failure to complete many of the ambitious projects he initiated. His legacy is quite modest compared to the promise entailed in his formidable gifts as painter, sculptor, architect, military engineer, and inventor. He did not leave a single complete statue, machine, or book despite sketches and plans for many exciting and innovative constructions. Thousands of pages of notes, full of ingenious plans and projects, sketches and hypotheses, went unread and unpublished for centuries. Indeed, the secretive left-handed artist kept his notebooks in a kind of code. It is tempting to speculate that if Leonardo had published extensively or completed his ambitious studies, he might have revolutionized several scientific disciplines, but that which is unknown, disorganized, and incomplete cannot be considered a contribution to science. Nevertheless, even his unfinished work indicates the range of ideas and dreams possible in the context of the fifteenth century.

Observing the workings of nature in fine detail and grand design, Leonardo's mercurial mind darted from astronomy to anatomy, from music to machines. He thought about the movement of the earth, the nature of sound and light, the use of rings to determine the age of trees, and the nature of fossil shells. Studies of the superficial anatomy of the human body inexorably led to an exploration of general anatomy, comparative anatomy, and physiological experiments. Through dissection and experimentation Leonardo believed he would uncover the mechanisms that governed movement and even life itself. Leonardo began his dissections as an artist, but continued as a scientist concerned with anatomy not only as a means to an end, but as a study worth doing as an end itself. He attended public anatomies whenever possible and received permission to study cadavers at a hospital in Florence. He may have dissected as many as 30 human bodies including a 7-month fetus and a very elderly man.

Not all of Leonardo's methods are known, but he seems to have been the first anatomist to make wax casts of the ventricles of the brain by injecting hot wax into these hollow areas. When the wax solidified it was possible to do careful dissections of the delicate tissues. Leonardo used serial sections and developed other techniques for studies of soft tissues such as the eye. To facilitate dissection, he placed the eye in the white of an egg and heated it until the albumen coagulated. He carried out many studies of the muscles and even constructed models to study the mechanism of action of the muscles and heart valves. Various vivisections gave him insight into the heartbeat. For example, he drilled through the thoracic wall of a pig and, keeping the incision open with pins, observed the motion of the heart. Although he realized that the heart was actually a very powerful muscle, he generally accepted Galen's views on the movement and distribution of the blood, including the imaginary pores in the septum. Leonardo's dissections of insects, fish, frogs, horses, dogs, cats, birds, and so forth represent the first significant efforts since Aristotle and Galen to seriously exploit the advantages of comparative anatomy. Like so many of his projects, Leonardo's great book on human anatomy was left unfinished. When he died, his manuscripts were scattered among various libraries, and some were probably lost.

True anatomical illustrations appeared in the sixteenth century. Of course older books had included interesting pictures, but these were generally formalized depictions of autopsies, wound-men, and astrological figures. Illustrations in anatomy texts were generally more distracting than informative; while cheerful-looking people held up flaps of skin to display internal organs, the text rambled on about other matters.

Renaissance artists and anatomists were not the first Europeans to revive the Alexandrian practice of human dissection, since postmortems had been a component of medical education in medieval Italian universities and autopsies were occasionally conducted to investigate suspicious deaths and epidemic diseases. Nevertheless, the opportunity to conduct or observe human dissections remained quite rare. When public anatomies were performed, the stylized ceremony was hardly likely to produce any new insights or inspire the typical student to challenge established authorities or engage in original research. The professor, trained in the scholastic pattern, would not have conducted the dissection himself. A barber-surgeon or technician did the actual autopsy while the professor explicated passages from Galen. If the reader and technician were well synchronized, the students would see the organs as the professor described them, but even this was unlikely. Presumably, Galen himself would have objected to this procedure; he had found that he could not entrust even the task of skinning his Barbary apes to an assistant without losing valuable information. The tools used by surgeons and anatomists were quite primitive; the technician at an anatomy opened the abdomen and chest cavity with only a knife. After a brief display of the mangled internal organs, the technician was supposed to demonstrate the muscles, nerves, and blood vessels. But this was generally too difficult for the technician and too boring for the audience. In any case, most members of the audience preferred scholastic disputations between the Galenists of the medical faculty and the Aristotelian professors of philosophy to the autopsy itself. Medical students knew that examinations and dissertation were based on authoritative texts and attended dissections primarily as a means of confirming accepted ideas and preparing for examinations. Like students in a modern laboratory course performing a typical "cookbook" experiment, medieval and Renaissance students were engaged in acquiring standard techniques rather than a search for novel observations.

Even if Renaissance anatomists dissected only to supplement their studies of the newly restored Galenic anatomical texts, their efforts were confounded by many difficulties. Probably the greatest problem was the shortage of bodies, but anatomists also lacked reliable guides to direct their efforts and tools with which to exploit the available materials. Furthermore, any observations they might make would be most difficult to describe to other anatomists because there was no standard nomenclature for the parts of the body. Descriptions were issued in a jumble of Greek, Latin, Arabic, and vernacular terms. The same structure might be referred to by different names or the same term might be applied to different structures. By 1500, Galen's *On the Use of the Parts* was available in editions suitable for use by students. Indeed, assisting Johannes Guinter (1505–1574) in preparing his text *Anatomical Institutions According to the Opinion of Galen for Students of Medicine* was an important step in the scientific awakening of Andreas Vesalius (1514–1564), the reformer of anatomy.

ANDREAS VESALIUS ON THE FABRIC OF THE HUMAN BODY

During the year 1543 two books appeared that challenged long-held convictions about the nature of the heavens and the human body. Just as *De revolutionibus* by Nicolaus Copernicus (1472–1543) transformed ideas about the movements of the earth and the heavens, *On the Fabric of the Human Body* by Andreas Vesalius ultimately revolutionized Western conventions concerning the nature of the human body. Vesalius regarded his massive treatise as the first advance in anatomy since the time of Galen. Indeed, when the young Vesalius began his study of human anatomy, the authority of the ancient anatomist had grown so weighty that anatomists tended to explain away discrepancies between his descriptions and their own observations in terms of specific abnormalities or general changes in the human body that had occurred since the time of Galen. For example, Jacobus Sylvius (1478–1555) defended Galen's description of the human sternum as made up of seven bones on the grounds that it was possible that in a more heroic age the chest might have contained more bones than later degenerate specimens.

According to a horoscope cast by Girolamo Cardano, Vesalius was born at 5:45 A.M. on December 31, 1514, in Brussels, Belgium. His father was imperial pharmacist to Charles V and often accompanied the emperor on his travels. When quite young, Vesalius began to teach himself anatomy by dissecting mice, moles, and other small animals. Although he studied at both Paris and Louvain, institutions noted for their extreme conservatism, his innate curiosity overcame the potentially stifling effect of education. As a medical student at the University of Paris, Vesalius served as assistant to Jacobus Sylvius, a dominant influence in Parisian anatomy and a staunch Galenist. Sylvius was preparing a great anatomical textbook that was actually a commentary on Galen. While carrying out dissections directed by Sylvius, Vesalius became increasingly critical of his mentor and other medical professors who merely studied Galen and lectured about things they had never investigated for themselves. Later Vesalius described Sylvius as a bad-tempered man more at home wielding his knife at the banquet table than in the dissecting room. Pursuing his own interests, Vesalius became so skillful that he was called upon to perform public dissections. The outbreak of war forced him to leave Paris without a degree and return to Louvain in 1536.

After a brief stay in Louvain, Vesalius enrolled in the medical school of the University of Padua. In December 1537 he received his degree and an appointment as lecturer-demonstrator in anatomy and surgery. Because Padua was a relatively enlightened institution, Vesalius was able to institute many reforms in the teaching of anatomy. His lecture-demonstrations attracted hundreds of observers. Never one to spare the feelings of his more reactionary colleagues, Vesalius told his students that they could learn more at a butcher shop than from arrogant professors who knew nothing more than the words of Galen, a man who had never had the opportunity to dissect human beings.

Obtaining human bodies was a major problem for all anatomists and medical schools. Since medical students were required to observe human anatomies, teachers sometimes encouraged unorthodox means of obtaining the necessary material. Medical students often obtained at least part of their education in graveyards, snatching bones out of the teeth of savage dogs. In several autobiographical passages Vesalius described the risks he had taken to obtain bodies. On one occasion he stole the bones of a criminal left on the gallows, sneaking into town with some of the bones under his coat over a period of several days and hiding the skeleton under his bed.

The anatomies conducted by Vesalius occupied him from morning to night for 3 weeks at a time. To minimize the problem of putrefaction, anatomies were done in the winter. A judge in Padua's criminal court became so interested in Vesalius's work that he thoughtfully set the time of execution

The title page of *On the Fabric of the Human Body* shows Andreas Vesalius, surrounded by excited observers, performing a dissection in the crowded anatomical theater at Padua

of convicted felons to suit the anatomist's schedule. Several bodies were dissected at the same time so that different parts of the body could be studied and their relationship examined. Large diagrams were prepared for the guidance of students. The anatomies began with the skeleton and then proceeded to the muscles, blood vessels, and nerves. Another cadaver would be used for the demonstration of the organs of the abdomen, chest, and brain. To improve the technical aspects of dissection, Vesalius introduced many new tools, some of his own design and some based on those used by various artisans he had consulted. Human dissection as performed by Vesalius in the crowded anatomical theater of Padua was a dramatic as well as an enlightening event.

In 1540, as a mark of his independence from Galen, Vesalius conducted a dramatic demonstration in which he assembled the bones of an ape and those of a man in order to show that Galen had made hundreds of errors with respect to human anatomy. Finally, in 1543 Vesalius published his great anatomical treatise *De humani corporis fabrica* (*The Fabric of the Human Body*) and a shorter text, known as the *Epitome*, which was intended to serve students. The *Epitome* even contained pictures rather like paper dolls so that the reader could cut out illustrations of the internal organs and reassemble them. About 250 blocks had been painstakingly prepared for incorporation into the text so that words and illustrations would complement and clarify each other. Both texts were extensively plagiarized and disseminated in editions so carelessly prepared that they caused Vesalius considerable pain and embarrassment. Of course the *Fabrica* infuriated Galen's still loyal followers, who denounced the text as "filth and sewage" and renamed its author "Vesanus" (madman). Conjuring up images of Vesalian anatomists as the "Lutherans of Physic," outraged Galenists warned that the heresy of such medical innovators was as dangerous as Martin Luther's (1483–1546) effect on religion. Ironically, many critics attacked the *Fabrica* on the grounds that the illustrations might be beautiful as works of art, but were actually false and misleading and would dissuade students from participating in dissections. Of course Vesalius himself was the supreme champion of anatomical research and throughout his text emphasized the importance of performing dissections. Careful directions are included for even the most basic aspects of preparing a corpse for dissection.

Opposition to his work was so intense that Vesalius decided to abandon research and the academic world. Following in the footsteps of his father, he became court physician to Emperor Charles V, Holy Roman emperor and king of Spain, to whom he had dedicated the *Fabrica*. Once in service to

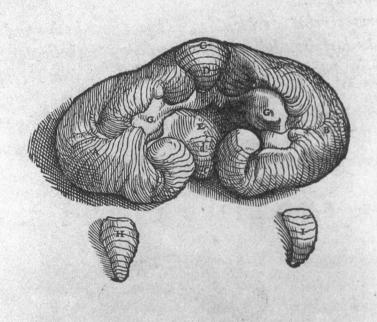

Interior view of the cerebellum as depicted in *De humani corporis fabrica*, 1543

the emperor, Vesalius had little opportunity for research or writing. The patronage of kings, popes, and wealthy noblemen was as important to Renaissance researchers as the patronage of government agencies and private foundations is to modern scientists, but imperial service could be almost as unpleasant as the stormy academic world. Royal patrons could be difficult and demanding patients. Charles V was a particular challenge to the skill of his physicians, with his bad combination of gout, asthma, gargantuan appetites, and predilection for quack remedies. When the University of Pisa offered Vesalius the Chair of Anatomy, Charles refused to release him. In 1556 Charles abdicated in order to retire to a monastery. Vesalius was transferred to the service of Philip II, the son of Charles V and successor to the throne. During this period Vesalius was sent to several other royal courts. When Henry II of France was injured while jousting, Vesalius and the famous French barber-surgeon Ambroise Paré (1517–1590) were among the physicians assembled. After experiments on the heads of condemned criminals, Paré and Vesalius agreed that the wound was fatal. Unhappy with his role in service to Philip II, Vesalius hoped to return to the academic world and embarked on a pilgrimage to the Holy Land as a means of escaping from his employment. The fact of his death on October 15, 1564 during the return voyage from Jerusalem is known, but the exact cause and place of death are uncertain.

Given the power of the Galenic tradition, it is remarkable that Vesalius, a man steeped in conservative academic scholarship, confronted and rejected the authority of Galen that generations of scholars and physicians had accepted virtually without question. If we abandon modern assumptions about the nature of progress and the evolution of science, it is proper to ask what factors forced Vesalius to demand that henceforth anatomists study only the "completely trustworthy books of man." On several occasions Vesalius attributed his disillusionment with Galen to his "discovery" that Galen had never dissected the human body, but Galen had confessed to this deficiency. Moreover, Vesalius was a scholar who knew the works of Galen better than most of his critics; frequently he pointed out that when Galen made a new discovery he had been willing to call attention to previous errors. It is more likely that practical problems concerning venesection forced Vesalius to question Galenic dogma. As the texts of Hippocrates and Galen became available in their pristine forms, medical humanists and physicians engaged in violent controversies about the proper means of performing phlebotomies. When Hippocratic and Galenic texts contradicted each other, the blame for obscurity and confusion could no longer be attributed solely to corrupt medieval influences. How then should physicians determine the proper site for venesection, the amount of blood that should be taken, and how often the procedure should be repeated? When Vesalius began to ask himself whether such questions could be answered by means of facts established by anatomical investigations, the learned debates among his colleagues seemed to be nothing more than "horsefeathers and trifles."

While the *Fabrica* retained some errors reflecting the survival of Galenic traditions, Vesalius came close to this goal of describing the human body as it really is, without deference to ancient authorities. Vesalius opened the *Fabrica* with a preface that amounts to a defense of science and a warning about the dangers that result from obstacles to the arts and sciences. Along with a liberal dose of flattery for Charles V and the University of Padua, Vesalius praised the Senate of Venice for their liberality towards learning. Some of his arguments could well grace a modern research grant proposal. The first book of the *Fabrica* is devoted to the skeleton supporting the whole body and the second to the muscles. The third book describes the vascular system. The fourth book was devoted to the description of the nervous system. The abdominal viscera and generative organs were described in the fifth book; the sixth book described the thoracic viscera. In book seven, Vesalius described the brain, pituitary gland, and the eye. A concluding chapter on animal vivisection provided students with exercises to train their hands and senses for surgery and the closure of wounds.

Still, it must be admitted that in contrast to his success in anatomy, Vesalius did not establish any physiological concepts or methods that were fundamentally different from those of Aristotle and Galen. Although Vesalius gave an exhaustive description of the structure of the heart, arteries, and veins, he generally followed Galenic concepts in explaining the function of these organs. He searched diligently for the Galenic pores in the septum of the heart but, unable to find them, concluded that they either did not exist or were invisible. In the second edition of the *Fabrica* he explicitly denied that any such channels could be found to completely penetrate the septum, but rather than using this as a way to question Galenic dogma, he merely called the reader's attention to this mystery.

Despite the rapid dissemination of the *Fabrica* and a plethora of unauthorized derivative versions, many anatomists continued to see the human body largely in terms of Galenic structures. Partly, this was because access to human cadavers remained limited and anatomists often used the same animals Galen had dissected in order to clarify their studies of particular organs. Even Vesalius admitted that when he began performing public anatomies, he always kept the head of an ox or lamb handy to demonstrate Galen's *rete mirabile*, a network of vessels which is found at the base of the

ιotum perdat, illumῷ denuò occlufo uentriculo
ἀtionem quam me defcripturum paulo antè pol
ι: quanquam uocis occafione fuem accipere ma-
utcunῷ illum afficias, fubinde neque latrat, neῷ
m interdum expendere nequis. Primùm igitur
:dem, & liberum corporis truncum porrigat, af-
·madmodum hîc modò interiecta tabella propo-

ιgata, ac dein cuipiam afferis anulo, aut foramini,
:et collum exporrectum, caputῷ immotum fit, &
                              Kk   interim

A depiction of a pig laid out on a dissecting table

brain in cattle but not in humans. Galenic theory required the presence of this "wonderful net" because it allegedly produced animal spirits. After he freed himself from Galen's influence, Vesalius openly declared this network was not present in humans and ridiculed his own youthful credulity. Nevertheless, although Vesalius and his contemporaries were clearly capable of accepting originality in terms of observed and isolated facts, they found it more difficult to set themselves free of Aristotelian and Galenic theories. It was possible to find a semi-Galenic alternative locus for the generation of the animal spirits by assuming that Galen actually meant to say that only part of the process that generated the animal spirits occurred in the *rete mirabile*. If the final modification involved the brain and its ventricles, the Galenic function of the *rete mirabile* could be assigned to the cerebral arteries.

Long after anatomical research was recognized as the cornerstone of Western medical science, anatomists were often forced into methods of obtaining bodies that were just as dangerous as those employed by the young Vesalius. Even if anatomists were fortunate enough to be awarded the bodies of criminals whose crimes were considered vicious enough to merit a sentence of "death and dissection," they faced the prospect of succumbing to the "putrid miasma" of the dissecting rooms; that is, the smallest cut could become infected and cause a fatal blood poisoning. Medical schools did not officially assist or encourage body snatching, but some were known to provide free tickets to the dissecting rooms for enterprising students. Rumors of bizarre experiments allegedly carried out by medical scientists reinforced the widespread fear of those who carried out dissections and perhaps worse atrocities on human bodies.

THE RENAISSANCE, NATURAL MAGIC, AND ALCHEMY

Even though the Renaissance has been characterized as the age of the rebirth of art and science, it was also an era in which superstition, mysticism, and the occult sciences flourished. While the ways in which science, superstition, mysticism, and magic were linked underwent profound changes, the persistent appeal of the occult sciences is reflected in the fact that the name of the sixteenth-century astrologer and physician Nostradamus (Michel de Notredame, 1503–1566) is likely to be better known today than those of his remarkable contemporaries Copernicus, Vesalius, or Paracelsus. Many varieties of mysticism, number magic, astrology, as well as neo-Platonism and Hermeticism challenged the authority of Aristotelian and Galenic doctrines. Natural magic and the occult sciences also affected medical philosophy and the nascent life sciences. Not all Renaissance natural philosophers and physicians accepted the primacy of anatomical studies. Indeed, some naturalists totally rejected anatomical research as irrelevant to understanding health and disease, life and death, and immersed themselves in the ancient, obscure, and secretive philosophies and techniques of alchemy and natural magic.

While chemistry as we know it today may be regarded as a young science, founded upon the atomic theory of John Dalton (1766–1844), chemical technologies have roots that go back to the earliest civilizations in their work with glass, metals, cements, perfumes, medicines, and alcoholic beverages. Ancient descriptions of chemical manipulations suggest that elaborate rituals were used to standardize processes which were but vaguely understood. Metallurgy was one of the most important roots of chemistry because it included the smelting of ores, interpreted as the transformation of ore to metal. Such technological achievements as making bronze, brass, and iron and purifying gold and silver were of obvious economic importance, as well as being a source of great power to the societies that possessed them.

The alchemists, who absorbed and transformed protochemcial techniques and philosophy, are often dismissed as mystics, quacks, or fools. Alchemists have been praised as pioneers of modern chemistry and damned as charlatans. Nevertheless, alchemists have a special place in the history of biology, for they were the first to attempt chemical analyses of organic materials. They did not do so as proto-biochemists, but because they believed that the whole universe was alive and did not limit their investigations to the inorganic world. While their experiments may have been based on complex, obscure, and mystical religious and philosophical theories, their techniques were often quite sophisticated improvements of ancient methods.

Alchemy did not originate as an *empirical* science, but as a *sacred* science concerned with manipulating the passions, marriages, growth, death, and transmutations of matter. Alchemical philosophy was shared by many ancient cultures, including Babylonia, China, India, Greece, and the Islamic world. According to the most fundamental alchemical traditions, the seeds of noble metals are contained in the base metals; if the right conditions and methods were applied, these seeds could be encouraged to grow just as the child grew in the womb. Alchemy was more than an attempt to transform base metals into gold; it encompassed a broad range of doctrines and practices, including the search for special elixirs of health, longevity, and immortality. Alchemists believed that beyond the known gross qualities of common crude materials there must be other subtle and hidden qualities. In their search for these occult properties, the alchemists sometimes discovered or produced novel substances. For example, in trying to distill off the essence of wine, they discovered *aqua ardens* (strong liquors), which were transformed into medicinal liquors and cordials.

The search for the elixirs of life was not exclusively confined to Chinese alchemy. The European alchemist most associated with the quest for better living through chemistry was Philippus Aureolus Theophrastus Bombastus von Hohenheim (1493–1541). Fortunately, he is generally remembered as Paracelsus, presumably indicating one who was greater than Celsus. The Paracelsians, his seventeenth-century followers, attributed the birth of a revolutionary new chemical or "spagyric" system of therapeutics to him. (Spagyric is a term derived from the Greek words meaning "to separate" and "to assemble.") Little is known of his early life or how he came to be called Paracelsus, but legends and speculation abound. His father, Wilhelm von Hohenheim, was a physician at Einsiedeln, a small town near Zurich; his mother was a nurse at the monastery hospital. She seems to have suffered from some form of insanity and eventually committed suicide.

After leaving home at the age of 14, and visiting various universities, Paracelsus rejected the structured and scholastic atmosphere of academia and turned to the study of alchemy. He learned the art, philosophy, and legends of alchemy and metallurgy from the famous alchemist and abbot Johannes Trithemius (1462–1516), performed an apprenticeship at the mines of the Tyrol, and traveled throughout Germany, Spain, France, and possibly as far as Moscow and Alexandria in search of the secrets of alchemists, astrologers, magicians, and peasants. His journeys were essential, he explained, because Nature was the divine book of Creation that the seeker of knowledge must tread with his feet. Although there is no evidence that he ever received a formal degree, he awarded himself the title of "Double Doctor," presumably for honors conferred on him by God and Nature, and he apparently enjoyed a successful, if always precarious, private practice. He was known for dramatic cures based on simple, inexpensive medicines. When celebrating his own methods, Paracelsus ridiculed pharmacists and physicians as nothing more than officially approved asses. Worse yet, he took away their business by undercutting their prices and curing their disillusioned patients.

In 1526 Paracelsus returned to Switzerland and, through the intervention of influential patients, was appointed Professor of Medicine and City Physician of Basel. At the University, Paracelsus demonstrated his talent for staging what would now be called media events when he initiated his lectureship by burning the works of Avicenna and Galen to show his contempt for ancient dogma and the curriculum favored by his outraged colleagues. Wearing the alchemist's leather apron rather than academic robes, he also broke

Paracelsus surrounded by the symbols, texts, instruments, and apparatus appropriate to alchemy and astrology

with tradition by lecturing in the vernacular rather than in Latin. Not surprisingly, he was forced to leave Basel only 2 years later. After many similar episodes, he was invited to settle in Salzburg by the Archbishop Duke Ernest of Bavaria, a man deeply interested in alchemy. Shortly afterwards, when only 48 years of age, Paracelsus died suddenly in a mysterious but certainly unnatural fashion. His friends said that he was the victim of a deliberate hostile attack, but his enemies said he met with an accident while drunk. Despite his turbulent life, Paracelsus was a prolific writer. Very few of his books were published during his lifetime, but more than 300 works (probably including many forgeries) were later attributed to him.

Paracelsus has been called everything from genius to quack, from founder of medicinal chemistry to a bizarre footnote to the history of chemistry. Presumably, he was a combination of magician and scientist, healer and fraud. He claimed that with natural magic and alchemy, human beings would one day see beyond the mountains, hear across the oceans, divine the future, make gold, cure all diseases, and even create life. After his death myths concerning his activities and accomplishments grew even more extravagant. Conrad Gesner called the followers of Paracelsus druids and magicians.

For the Paracelsians, *natural magic* was the key to understanding nature and her laws. While natural magic was essentially an occult science, allied to alchemy and astrology, it was also an experimental art with practical goals. Alchemists contended that life was a chemical process and thus that the problems of physiology and pathology could be explained in chemical terms; disease must be the result of a defect in body chemistry. Such concepts might be dismissed as too obvious for comment today, but they were considered strange and obscure in the sixteenth and seventeenth centuries. Attempting to discover the most exotic secrets of nature while creating alternatives for the powerful Aristotelian-Galenic synthesis, the Paracelsians immersed themselves in Christian neo-Platonic and Hermetic philosophy. Alchemists might challenge the old Greek elements and humors, but the three chemical principles of Paracelsus—*salt*, *sulfur*, and *mercury*—were equally ambiguous. According to Paracelsus salt was the principle of solidity, sulfur the principle of inflammability and malleability, whereas mercury was the principle of fluidity, density, and that which was metallic in nature.

All physiological processes, according to Paracelsian logic, were fundamentally chemical transformations governed by the *archeus* or internal alchemist of the body. Disease was the result of some malfunction of the archeus and death was due to its final loss. Dispensing with traditional humoral explanations, Paracelsus argued that disease processes should be analyzed in accordance with the methods used to analyze chemicals. While often obscure and inconsistent, some Paracelsian categories were peculiarly appropriate for metabolic diseases, dietary disorders, and certain occupational diseases. For example, gout and gouty arthritis were classified as "tartaric diseases" based on an analogy with the materials that precipitate out of wine casks. Because gout involves the abnormal local deposition of a metabolic product called uric acid, it is indeed a condition where body chemistry has gone wrong.

In terms of medical philosophy, alchemy could provide new chemical analogies for physiological functions; in therapeutics, alchemists could use the art to prepare new drugs. In contrast to the prevailing preference for polypharmacy, which called for the use of complex mixtures of substances, Paracelsus suggested that specific remedies should be designed for specific diseases in accordance with their specific causes. Orthodox physicians and pharmacists preferred to prescribe complex and expensive traditional remedies which were generally composed of herbs but might also contain exotic ingredients such as viper's flesh, mummy powder, and unicorn horn.

In purifying drugs and poisons, alchemists gained an appreciation of how tiny quantities of chemicals could exert great effects on the body. Their search for valuable remedies was, however, often guided by astrological concepts such as the correspondence between the seven planets, the seven metals, and the parts of the body. The doctrine of signatures also provided clues in the search for new drugs. According to this concept natural objects were stamped by the Creator with signs, which could be interpreted by a true physician, that revealed their usefulness to human beings. Paracelsians believed that violent diseases required violent remedies, especially those based on metals and their salts. Fortunately, many of the more toxic minerals were used in preparations that caused the patient to vomit so quickly that there was no time to absorb a lethal dose. Two very important drugs associated with Paracelsus were laudanum and ethyl ether, or "sweet vitriol," which might have been produced from alcohol and sulfuric acid. Although Paracelsus recognized the soporific properties of ether through tests on humans and chickens, it was not used as a surgical anesthetic agent until the 1840s. Such examples can make the work of Paracelsus seem remarkably prescient, but, just as the products of his chemical manipulations were likely to be far from pure, safe, and effective, his writings were far from clear; it is easy to read too much into the Paracelsian literature and confuse obscurity with profundity. Indeed, scholars still argue about whether it is possible to find a structured system of thought in the writings of Paracelsus and the Paracelsians. Still, the alchemists called attention to the need to emphasize the study of function rather than form alone. Attempts to deal with the chemistry of life, however, were less successful than what might be called studies of animated anatomy.

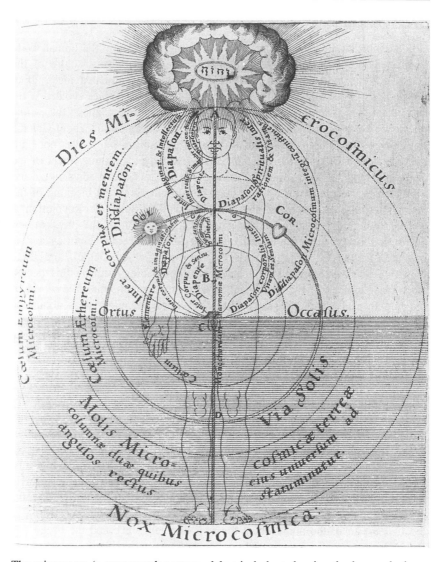

The microcosm (a seventeenth-century alchemical chart showing the human body as world soul)

ANATOMY AND PHYSIOLOGY

The attempts of Renaissance anatomists and artists to study the human body without deference to Galen revolutionized ideas about the structure of the body, but the theories of Aristotle and Galen still dominated ideas about physiological functions and the nature of life. Even Vesalius avoided a direct attack on Galenic physiology and was rather vague about the whole question of the distribution of the blood and spirits. Questions about Galenic physiology were raised by sixteenth-century anatomists, but it was not until the seventeenth century that William Harvey's study of the movement of the heart and blood produced a significant challenge to Galenic physiology and ultimately revolutionized the way scientists thought about the meaning of the heartbeat, pulse, and movement of the blood. The Scientific Revolution is usually thought of in terms of the great transformation of the physical sciences that occurred during the sixteenth and seventeenth centuries and is primarily associated with Nicolaus Copernicus, Galileo Galilei, and Isaac Newton. But developments in anatomy and physiology during this period were changing ideas, about the nature of the microcosm, the little world of the human body, as much as astronomers were changing ideas about the nature of the macrocosm. The challenge to ancient ideas about the human body was as threatening to ancient systems of thought as that engendered by removing the earth from its place at the center of the universe.

Perhaps it is possible to imagine a people without an interest in astronomy, but it is impossible to imagine any uninterested in their own life's blood. Ancient and even modern emotional responses to blood involve images far removed from the actual physiological role of this remarkable liquid tissue. Blood has been used in religious rituals, charms, amulets, medicines, and, of course, in horror films. Many ancient cultures shared the belief that traits such as youth, courage, virility, and character could be transmitted through blood. Scientists and philosophers also treated the blood as a most powerful fluid. Hippocrates and Aristotle shared the belief that the movement of the blood was fundamental to life. For Aristotle, the heart was the first organ of the body, being the seat of intelligence and the organ that added the innate heat to the blood. Pulsations in the blood system were thought to be the result of a kind of boiling up that occurred when the blood in the heart met the pneuma drawn in by respiration. Galen demonstrated that both arteries and veins carry blood and attributed the bright red color of blood in the arteries and left ventricle to the presence of pneuma or vital spirits. Vesalius was unable to identify the pores in the septum that Galen's scheme required to get blood from the right to the left side of the heart, but for the most part Vesalius was rather vague about the whole question of the

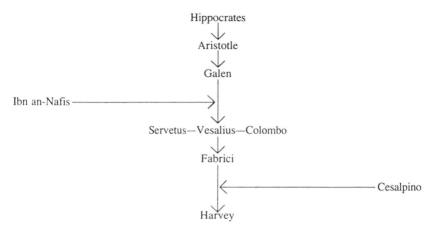

Discovery of the minor and major circulation

distribution of the blood and spirits. Other sixteenth-century anatomists, however, were able to confront the issue of the invisible septal pores and suggest an alternative pathway from the right heart to the left heart, which is known as the pulmonary or minor circulation.

Michael Servetus (ca. 1511–1553), the first European to do this, was a man who spent his whole life struggling against dogmatism, conformity, and orthodox religion; he died a martyr to his principles. Servetus so antagonized both Protestant and Catholic authorities that he was burned alive by order of a Protestant court and in effigy by the Catholic Inquisition. In challenging the theological orthodoxies of his day, Servetus also challenged Galenic assumptions about the movement of the blood. Born in Tudela, Spain, Servetus studied law at Toulouse and served for a time as secretary to Holy Roman Emperor Charles V. Traveling almost constantly, he supported himself by writing, editing, and translating while he studied at various universities. His youthful work *On the Errors of the Trinity* (1531) convinced both Protestant and Catholic censors that Servetus was a heretic who did not deserve to live. To avoid his enemies, Servetus established a new identity as "Michael Villanovanus." By 1536 he was studying geography, astronomy, and medicine at the University of Paris and, like Vesalius, serving as assistant to Johann Guinter of Andernach. At the time both Servetus and Vesalius were anatomical novices and had barely begun to question Galenic dogma. Eventually, faced with the threat of excommunication for indulging in the forbidden area of judicial astrology (essentially fortunetelling), Servetus was forced to leave Paris without a medical degree and resume his underground life.

To support himself, Servetus practiced medicine and wrote, edited, and translated various books including one about pharmaceutical products called *On Syrups* and a new edition of the *Geographia* of Ptolemy of Alexandria. Even when simply editing classic texts, Servetus could not resist adding his own opinions. While describing France, Servetus described the miraculous ceremony of the Royal Touch, in which the king of France was supposed to cure victims of scrofula. "I myself have seen the king touch many attacked by this ailment," Servetus wrote, "but I have never seen any cured."

While living in Vienna, Servetus took up a peculiar correspondence with John Calvin (1509–1564), the French Protestant reformer in Geneva. Ostensibly asking for information, Servetus sent his annotated copy of Calvin's *Institutiones* to Geneva along with an advance copy of his own radical treatise, *On the Restoration of Christianity* (1553). Calvin responded by denouncing the author of the *Restoration* to the Catholic Inquisition and sending pages torn from the book as evidence. Servetus was arrested, but managed to escape from prison before he was tried, convicted, and condemned to be burned in a "slow fire until reduced to ashes" along with his book. Within 4 months of his escape from death at the hands of the Catholic Inquisition, Servetus was captured in Geneva. Once again he was condemned to be burned at the stake along with his book. Attempts to mitigate the sentence to death by the sword or immolation "with mercy" (meaning strangulation before the burning began) were unsuccessful. A monument to the unfortunate heretic was erected by the Calvinist congregation of Geneva in 1903.

In the *Restoration*, a 700-page theological treatise exploring the ramifications of his view of the relationship of God to the world and to human life, Servetus included a brief account of the pulmonary circulation. Anatomy and physiology fit into this text because, although his primary inspiration was religious rather than scientific, Servetus drew on ideas from law, politics, astronomy, physics, and medicine to substantiate his theological arguments. To understand the relationship between God and humanity and to know the Holy Spirit, Servetus argued, one must understand the spirit of man. This required exact knowledge of the human body—its structures, functions, and its moving and nonmaterial components, especially the blood and the spirit, for as stated in Leviticus "the life of the flesh is in the blood." In his discussions of human anatomy, Servetus mixed evidence from human dissection, passages from Galen, and his own theological concepts. Direct observation suggested that the septum did not have the pores required for Galen's

Michael Servetus (in the background the ill-fated theologian is shown being burned at the stake)

scheme. If these pores did not exist, then not much blood could pass from the right to the left heart. Galen's doctrines were also called into question by simple considerations about the structure of the heart and the attached blood vessels; in particular, anatomists should ask why the pulmonary artery was so large and why so much blood was forcefully expelled from the heart to the lungs. Certainly more blood was sent to the lungs than the quantity needed for their own nourishment. The blood must, therefore, go from the right side of the heart to the lungs for aeration as well as for the expulsion of sooty vapors; most of the blood then returned to the left side of the heart via the pulmonary vein. According to Galen, aeration was the function of the left ventricle, but it was during passage through the lungs that the color of the blood changed. Galen was led into error, Servetus explained, because he misunderstood the function of the lungs. Galen did not recognize the continuous flow of blood through the lungs, but assumed that if any blood returned to the heart from the lungs it was due to leakage.

Both biblical arguments and physiological observations, Servetus contended, proved that the soul is to be found in the tides of the blood rather than confined to the heart, liver, or brain. Satisfied that he had reconciled physiology and theology as to the unity of the spirit, Servetus did not consider the systemic or major circulation of the blood, nor did he absolutely preclude the possibility that some blood might seep through the septum. Ibn an-Nafis had been even more forceful in rejecting the existence of passages through the septum of the heart, but there is no compelling evidence that Servetus was aware of the work of the thirteenth-century physician. Having established his main point, Servetus generally accepted other aspects of the Galenic system.

Since only three copies of the *Restoration* seem to have survived the flames, it is unlikely that Servetus's discovery of the pulmonary circulation was better known to anatomists and scholars than that of Ibn an-Nafis. Even if anatomists were familiar with the work of Servetus, they were unlikely to admit any association with its ill-fated author. Still, it must be admitted that every age has its own invisible networks for the dissemination of ideas, especially those considered dangerous and subversive. If the story of Servetus was transmitted in whispers among his contemporaries, it might have said more about the dangers that faced incautious and unorthodox thinkers than it did about the errors of Galenic physiology. In any case, a more accessible account of the pulmonary circulation was also published in the 1550s by the well-known anatomist Realdo Colombo (Realdus Columbus; ca. 1510–1559).

Colombo, son of an apothecary, served as apprentice to a surgeon for 7 years before studying medicine and anatomy at the University of Padua.

According to the university records for 1538 Colombo was an outstanding student of surgery. When Vesalius left Padua to supervise the publication of the *Fabrica*, Realdo Colombo was appointed as a temporary substitute. Eventually Colombo succeeded Vesalius as professor of surgery and anatomy and remained at the University of Padua until 1545 when he was offered a professorship at Pisa. Three years later, at the invitation of Pope Pius IV, he accepted the chair of anatomy at the University at Rome.

Colombo and Vesalius seem to have had a cordial collegial relationship before Colombo assumed his professorship and distinguished himself as one of the most vociferous critics of the *Fabrica*. In response, Vesalius called his rival a scoundrel, an ignoramus, and an "uncultivated smattering" whose general education was as deficient as his mastery of Latin. Later the anatomist Gabriello Fallopio (1523–1562) added to the attack on Colombo by accusing him of plagiarizing the discoveries of other anatomists.

From the time of his first major public anatomical lectures to his death in 1559, Colombo boasted about his skills in dissection and vivisection and his plans for the publication of a major illustrated anatomical treatise that would detail the errors made by Vesalius and supersede the *Fabrica*. This ambitious enterprise, which was to involve Michelangelo as the illustrator, never materialized, and Colombo actually left only one medical treatise, *De re anatomica*, which was published by his heirs in 1559. Lacking illustrations, this posthumous work has been criticized as an obvious and inferior imitation of the *Fabrica*. While Colombo apparently exaggerated the number of dissections he had performed and attempted to take credit for the discoveries of others, his description of the pulmonary circulation was the first clear-cut statement about the system made by a prominent European anatomist and thus was likely to reach a wide audience. It is possible that Colombo had been demonstrating the pulmonary circulation as early as 1545.

Unlike Servetus and Ibn an-Nafis, Colombo made it clear that his ideas on the pulmonary circulation were based on clinical observations, dissections, and experiments on animals. Colombo described the anatomy of the heart in detail and corrected many ancient errors. Praising his own resourcefulness and originality, he claimed to be the first to discover the role of the lungs in the preparation and generation of the vital spirits. Denouncing other anatomists for erroneously assuming the existence of a direct path from the right to the left ventricle, Colombo urged his readers to confirm his conclusions. The blood was carried from the right side of the heart by the "artery-like vein" (pulmonary artery) to the lung where it was rarefied and mixed with air. The "vein-like artery" (pulmonary vein) brought the mixture of blood and air back to the left ventricle of the heart to be distributed to all parts of the body via the arteries. Despite his glowing praise for his own originality and daring in revealing this error in the dogmas of Galen, Colombo's general views on the functions of the heart, lungs, and blood were rather conservative. Even his claim to priority has been questioned on several accounts. In 1546 Servetus sent a manuscript copy of the *Restoration* to a doctor in Padua. If Colombo was not aware of this manuscript, he might have learned about the *Restoration* when he settled in Rome where the printed copies survived. Even if Colombo had begun his study of the pulmonary circulation as early as 1545, his ideas might have been stimulated or confirmed by learning about the speculations of Servetus.

While the role of Ibn an-Nafis and Michael Servetus in the series of events that led to the discovery of the circulation of the blood is ambiguous, some historians have argued that Andrea Cesalpino (1519–1603) should be honored as the discoverer of both the minor and major circulation. Cesalpino was a very learned man who combined a great reverence for the ancients, especially Aristotle, with an appreciation of more modern aspects of natural history and medicine. From 1555 to 1592 he was professor of medicine and botany at Pisa. In 1592 he was called to Rome to serve as physician to Pope Clement VIII and as professor at the Sapienza University. Although his main scientific interest was botanical classification, Cesalpino also wrote a number of books pertaining to anatomy and practical medicine. As indicated by his *Quaestionum peripateticarum*, his scientific and medical work was firmly grounded on an Aristotelian foundation. Opposed to Galenic doctrines that conflicted with the views of Aristotle, Cesalpino was eager to support Aristotle's claim that the heart was the most important organ of the body. For the Galenic system to work, Cesalpino noted, the lungs and the heart must expand and contract at the same time. Yet it was obvious that we are able to regulate our breathing by our will, although we cannot control the heartbeat. Similarly, physicians knew that the pulses and the respiration might be fast or slow, strong or weak, and that such changes in the respiration need not correspond to changes in the pulses. In a discussion of bloodletting, Cesalpino called attention to the well-known fact that veins swell on the far side of the ligature. This observation could have led him to an exploration of the idea that the direction of blood flow is from the arteries to the veins and from the veins to the heart. But preoccupied with Aristotelian ideas about the primacy of the heart and the nature of the innate heat, Cesalpino entered into a digression on Aristotle's concept of the nature of sleep and the movement of animal heat during sleeping and waking states.

Like Aristotle, Cesalpino thought that the primary organ of the body was the heart, which he lovingly described as a fountain from which four great veins irrigated the body "like the four rivers that flow out from Paradise." Cesalpino had a remarkable gift for infusing ambiguous passages with words like *circulation* and *capillary vessels* that sound remarkably modern, at least in translation. If the scattered references to the movements of the heart and blood in his writings are brought together and arranged in appropriate patterns, a case can be made for awarding him a place in the history of the discovery of the circulation of the blood, but such an exercise also demonstrates the difficulty of establishing the relationship between observations, meaning and context, theory formation, and the nature of a scientific discovery.

In calling attention to the way in which observations made in the course of bloodletting could lead to insights into the workings of the body, Cesalpino was not alone. Several other sixteenth-century anatomists were led to the investigation of the valves of the veins from such considerations. When a ligature was tied around the arm in preparation for bleeding, little knots or swellings, which correspond to the valves of the veins, can be seen along the course of the veins. Studies of the venous valves play an important role in the history of the discovery of the circulation because William Harvey (1578–1657) suggested that thinking about the purpose of these structures led him to wonder whether the blood might travel in a circle. Harvey's teacher at the University of Padua, Girolamo Fabrici (Fabricius; ca. 1537–1619) probably began studying the structure, distribution, and function of the venous valves in 1574. Three years after the death of Fallopio, Fabrici was appointed to the chair of anatomy and surgery. Like his predecessor, Fabrici was much less interested in teaching than in pursuing his own research in anatomy and embryology. Sometimes he disappeared before completing his courses, and when he was unable to avoid his teaching obligations, he angered students who had come to Padua from all over Europe by his obvious display of boredom and indifference. Presumably, then as now, students thought it more natural for the teacher to be enthusiastic and the students to be bored and apathetic. After serving the university for 50 years, Fabrici claimed that illness made his retirement necessary. Poor health did not prevent him from continuing to work; he published eight new works and revised several previous books during his retirement.

In 1603 Fabrici published a brief work called *On the Valves in the Veins*. Other anatomists had observed the peculiar membranous structures in the valves, but this was the first complete account of their structure, position, and distribution throughout the entire venous system. According to Fabrici, nature had fashioned the valves to oppose the flow of blood *from* the heart *towards* the periphery. Comparing these structures to the floodgates of a millpond, Fabrici suggested that their function was to regulate the volume of blood distributed to the parts of the body so that each could obtain its proper nourishment. The arteries did not need such structures because the problem of equitable distribution had been solved by means of the pulsations of the thick arterial walls. In his anatomy courses and his public lectures, Fabrici frequently demonstrated the action of the venous valves. This is easily done by tying a ligature around the arm of a volunteer. Because the ligature blocks the flow of venous blood, little knots can soon be seen along the course of the engorged veins. Dissection of the veins in a cadaver suggested that the membranous structures distributed along the course of the veins corresponded to these swellings. A simple experiment involving an attempt to push the blood past these knots clearly demonstrates that the valves oppose the flow of blood. Intrigued by such experiments, Harvey realized that the venous valves did not retard the flow of blood from the heart to the periphery but directed the blood *towards* the heart.

William Harvey, a quiet and in many ways conservative man, was the eldest of the seven sons of Thomas Harvey and the only member of this close-knit family of farmers and merchants to become a scientist. After receiving the Bachelor of Arts from Caius College, Cambridge, Harvey stud-

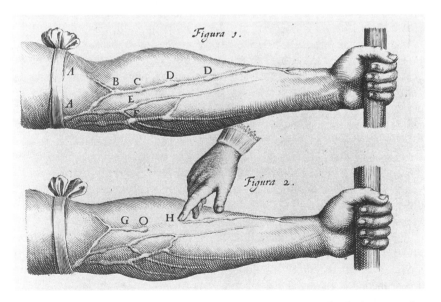

William Harvey's demonstration of the circulation of the blood: a simple demonstration of the valves of the veins provides evidence of the direction of blood flow

ied medicine at the University of Padua. In 1602 Harvey received the degree of Doctor of Medicine and returned to England. He soon established a successful practice and married Elizabeth Browne, daughter of Lancelot Browne (d. 1605), physician to Elizabeth and James I. In short order Harvey was elected Fellow of the College of Physicians (1607), appointed physician to St. Bartholomew's Hospital (1609), Lumleian Lecturer for the College of Physicians (1615–1656), and physician extraordinary to James I and Charles I. In 1631 Harvey was promoted to physician-in-ordinary to Charles I and became the king's senior physician-in-ordinary in 1639. (Peculiarities of nomenclature meant that the title *ordinary* ranked higher than *extraordinary* in the royal medical hierarchy.) Despite considerable personal risk and sacrifice, Harvey remained the faithful friend of Charles I throughout the Civil War between the followers of King Charles I and the Parliamentary forces under Oliver Cromwell (1599–1658). When seen against the political and social turmoil of the age in which he lived, Harvey's achievements are particularly remarkable. As physician to the king, Harvey frequently traveled with the court to continental Europe, where he is believed to have discussed his discoveries and theories with other physicians. During the war, Harvey's house in London was attacked and many of his manuscripts and collections were destroyed. When the court moved from London to Oxford, Harvey obtained a professorship there. His remarkable temperament is illustrated by the story of how at the battle of Edgehill he sat under a hedge just beyond the battlefield reading a book. After Cromwell's followers prevailed and King Charles was publicly beheaded in 1649, Harvey retired to the countryside to live with his brothers. Suffering from gout, depressed, and often in pain, Harvey dosed himself with peculiar remedies and large quantities of opium. There is suggestive evidence that he made several suicide attempts. In his will, Harvey, who had no children, left his estate to Oxford University, which has an annual festival dedicated to him.

Despite the reform of anatomy accomplished by Andreas Vesalius at the University of Padua, medical teaching and medical philosophy were still dominated by Galenic theories at the time Harvey first began to think about the circulation of the blood. Long before 1628 when he assembled his evidence and published *De motu cordis* (*Anatomical Treatise on the Movement of the Heart and Blood*), Harvey seems to have arrived at an understanding of the motion of the heart and blood. The notes for his first Lumleian Lecture show that by 1616 he was performing demonstrations and conducting experiments to show that blood passes from the arteries into the veins. He concluded that the beat of the heart impels a continuous circular motion to the blood. Although it was possible to discuss innovative ideas as the

For J. Hinton at the King's Arms in Newgate Street.

William Harvey

Lumleian lecturer, Harvey was reluctant to actually publish his novel and unprecedented views for fear of having "mankind at large for my enemies."

In thinking about the function of the heart and the blood vessels, Harvey came to reject Galen's scheme while moving closer to Aristotle's idea that the heart is the most important organ in the body. Like Aristotle, whom he greatly admired, Harvey asked seemingly simple but truly profound questions about the nature and purpose of things in order to understand their final causes. Harvey wanted to know why the two structurally similar ventricles of the right and left heart should have such different functions as the control of the flow of nutritive venous blood and the distribution of the vital spirits in the arterial blood. If arterial blood and venous blood were fundamentally distinct, why did they differ only in color while they appeared to be identical in taste, odor, and other observable characteristics? Why should the artery-like vein nourish only the lungs, while the vein-like artery had to nourish the whole body? Why should the lungs need such an inordinate amount of nourishment? Why did the right ventricle have to move in addition to the movement of the lungs? But the unique question that Harvey asked was one that is astonishingly simple, even child-like. Furthermore, it is a question that lent itself to experimental and quantitative answers. That question was: *How much blood is sent through the body with each beat of the heart?* Indeed, when reading *De motu cordis*, one can hardly avoid the thought that, at least *in principle*, the experiments by which Harvey answered this question could have been performed hundreds of years before. That is, Harvey's observations did not require the use of any materials or instruments that could not have been essentially duplicated by Hippocrates, Aristotle, Galen, or any of the thousands of students who had dozed through lectures on the whole of anatomy.

In his first chapter, "Difficulties in Experiment," Harvey dismissed the work of anatomists who would study only one species, man, and that one only when dead. Obviously, the proper use of comparative anatomy and vivisection of various animals could provide valuable evidence of vital functions. He admitted that when he first tried to understand the movement of the heart, its motion seemed so rapid and complex that he thought only God could comprehend it. But Harvey saw the advantages of using cold-blooded animals and mammals brought close to death by drugs or exsanguination as model systems in which the heartbeat was slower and easier to follow. Thus, the workings of the heart and arteries could be thought of as analogous to the action of wheel upon wheel in a complex machine or the mechanism of a gun in which one could not see the rapid action of trigger, flint, steel, spark, and powder but could logically separate out each step.

De motu cordis (courtesy of the National Library of Medicine)

With arguments based on dissection, vivisection, clinical experience, and the works of Aristotle and Galen, Harvey proved that in the adult all the blood must go through the lungs to get from the right to the left side of the heart. Acknowledging the work of his predecessors, Harvey clarified various aspects of the pulmonary circulation. He proved that the heart is muscular and that its important movement is the contraction. Previously, anatomists had taught that the heart was not a muscle and that dilation was the important function; movement had been ascribed to the chest and lungs rather than to the heart itself. Most of all, Harvey proved that it was the beat of the heart that caused a continuous circular motion of the blood: from the heart into the arteries and from the veins back to the heart. In support of the novel idea of a continuous circulation of the blood, Harvey turned to quantitative considerations. Although he did experiments aimed at getting an accurate measurement of the quantity of blood put out by the heart with each beat, he emphasized that exact measurement was unnecessary. Even the most cursory calculation would prove that the amount of blood pumped out of the heart per hour was so great that it exceeded the weight of the entire body. That is, if the heart pumps out 2 ounces of blood with each beat and beats about 70 times per minute, the heart must expel about 600 pounds per hour. Such a quantity is three times the weight of even a very large man. Similar calculations were carried out for various animals, such as sheep and dogs. In all cases the amount of blood pumped out by the heart in even a half hour is more blood than is contained in the whole animal. One could exsanguinate experimental animals, as a butcher exsanguinates an ox, and easily prove this. Blood must, therefore, move in a circle through the body. The movement of blood must be continuous, rather than a one-way trip of blood constantly synthesized from food by the liver. It seems to follow that the blood in the body of a living animal is impelled to move in a circle and is kept in a state of ceaseless motion. The purpose of the motion and contraction of the heart was to impart this circular motion to the blood.

Once this basic principle was grasped, many observations fell neatly into place. Every butcher knew, for example, that the heart must still be beating to drain out the blood of a slaughtered animal. Under these circumstances the butcher could perform this task in about one quarter of an hour by opening an artery. With Harvey's theory scientists could finally explain what every butcher already knew. Experience gained through phlebotomy and observations on patients, such as the throbbing of an arterial aneurysm and the pulses of the wrists, temples, and neck, similarly supported the central thesis of *De motu cordis*. Knowledge of the continuous circulation of the blood explained many puzzling clinical observations. Circulation explained why poisons or infections at one site could affect the whole system, as could bites from snakes and rabid animals. Harvey even saw in the work an explanation for the way externally applied medicines could be absorbed and distributed throughout the system. This included such quaint therapy as applying garlic to the soles of the feet.

It is all too easy to take Harvey out of his seventeenth-century context and make him appear more modern in outlook and approach than is really appropriate. Moreover, his methods and conclusions appear so reasonable and compelling to the modern reader that it might be assumed that his contemporaries must have reacted similarly and that *De motu cordis* should have dealt an immediate death blow to the Galenic system. The kind of evidence that appears unequivocal today, however, did not necessarily appeal to Harvey's contemporaries. Moreover, although Harvey had discovered virtually all that could be known about the anatomical and mechanical aspects of the circulatory system, many questions remained unanswered. In the absence of the microscope, Harvey could not see the capillaries that join the arteries and veins; he had to postulate hypothetical anastomoses or pores in the flesh to perform this function. Without any information about the chemical nature of the differences between venous and arterial blood, the relationship between respiration and the circulation remained obscure and Harvey could provide no substitutes for the Galenic spirits and the innate heat. Harvey acknowledged that his theory did not explain whether or not the heart did more than distribute the blood, by adding heat, spirit, perfection, or local motion. Demonstrating a remarkable degree of modesty and restraint, he admitted that such questions were beyond the range of his observational and experimental methods. A mechanical explanation of the movement of the heart and blood simply could not fully replace the Galenic system that had so completely, if incorrectly, incorporated explanations for the *purpose* of the heart, blood, lungs, liver, veins, arteries, and spirits.

Indeed, the new Harveian theory of the circulation seemed to raise more questions than it answered. How could the new system explain how the parts of the body secured their proper nourishment if the blood was not continuously consumed? If the venous blood was not continuously synthesized by the liver, what was the function of this large and bloody organ? If the blood moved in a continuous circle, how did the body distribute the vital spirits and the innate heat? If the vital spirit was not produced in the lungs or the left ventricle, what was the function of respiration? If the whole mass of the blood was constantly recirculated, what was the difference between the arterial and venous blood? What principles would guide medical practice if the great Galenic synthesis was sacrificed for the theory of the circulation? Certainly Harvey's work did not lead to the rejection of venesection as a major therapeutic tool. Instead, ideas about the circulation provoked new arguments about the selection of appropriate sites for venesection and seemed to stimulate interest in bloodletting and other forms of depletion therapy. Even Harvey did not worry about the compatibility, or incompatibility, of venesection with the continuous circulation of the blood, nor does he seem to have considered the possibility of therapeutic blood transfusions. By 1660, however, his followers were engaged in attempts to transfuse and infuse many curious substances into animal and human veins. Seventeenth-century physiologists and physicians, quite unaware of the dangers of such experiments, had great hopes for the efficacy of blood transfusion. If the macrocosm and microcosm could be explained in terms of four elements and four humors, there was no compelling reason to expect major incompatibilities between the blood of different individuals, or even different species. The first significant experiments on blood transfusion were performed by Christopher Wren, Richard Lower, and Robert Boyle in England, and by Jean Denis in Paris. Wren and his colleagues began rather cautiously with a series of experiments on animals, but Denis proceeded immediately to experiments on humans. After the deaths of several patients and a great deal of sordid publicity, Denis and his English counterparts abandoned this highly experimental form of therapy.

Ironically, some defenders of Galen attacked Harvey for assuming that the results of experiments on animals could be applied to human anatomy and physiology. On the other hand, French botanists thought that the discovery of the circulation of the blood could be used to explain the circulation of sap; the analogy was intriguing, but they soon realized that the facts of plant anatomy and physiology were not compatible with models based on the circulation of the blood in animals. Several distinguished scientists accepted the general idea of the circulation but disagreed with particular aspects of Harvey's ideas, such as his estimate of the speed of the movement of the blood or the amount of blood driven into the arteries by the contraction of the ventricles. For example, the French physician Jean Riolan (1571–1657) suggested that the blood made the complete circuit through the body only two or three times a day. Not all the blood returned to the heart, because the ebb and flow of the blood in the course of its leisurely journey allowed the parts of the body to consume some of the nourishing blood. Rejecting the idea that it was the action of the heart that propelled the circulation, Riolan argued that it was actually the blood that kept the heart in motion. Like Galen, Riolan believed that the liver was the most important organ; the role of the heart, he argued was to warm the blood and restore the spirits.

Harvey took these arguments seriously enough to publish a refutation of Riolan's views in 1649.

Sure of the truth of his own work, but well aware of its revolutionary nature, Harvey predicted that only the new generation of scientists would understand and accept it. Illness, the infirmities of age, and the loss of many of his precious manuscripts during the Civil War prevented Harvey from accomplishing all his goals, but he did live to see the beginnings of a new experimental physiology. Ideas, methods, and questions inspired by the work of William Harvey provided the young scientists who eventually founded the Royal Society with a new research program for establishing the workings of the human body and developing a new experimental natural science.

SANTORIO SANTORIO AND THE QUANTITATIVE METHOD

It is easy to attribute much of Harvey's success to his brilliant use of the quantitative method, but not all seventeenth-century attempts to apply quantitative methods to the study of physiological problems were as successful. Although Santorio Santorio (Sanctorius; 1561–1636) has been called the founder of quantitative experimental physiology, his years of painstaking measurements primarily demonstrate that the quantitative methods available to him could not solve the fundamental problem he was trying to solve. After graduating from the University of Padua in 1582, Santorio practiced medicine in Venice. In 1611 he was appointed to the chair of theoretical medicine at Padua where he remained until his resignation in 1629, 5 years after his students charged him with negligence on the grounds that he was evading his teaching obligations in order to maintain his private practice. Freed from his duties at the university, he moved to Venice and devoted the rest of his life to his medical practice and research on the phenomenon known as *insensible perspiration*. Many physicians and scientists thought that in addition to the respiration that took place through the lungs, there was another form of respiration that involved the release of imperceptible vapors through the skin.

Santorio believed that he could study the mechanism of insensible perspiration by the process of exact measurement. For much of his experimental career he sat on a special balance of his own design, which consisted of a chair suspended from a steelyard, patiently weighing himself before and after eating, drinking, sleeping, resting, and exercising. In 1614 he published a book of aphorisms based on his work called *Ars de medicina statica aphorismi* (*Medical Statics*). The book went into at least 32 editions and was translated into several languages.

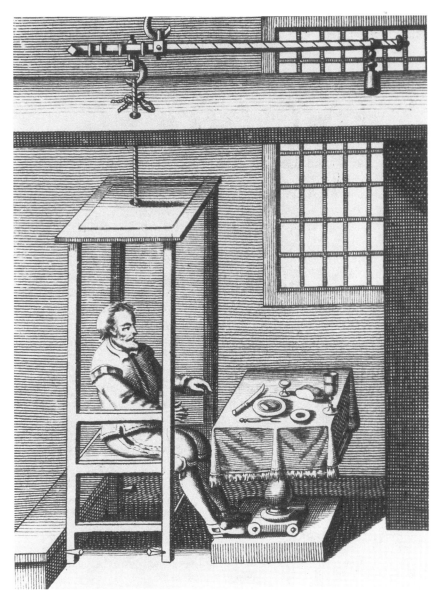

Santorio Santorio in his weighing chair

Although the *Medical Statics* provided only vague indications of experimental methods, Santorio claimed to have created a totally new field of exact measurement in medicine. Based on his measurement of the amount of food and drink, and the sensible and insensible excreta of the body, Santorio provided deductions as to what changes had taken place in the body under various conditions. In Santorio's opinion, these experiments had some practical applications for health and medical practice. A man could take his meals in the balance and know when he had eaten just the right amount of food. Moreover, he claimed that physicians needed to understand the quantitative aspects of insensible perspiration in order to treat their patients. Critics responded that even if Santorio could provide *quantitative* measurements of the insensible perspiration, it was only the *qualitative* nature of the invisible vapors that was related to pathological phenomena.

Similar quantitative experiments being carried out in the nineteenth century were subjected to scathing criticism by the great French physiologist Claude Bernard (1813–1878). For example, a study of nutrition reviewed by Bernard had consisted of a balance sheet of all the substances taken into a cat's body and those excreted during 8 days of feeding and 19 days of fasting. Kittens born on the seventeenth day were included in the calculations as excreta. Bernard ridiculed such experiments as attempts to understand what occurred in a house by measuring who went in the door and what came out the chimney.

The *Medical Statics* is Santorio's most famous book, but it is so terse that it fails to reflect Santorio's motives for adopting the quantitative method or his interest in new measuring instruments such as the thermometer and pulse clock as ways in which to establish greater certainty for practitioners of what was fundamentally Galenic medicine. While no graduate or professor of the University of Padua could ignore the errors of Galenic anatomy, as long as a scientist was also guided by reason and experience, Santorio saw no reason to reject Galen as a medical authority. After all, it was Galen who taught anatomists to devote themselves to experimentation and observation.

SUGGESTED READINGS

Ambrose, Elizabeth Ann (1992). Cosmos, Antropos and Theos: Dimensions of the Paracelsian Universe. *Cauda Pavonis: Studies in Hermeticism* (New Series) *11:* 1–7.

Arcieri, G. P. (1945). *The Circulation of the Blood and Andreas Cesalpino of Arezzo.* New York: S. F. Vanni, Publishers.

Bainton, R. H. (1960). *Hunted Heretic: The Life and Death of Michael Servetus, 1511–1553.* Boston: Beacon.

Boas, M. (1962). *The Scientific Renaissance, 1450–1630.* New York: Harper and Row.

Bonelli, M. L. R., and Shea, W. R. (1975). *Reason, Experiment, and Mysticism in the Scientific Revolution.* New York: Science History.

Bylebyl, J. J., ed. (1978). *William Harvey and His Age. The Professional and Social Context of the Discovery of the Circulation.* Baltimore, MD: Johns Hopkins Press.

Castiglioni, Arturo (1931). The Life and Work of Santorio Santorio (1561–1636). Trans. by Emilie Recht. *Medical Life* n.s.*38* (12): 729–785.

Choulant, J. L. (1962). *History and Bibliography of Anatomical Illustration.* Trans. and annotated by Mortimer Frank, repr. with additional essays. New York: Hafner.

Cipolla, C. M. (1969). *Literacy and Development in the West.* Baltimore, MD: Penguin Books.

Cohen, I. B. (1985). *Revolution in Science.* Cambridge, MA: Harvard University Press.

Davis, A. B. (1973). *Circulation Physiology and Medical Chemistry in England 1650–80.* Lawrence, KS: Coronado Press.

Debus, A. G., ed. (1972). *Science, Medicine and Society in the Renaissance.* 2 vols. New York: Neale Watson Academic Publications. 2 volumes.

Debus, A. G. (1978). *Man and Nature in the Renaissance.* New York: Cambridge University Press.

Debus, A. G. (1986). *Chemistry, Alchemy, and the New Philosophy, 1500–1700.* London: Variorum.

Eisenstein, E. (1983). *The Printing Revolution in Early Modern Europe.* New York: Cambridge University Press.

Fabrizzi, Girolamo (1603). *De venarum ostiolis. By Fabricius ab Aquapendente.* Facsimile, with English trans., ed. by K. J. Franklin. Springfield, IL: Charles C. Thomas, 1933.

Fevre, L., and Martin, H. J. (1984). *The Coming of the Book. The Impact of Printing, 1450–1800.* Trans. David Gerard. London: Verso Editions.

Fishman, A. P., and Dickinson, W. R., eds. (1964). *Circulation of the Blood. Men and Ideas.* New York: Oxford University Press.

Frank, R. G., Jr. (1980). *Harvey and the Oxford Physiologists. Scientific Ideas and Social Interactions.* Berkeley, CA: Univeristy of California Press.

French, R., and Wear, A., eds. (1989). *The Medical Revolution of the Seventeenth Century.* New York: Cambridge University Press.

Fulton, J. F. (1953). *Michael Servetus, Humanist and Martyr.* New York: Reicher.

Gjertsen, D. (1984). *The Classics of Science. A Study of Twelve Enduring Scientific Works.* New York: Lilian Barber Press, Inc.

Grendler, Paul F. (1989). *Schooling in Renaissance Italy. Literacy and Learning, 1300–1600.* Baltimore, MD: Johns Hopkins Press.

Hackett, Earle (1973). *Blood, The Paramount Humor.* London: Jonathan Cape, Ltd.

Harris, Charles R. S. (1973). *The Heart and the Vascular System in Ancient Greek Medicine from Alcmaeon to Galen.* Oxford: Clarendon Press.

Hartmann, F. (1973). *The Prophecies of Paracelsus*. New York: Rudolf Steiner Publications.

Harvey, W. (1628). *Exercitatio anatomica de motu cordis et sanguinis in animalibus*. Translated by C. D. Leake (1930). Springfield, IL: Charles C. Thomas.

Harvey, W. (1961). *Lectures on the Whole of Anatomy*. Annotated translation by C. D. O'Malley, F. N. L. Poynter, and K. F. Russell. Berkeley, CA: University of California Press.

Harvey, William (1976). *An Anatomical Disputation Concerning the Movement of the Heart and Blood in Living Creatures*. Trans. by G. Whitteridge. Oxford: Blackwell.

Keele, K. D. (1965). *William Harvey: The Man, the Physician and the Scientist*. London: Nelson.

Keele, Kenneth D. (1983). *Leonardo da Vinci's Elements of the Science of Man*. New York: Academic Press.

Kerrigan, W., and Braden, G. (1989). *The Idea of the Renaissance*. Baltimore, MD: Johns Hopkins University Press.

Keynes, G. L. (1966). *The Life of William Harvey*. Oxford: Clarendon.

Leonardo da Vinci (1952). *Leonardo da Vinci on the Human Body*. Translations, text and introduction by Charles D. O'Malley and J. B. de C. M. Saunders. New York: Schuman. (Reprint, New York: Dover Publications, 1983.)

Lind, L. R. (1974). *Studies in Pre-Vesalian Anatomy: Biography, Translations, Documents*. Philadelphia, PA: American Philosophical Society.

Lindeboom, G.A. (1975). *Andreas Vesalius and His Opus Magnum: A Biographical Sketch and an Introduction to the Fabrica*. Netherlands, Nieuwenkijk: de Forel.

MacCurdy, Edward (1956). *The Notebooks of Leonardo da Vinci*. London: Jonathan Cape.

McKnight, Stephen A. (1989). *Sacralizing the Secular: The Renaissance Origins of Modernity*. Baton Rouge, LA: Louisiana State University Press.

McVaugh, Michael, and Siraisi, Nancy G., eds. (1990). *Renaissance Medical Learning. Evolution of a Tradition. Osiris*, 2nd series, volume 6.

O'Malley, C. D. (1953). *Michael Servetus. A Translation of His Geographical, Medical and Astrological Writings with Introductions and Notes*. Philadelphia, PA: American Philosophical Society.

O'Malley, C. D. (1964). *Andreas Vesalius of Brussels, 1514–1564*. Berkeley, CA: University of California Press.

Pagel, W. (1958). *Paracelsus: An Introduction to Philosophical Medicine in the Era of the Renaissance*. Basel: Karger.

Pagel, W. (1976). *New Light on William Harvey*. Basel: S. Karger.

Pagel, Walter (1984). *The Smiting Spleen. Paracelsianism in Storm and Stress*. Basel: Karger.

Paracelsus (1941). *Four Treatises of Theophrastus von Hohenheim Called Paracelsus*. Baltimore, MD: Johns Hopkins University Press.

Paracelsus (1951). *Paracelsus: Selected Writings*. Intro. by J. Jacobi; Trans. by N. Guterman. New York: Pantheon Books.

Richardson, Ruth (1987). *Death, Dissection and the Destitute*. London: Routledge and Kegan Paul.

Sanctorius (1720). *Medicina Statica*: Being the Aphorisms of Sanctorius, Trans. into English with large Explanations. 2nd ed. To which is added Dr. Keil's *Medica Statica Britannica* by John Quincy, MD. London: Printed for W. & J. Newton in Little-Britain.

Sarton, G. (1957). *Six Wings: Men of Science in the Renaissance*. Bloomington, IN: Indiana University Press.

Saunders, J. B. de C. M., and O'Malley, C. D. (1947). *Andreas Vesalius Bruxellensis: The Bloodletting Letter of 1539*. New York: Henry Schuman.

Saunders, J. B. (1978). The history of the venous valves. *Developments in Cardiovascular Medicine*. Ed. by C. J. Dickinson and J. Marks. Lancaster, England: MTP Press, pp. 335–51.

Shumaker, W. (1972). *The Occult Sciences in the Renaissance*. Berkeley, CA: University of California Press.

Sigerist, H. E. (1933). *The Great Doctors: A Biographical History of Medicine*. Trans. by E. and C. Paul, 1958. Garden City, NY: Doubleday.

Singer, C. (1957). *A Short History of Anatomy and Physiology From the Greeks to Harvey*. New York: Dover.

Singleton, C., ed. (1968). *Art, Science, and History in the Renaissance*. Baltimore, MD: Johns Hopkins University Press.

Vesalius, Andreas (1969). *The Epitome of Andreas Vesalius*. Trans. from the Latin, with introduction and anatomical notes by L. R. Lind and C. W. Asling. Cambridge, MA: MIT Press.

Vickers, Brian, ed. (1984). *Occult and Scientific Mentalities in the Renaissance*. Cambridge, England: Cambridge University Press.

Wear, Andrew (1981). Galen in the Renaissance. In *Galen: Problems and Prospects*. Ed. by Vivian Nutton. London: The Wellcome Institute for the History of Medicine.

Westman, Robert S. (1977). *Hermeticism and the Scientific Revolution*. Los Angeles, CA: Clark Library.

Whitteridge, G. (1971). *William Harvey and the Circulation of the Blood*. New York: American Elsevier.

Wightman, W. P. D. (1962). *Science and the Renaissance*. 2 vols. Edinburgh: Aberdeen University.

Willus, F. A., and Keys, T. E., eds. (1961). *Classics of Cardiology*. 2 vols. New York: Dover.

4

THE FOUNDATIONS OF MODERN SCIENCE: INSTITUTIONS AND INSTRUMENTS

Looking back to the seventeenth century from the vantage point of 1787 as he prepared the second edition of his *Critique of Pure Reason*, the German philosopher Immanuel Kant saw it as the end of centuries of darkness. Suddenly, it seemed, a new light had burst upon natural philosophers and illuminated the true methods of scientific inquiry. Certainly it is possible to see Kant's bright light reflected in the work of such giants of this age as Galileo, Newton, and Harvey. And yet certainly deeper, darker, and more fundamental forces were at work during the age that witnessed the construction of modern experimental science. Perhaps there has never been any period in which prevailing concepts of the earth, the heavens, and the nature of human beings were more radically transformed than the seventeenth century, a time when European society was torn by the political, social, and spiritual turmoil that accompanied the shift from an agrarian society towards an increasingly urban and industrial world. The eternal and unchanging heavens had been radically rearranged by Galileo, and because of the work of William Harvey the blood was now known to travel continuously in a circle around the body, just as the earth moved in its orbit around the sun. If the work of Galileo and Harvey did indeed raise questions that could not be answered by the texts and methods that had served as the ultimate source of authority for hundreds of years, what kinds of individuals and what kinds of institutions could train a new generation of investigators?

Given the essentially conservative role played by the university, it might not be surprising that few of the leading participants in the scientific revolution held university appointments. Before the sciences emerged as specialized professions, the individuals drawn to the study of questions that now fall within the domain of the life sciences were associated with various institutions and occupations. The pre-Socratic philosophers, for example, were independent scholars, unattached to any formal educational institutions, whereas the Museum of Alexandria provided a setting and salary that encouraged research. Scholarly activity in the Middle Ages was generally confined to the monasteries until the universities began to assume this role, but the sciences did not occupy a central role in the curriculum. Many European universities, moreover, were dominated by religious orders, and virtually all remained committed to Aristotelian thought. During the Renaissance, biological research was carried out by artists, scholars, and physicians, some of whom were associated with major universities, while others earned their livelihood in private practice or in service to wealthy nobles and kings. The universities of the seventeenth century did not necessarily prevent the expression of new ideas, but the intellectual atmosphere was such that a scientist of Isaac Newton's stature could present his discoveries in a university lecture hall without creating a ripple of excitement or interest.

By the seventeenth century, membership in a novel institution called the *scientific society*, or *academy*, became the sign and symbol of the new scientific movement. The scientific society is in some sense a descendent of the Pythagorean brotherhood, Plato's Academy, and Aristotle's Lyceum, but with a new purpose and mission. Practical and experimental goals superseded abstract and philosophical pursuits. Eminent philosophers of science, such as René Descartes (1596–1650) and Francis Bacon (1561–1639), provided the philosophical rationale for the new institution, while Harvey, Galileo, and others provided the models and methods that inspired the academicians. New instruments, such as the telescope, thermometer, air pump, and microscope, as well as new ideas, institutions, journals, and other means of communication made it possible for scientists to ask new questions and design new ways of finding satisfactory answers.

The intimate association between the new experimental science and the academy rather than the university raises an obvious question: Why did the seventeenth century university fail to serve as the incubator of the new science? Clearly, the university and the scientific society differed fundamentally in structure, function, and philosophy; the pedagogical purposes of the university encouraged formal methods, based on teacher-student relationships, that may become almost adversarial and are not conducive to innovations. On the whole, universities, despite the achievements of certain individuals, continued to stress the ancient, respectable subjects where authoritative texts and recognized scholarship existed. Universities generally regarded their mission as the preservation and transmission of the wisdom of the past to the new generation rather than the encouragement of novel ideas and techniques.

Unlike the university, the academy served as an informal setting where talented and enthusiastic individuals interacted as friends and equals. Demonstrations and experiments stimulated new enquiries by bringing together a critical mass of intellectuals in a stimulating environment. The exchange of information was enhanced by the regular publication of journals and the transactions of the societies. Competition for awards and prizes encouraged the search for new facts and ideas and led to study and critical appraisal. Academies that succeeded in obtaining government charters, official protection, and wealthy sponsors could serve as safe places for the search for new facts and theories.

While the most famous and durable scientific societies were the Royal Society of London and the Royal Academy of Sciences of Paris. Italy, which had been the heart of the Renaissance, was also the site of the earliest academies. The short-lived *Accademia Secretorium Naturae* was established in the 1560s and met at the house of its leader, Giambattista della Porta (1538–1615), in Naples. This group took the important step of requiring potential members to present a new fact in natural science as a condition for admission. A group called the *Accademia dei Lincei* (Academy of the Lynx Eye) met in Rome from 1603 to 1630 under the patronage of Duke Federigo Cesi. Distinguished members of the Academy included Galileo, della Porta, and Nicolas Fabri de Peiresc, a Frenchman interested in Galileo's work and the exchange of scientific information throughout Europe. Galileo's microscope was developed with the support of this academy, as were other important instruments and publications. The condemnation of the Copernican theory caused considerable tension among members of the academy, which disbanded shortly after the death of its patron. The Florentine *Accademia del Cimento* (Academy of Experiments) encouraged the cooperative efforts of natural philosophers. The group included disciples of Galileo as well as Giovanni Borelli, a mathematician with a special interest in muscle physiology, and Francesco Redi, best known for his experiments on spontaneous generation. The members of the Academy assembled and tested an outstanding collection of scientific instruments as indicated by the descriptions published in their *Report of Experiments* in 1667; unfortunately, the Academy ended abruptly in that same year.

The early Italian societies emphasized experimental science and generally avoided speculative or controversial areas. This was not because the sciences had become well separated from philosophy and religion at this stage, but because natural philosophers, for their own protection, attempted to avoid potentially dangerous topics. While this may have stifled discussions of broad theoretical concerns, it did encourage experimentation, observation, and the refinement of specialized scientific instruments. Academies developed in many European countries and in the United States, but none of these matched the prestige and stability of the Royal Society of London and the French Academy of Sciences.

The political turmoil of seventeenth-century England inhibited the development of formal academies. Indeed, the informal association of natural philosophers held during the Civil War has been referred to as an "Invisible College." Although the Royal Society was not officially charted by Charles II until 1662, it grew out of informal networks established in the 1640s in London and Oxford by groups of natural philosophers who held meetings at the homes of participants, taverns, and Gresham College. After a lecture by Christopher Wren at Gresham College in November 1660, the assembled natural philosophers formally declared interest in establishing a college for the promotion of "Physico-Mathematical Experimental Learning." A chairman was elected and, of the many persons interested in science, 41 men were chosen for the official project. The king gave his verbal approval and, 2 years later, a formal charter was granted for the incorporation of the Royal Society for the Improvement of Natural Knowledge. The events that established the Society were recorded in great detail by Thomas Sprat in his *History of the Royal Society* (1667). The invaluable secretary of the Society, Henry Oldenburg, another zealous advocate of the experimental method, carried out a massive correspondence that linked the Royal Society with scientists and academies all over Europe. In many ways, Francis Bacon was the guiding spirit of the Royal Society.

FRANCIS BACON

In spite of his status as prophet, critic, and philosopher of the new experimental science, Francis Bacon made no direct contributions to scientific knowledge. Indeed, he did not distinguish himself as a judge of facts and fancies; he failed, for example, to appreciate the role of mathematics in scientific theory and the importance of the work of Copernicus and Galileo. He did, however, influence the philosophy and institutions that directed the course of science. Bacon believed that science could play a major role in increasing human knowledge, power, and control over nature. His impact on the sciences came about through his emphasis on defining the methodology of science, suggesting means of ensuring its application, and providing encouragement and direction for the new scientific enterprises he predicted.

Francis Bacon, son of Sir Nicholas Bacon, Keeper of the Great Seal, devoted much of his energy to the scramble for political power and wealth. Under King James I, Bacon rose to the rank of Lord High Chancellor. Despite the burden of his public career, Bacon energetically developed grandiose schemes for the review and analysis of all branches of human knowledge, planned an encyclopedia of the crafts and experimental facts, and proposed plans for educational and institutional means by which the sciences would be redirected and exploited to provide for human welfare, comfort, and prosperity. Despite the later popularity of Bacon's *Advancement of Learning*, William Harvey dismissed Bacon's texts with the remark, "He writes philosophy like a Lord Chancellor." But Thomas Jefferson's trinity of portraits of the world's greatest men consisted of Newton, Locke, and Bacon. Eventually, Bacon, who so earnestly, if ineffectually, struggled to fulfill his vision of a life in science dedicated to the improvement of human welfare, became the target of a twentieth-century wave of anti-rationalism and anti-science.

In time, the "Baconian method" became synonymous with the "scientific method." Bacon's method, also known as the *inductive method*, involves the exhaustive collection of particular instances or facts and the elimination of factors that do not invariably accompany the phenomenon under investigation. Rather than passively collect facts, the scientist must be actively involved in *putting the question to nature*. Recognizing the impossibility of absolute proof of inductive generalizations based on a finite number of observations, Bacon called attention to the power of "negative instances" that can falsify an induction. In other words, experimental results that contradict a general theory may reveal more about nature than another bit of data that appears to support the theory. This principle, known as *falsifiability,* is usually associated with the twentieth-century philosopher Karl Popper, but Bacon apparently recognized its significance. Scientists would analyze experience "as if by a machine" to arrive at true conclusions by proceeding from less to more general propositions. The result of applying this scientific method, Bacon assured his readers, would be a great new synthesis of all human knowledge, a true and lawful marriage between the empirical and the rational faculty.

Deeply suspicious of mathematics and pure deductive logic, which proceeds from the more to the less general by using intuitive thinking, Bacon was certain that valid hypotheses would be derived from the assembly and analysis of "Tables and Arrangements of Instances." Bacon rejected the scholasticism of the universities and launched open attacks on Aristotle and Plato; fact gathering and experiment, he insisted, must replace the sterile burden of deductive logic, which could never generate new scientific knowledge. The crafts, he argued, had developed because many individuals had contributed to their growth, but philosophy had been stifled in its progress because the accepted philosophical systems were the products of only one mind. Instead of adding to human understanding, followers of the great thinkers had surrendered their intelligence to past masters of philosophy. Theology had to be separated from science and philosophy, but natural philosophy must be reunited with the craft traditions that had nourished it during the time of the pre-Socratics. Among the ancient philosophers, Bacon had especially high regard for Democritus and Lucretius, the advocates of atomic theory. Under the direction of a new community of scientists, the union of natural philosophy with the craft traditions would produce understanding and control of nature along with valuable inventions to improve the human condition.

Science would be the basis of a new system of education, because the universities could not provide the technical and utilitarian training required to redirect science and technology. In his utopian novel, *The New Atlantis* (1627), Bacon described an imaginary state-supported research institution called "Salomon's House." The goal of the ideal research foundation was to assemble knowledge of the causes and secrets motions of all things, thus enlarging of the bounds of the "Human Empire," in order to effect "all things possible." Financial aid for scientific research was a high-priority item in Bacon's schemes. Properly supported, science could create a brotherhood of experts and an international government. Scientists and technologists would serve as the supervisors and reformers of a world made as beautiful as Eden before the fall of man. James I, who supposedly fell asleep reading Bacon's *Novum Organum*, ignored his requests for the support needed to carry out this grand project. To reassure the theologians who feared that science might subvert the authority of religion, Bacon portrayed science as the most faithful "handmaid" of religion and, except for the word of God, the "surest medicine" against superstition. When scientific knowledge had purged the mind of "fancies and vanities," it would be better able to submit to "the divine oracles" and "give to faith that which is faith's."

RENÉ DESCARTES

Like Bacon, the French philosopher René Descartes believed that a new science would lead to knowledge and inventions that would promote human welfare. Unlike Bacon, Descartes was a gifted mathematician, honored as the inventor of analytic geometry, and the advocate of a deductive, mathematical approach to the sciences. Like Bacon, Descartes rejected the conceit that science should be purely theoretical and abstract, with pretensions of total separation from the human condition and the world of practical problems. Cartesian science would allow human beings to master and possess nature's abundance and establish a new medical science capable of eliminating disease and the disabilities of old age.

While still a student at the Jesuit school of La Fleche, Descartes became interested in mathematics and dismissed the rest of the teachings as mere quibbling. Thereafter Descartes worked as an engineer, dabbled in astrology, and associated with mathematicians and natural philosophers. Apparently restless and dissatisfied with these activities, Descartes enlisted in the army of the Prince of Orange. During a lull in the warfare in the Netherlands, Descartes set off seeking diversion in Germany where he joined the Bavarian service. While sitting in a "stove" (actually a small, very warm room), he was overwhelmed with an emotional and intellectual experience that left him with a profound philosophical insight, which the 23-year-old philosopher immodestly described as the foundation of a "completely new science." Since then, many eminent scientists, philosophers, and historians have agreed that Descartes's work constituted a major scientific revolution despite the fact that there is no single great scientific principle or discovery associated with his name. Unlike the revolutionary discoveries associated with Newton and Harvey, many of Descartes's theories about the nature of matter and space soon became mere historical curiosities. Still, the Cartesian spirit is so highly esteemed in France that late twentieth-century French philosophers proclaim "Descartes, c'est la France."

In 1637 Descartes published his *Discourse on Method*, subtitled "For the correct use of reason and for seeking truth in the sciences," along with *Geometry*, *Dioptrics*, and *Meteorology*. Descartes had settled in Holland because he believed that his ideas were too unorthodox for France. During the 20 years he spent in Holland, he studied mathematics, optics, the theory of music, astronomy, meteorology, and anatomy. When Galileo was censored by the Inquisition for his support of the Copernican system, Descartes felt compelled to suppress his *Treatise of the World*, and probably destroyed some of his manuscripts, because the Copernican doctrine was so intimately connected with every part of his own work. Thereafter, fear of persecution for heresy influenced much of his behavior, but despite his habitual caution charges were filed against him. Queen Christina of Sweden offered Descartes a refuge and requested his services as a tutor. Ridiculed as a "hermaphrodite" because of her interest in science and mathematics, Queen Christina was blamed for Descartes's death, which occurred only 5 months after he joined her court in Stockholm.

Unlike Bacon, whose work he had read and criticized, Descartes placed *a priori* principles first and subordinated his observations and experimental findings to them. Nevertheless, he too had a grand scheme and task for natural philosophy. An examination of method was a primary part of his plan to use the mathematical method in developing a general mechanical model of the workings of nature. Although experimentation had a role in Descartes's system, it was a subordinate one. He believed that experiments should serve as illustrations of ideas that had been deduced from primary principles or should help decide between alternative possibilities when the consequences of intuitive deduction were ambiguous.

Descartes's method began with his own mind engaged in methodically confronting "clear and distinct ideas" with a form of skepticism so deep and profound that he was able to separate his search for truth from all past authorities. General and complex problems could thus be broken down into its simple components. Applying this method to science, philosophy, or any other rational inquiry, Descartes asserted, would not only resolve problems but would lead to the discovery of useful philosophical knowledge. It is impossible to leave Descartes without reference to his most famous axiom or first principle, "I think, therefore, I am." But he also said, "Give me motion and extension, and I will construct the world." So pervasive was Descartes's influence on science and philosophy that many contemporaries of Newton thought of themselves as the disciples of Descartes. Neither Descartes nor Bacon alone could have served as a complete guide for the development of science and technology. The great mathematician and physicist Christiaan Huygens (1629–1687) recognized this when he remarked that Descartes had ignored the role of experimentation, while Bacon had failed to appreciate the role of mathematics in scientific method. A synthesis of the two approaches was needed, or at least the admission that there is no one scientific method sufficient for posing and solving all possible problems. Mechanical fact finding, daydreams, and flashes of intuition have played a role in science, no matter what formal doctrine or method scientists professed to follow.

Many early participants in the Royal Society of London and the French Academy of Sciences saw themselves as builders of Salomon's House. Some of the members tried to carry out policies and projects suggested by Bacon, such as exhaustive studies of crafts and the compilation of catalogues and encyclopedias. Robert Hooke (1635–1703), the curator of experiments for the Society, reflected the Baconian principle that truth and utility are the same thing in his description of the "noble matters" pursued by the Royal Society as a means of improving industry, agriculture, commerce, and navigation. One of the most versatile, hard-working, and underpaid scientists of all time, Hooke performed demonstrations for the Society, worked as assistant to the wealthy Robert Boyle, and designed and improved many experimental instruments such as the air pump, balance, and watch spring. While pursuing the useful arts, manufactures, and inventions, members of the Royal Society pledged to avoid meddling with politics, morals, metaphysics, and divinity. To promote philosophical progress members pledged that they would avoid "Dogmatizing, and the espousal of any Hypothesis not sufficiently grounded and confirm'd by Experiments." Although the Royal Society did not emphasize biological studies, many of the inventions and experiments performed by members were of great importance in the development of the life sciences. Robert Boyle is best remembered as a reformer of chemistry, but his experiments on respiration and combustion, which used the air pump, were fundamental to studies of the physiology of respiration. Many experiments were conducted on the intravenous injection of food and drugs, as well as blood transfusions between man and beast.

Early histories of the scientific revolution analyzed the age, class, professions, and religious affiliation of the members of the Royal Society in great detail. Until the rise of modern science was reexamined from a feminist perspective, it was considered highly significant that 62% of the early members were Puritan, but the fact that 100% were male was considered too obvious for comment. Excluded from the formal academies of England and France, women explored the possibilities of creating academies such as that proposed by Margaret Cavendish in her play *The Female Academy* (1662). In the 1690s Mary Astell petitioned Queen Anne for funds to establish an educational institution for girls from noble families. Despite claims that the study of natural philosophy would enhance the moral virtue, modesty, and religious sensibilities of future wives and mothers, attempts to establish such institutions in England and France were generally unsuccessful.

Nevertheless, the exclusion of women from participation in the newly emerging experimental sciences was not inevitable during a period in which its institutional arrangements remained somewhat uncertain and ambiguous. Although the rise of modern science is closely associated with the scientific academies of the seventeenth century, the informal salons of aristocrat intellectuals, the courts of princes, the universities, and the workshops of artisans, also played a part. In seventeenth-century Europe natural philosophy was still a new and precarious enterprise; it lacked a secure base of support within the existing power structure, and its practitioners were not assured of any particular means of earning their keep. Those who were fortunate might find a wealthy and powerful patron willing to support artists, philosophers, or inventors. For ambitious young men, playing the role of intellectual ornament in the courts of Renaissance princes or the salons of the new urban aristocracy was a rather precarious existence. On the other hand, such setting provided a place in which wealthy and aristocratic women could participate in the arts and sciences as patrons and students when they found all other opportunities closed to them.

The medieval convent had provided a setting in which women such as Hildegard von Bingen might pursue learning, but the medieval universities set the precedent for excluding women. European universities were generally closed to women until the late nineteenth century; some did not admit women until well into the twentieth century. Outside the privileged world of the convent and university, guilds, household workshops, and the practice of medicine served as possible pathways to scientific participation for some women. But by the end of the eighteenth century women generally found themselves legally barred from the practice of medicine, surgery, and pharmacology. Similar trends occurred in other fields and institutions. Thus, as the prestige of scientific academies increased during the seventeenth and eighteenth centuries, and the trend towards the professionalization of science and medicine accelerated, the exclusion of women was also formalized. Women had been active participants in the Renaissance courts of Italy, aristocratic learned circles, and the salons that were the precursors of formal academies, but they were unwelcome in the new institutions, despite some controversy as to whether merit, regardless of sex, should be the primary criterion for admission.

Women could continue to engage in informal discussions of science at appropriate social gatherings and study with private tutors, but if they were too active in their pursuit of knowledge even the most privileged women might be subjected to censure and ridicule. Moreover, women who wrote about their scientific inquiries had to publish them anonymously. Women who were able to patronize the sciences but found themselves excluded from serious participation called attention to the lack of education and opportunities for women and warned that society would ultimately find it costly to lose the potential contributions that one half of the human race might have made to science, literature, and the arts.

THE ROYAL ACADEMY OF SCIENCES

Like the Royal Society of England, the Royal Academy of Sciences in France originated from a tradition of informal meetings and networks, but the two institutions developed quite differently thereafter. While the Royal Society of London lacked state funding and remained quite independent of government control, the French Academy became intimately linked to the state and so dependent on royal patronage that the academicians owed their appointments and salaries to the king. The French Academy helped to establish a new concept of science; it was the first institution in Western Europe to turn scientific inquiry from a hobby or avocation into a profession financed and controlled by the state. At a point in time when the word *scientist* had not yet been coined, the Academy served as a specialized niche in which amateurs evolved into professional scientists ostensibly committed to a new intellectual hierarchy based on experimental ingenuity, innovation, and accomplishments. Most of the scientific thinkers of this period supported themselves as physicians, teachers, clerics, or government officials or sought the patronage of wealthy nobles.

Early in the seventeenth century, various groups of intellectuals met privately in and around Paris to explore the new scientific theories then permeating Europe. Many individuals gravitated towards the fashionable private salons that competed for the time and attention of the most amusing, learned, and witty intellectuals and scientists. Private academies attempted to gain the support of wealthy courtiers, who established private laboratories and observatories. Wealthy patrons, however, were not reliable sources of support since they often had short attention spans and quickly turned to other sources of amusement after dabbling in the sciences. When the disadvantages of instability and chronic financial stress clearly outweighed the benefits of independence and spontaneity, the participants turned to the state for money and support.

On December 22, 1666, the Royal Academy of Sciences held its first official meeting in the private library of Louis XIV. Although the king was portrayed as the patron and sponsor of the Academy, he had no personal interest in science. State support for the arts and sciences had become part of France's quest for fame, honor, international recognition, and economic advancement. The Academy owed its official status to Jean-Baptiste Colbert, the king's chief advisor and minister of finance. Colbert saw the Academy as a way to glorify the king and expand the wealth, power, and prestige of France; practical advances in industry, commerce, medicine, public health, and warfare were also anticipated. During the next half-century the Academy grew into what might be characterized as a new organ of government. In exchange for state support, the Academy accepted regulations dealing with its duties and relations to king and state, membership and procedures. Luxuriously accommodated by the king at the Louvre, holding meetings and conducting experiments in the Royal Library, the academy was clearly a creature of the state. Academicians enjoyed many advantages at first. The academy had the resources to pay salaries, set national standards, establish regulatory power over various industries and inventions, and serve the government as a think tank on technical matters. Colbert generally allowed the academicians wide latitude in choosing their projects and governed the Academy only through general directives, but he clearly expected the Academy to perform useful functions such as analyzing the drinking water at Versailles, studying the fish found along the coasts of Brittany and Normandy, drawing up detailed and accurate maps of the tax district around Paris, finding better means of determining longitude and other navigation aids, studying the natural history of useful plants, and so forth.

The financial support granted by the king to the Academy of Sciences was not unprecedented in kind, but was remarkable in its extent. The level of funding awarded to the seventeenth-century Academy of Sciences was equivalent to the annual income of a wealthy French monastery. Support was provided both to individuals, generally in the form of lifetime pensions or lodgings in buildings belonging to the king, and to institutions in the form of subventions large enough for the construction of an observatory or chemical laboratory, and funds for equipment, research assistants, expeditions, and the costs of publishing research reports and catalogues. The contribution of academicians continued even after their deaths, thanks to the understanding that members should allow their colleagues to dissect their bodies. By the end of the seventeenth century, because of economic problems and political decisions, pensions generally declined or even disappeared. Members argued that pensions were essential to providing the leisure, peace of mind, and good morale needed for a cooperative, creative scientific community, but politicians had their own agenda.

During its first phase, Bacon's writings influenced the projects adopted by the Academy, which consisted primarily of assembling catalogues, histories, herbals, and encyclopedias. Interest in Baconian utilitarianism soon declined, and Descartes's philosophy of science became a major influence. Among the useful functions of the Academy were some that made it resemble manufacturing and trading companies, but the Academy had to demonstrate success in scientific research in order to survive and prosper as a vehicle of the new experimental science. This was difficult to do while balancing the demands of the state with an emerging professional identity.

Between 1666 and 1699 the Academy never had more than 34 or less than 19 members in any year. During this period about 60 men were awarded appointments as academicians, but those who actually worked on projects sponsored by the Academy at any one time rarely numbered more than 20. The seventeenth-century French Academy included a higher proportion of members trained as physicians, surgeons, or apothecaries than the Royal Society of England; this lent plausibility to the claim that researches sponsored by the Academy could provide information useful for medical practice and public health, as predicted by Descartes and Bacon. Despite much discussion of the virtues of cooperation among scientists, the Academy fostered the formation of a closed, elite, and exclusive community, which included as one of its principles the exclusion of women. In the 1830s three eminent and learned women had been proposed as member of the Académie Française, the precursor of the Academy of Science. But despite considerable controversy, Mademoiselle de Scudéry, Madame des Moulières, and Madame Dacier were not admitted. In 1910, the year before Marie Curie became the first person ever to win a second Nobel Prize, she was rejected by the Academy of Science. Curie's opponents argued that other women had been rejected by prestigious French academies and that the tradition should be maintained.

Unfortunately for the academicians, Colbert was replaced by the Marquis de Louvois, a man who believed that scientific research should be devoted to useful ends in the service of the king and the state. The reorganization of the Academy in 1699 affected the way in which members conducted research. Academicians were ordered to undertake a variety of practical and even trivial projects, and political appointments led to increased control over the Academy by men unsympathetic to the sciences. The Academy was expanded to incorporate 70 members in hierarchically arranged positions, salaried and controlled by the state.

Members of the Academy were envied and criticized; they were expected to referee disputes, solve technical problems, and serve as a symbol of progressive science. Tensions between the Academy and the increasingly restive French populace generated conflict and political pressures that eventually crippled the institution, but ultimately it was the French Revolution that destroyed the old system. Some of the academicians perished in the Reign of Terror, and others were forced to emigrate; eventually the institution was reorganized on a more egalitarian basis. After the Revolution, the National Institute of Arts and Sciences replaced the old Academy. The new institute was expected to serve as proof of egalitarian French cultural achievements and make useful discoveries. All members of the institute, salaried by the government, were expected to serve as government advisors and functionaries. Compared to the independent scientific and professional societies developing in other countries, the French Institute was a conservative force that tended to undermine the early promise of French science.

SCIENTIFIC SOCIETIES IN THE UNITED STATES

Even during the early colonial period, science was a subject of intense interest in America. The English colonies were particularly influenced by the Royal Society, and several Americans were elected fellows of the Society. In 1663 Governor John Winthrop was elected a fellow of the Society and became chief correspondent of the Royal Society in the Western World. The Reverend Cotton Mather maintained a correspondence with the Royal Society. His father, Increase Mather, had established a scientific society in Boston in 1683 modeled on the Royal Society. This later became the Boston Philosophical Society, but disputes over religion and politics caused its premature demise.

Among the founding fathers of the new republic, Benjamin Franklin and Thomas Jefferson were particularly active in promoting the development of science. Franklin established a group called the Junto, which met weekly to discuss natural history and philosophy. The Junto was the precursor of the American Philosophical Society, the oldest surviving learned society in the United States. The American Academy of Arts and Sciences, incorporated in Boston in 1780, was the second major learned society established in the United States. George Washington, Benjamin Franklin, Thomas Jefferson, Alexander Hamilton, James Madison, and John Adams were among its first members. Various specialized and regional societies formed and dissolved, but only the American Philosophical Society and the American Academy of Arts and Sciences evolved into permanent national learned societies. Engendered in an age permeated by Baconian utilitarianism, the American societies also contributed to the spirit of cultural nationalism. Like the Royal Society, the American academies remained essentially independent of university or government ties.

Government sponsorship of scientific societies and research is a quite recent development in the United States. Following World War I, various societies were grouped into national councils or institutes: the American Council on Education, the National Academy of Sciences, the National Research Council, the American Council of Learned Societies, and the Social Science Research Council. The roots of the National Academy of Sciences go back to two societies founded in the 1840s: the National Institution for the Promotion of Science and the American Society for Geologists and Naturalists (which became the American Association for the Advancement of Science). Scientists and government officials called for an academy to supplement the existing societies and serve as a guide for public policy concerning scientific matters. In 1863, Congress chartered the National Academy of Science to act as the official adviser to the federal government on scientific and technical questions. According to its charter the Academy must investigate and provide reports at the request of any department of the government. Although the government was expected to appropriate funds for the expenses involved in carrying out such studies, no funds were to be paid to the Academy for its services. Every year the National Academy of Sciences, the nation's most exclusive club for researchers, elects 60 new members. The process by which the candidates are selected is very secretive, but the Academy has repeatedly been attacked as an exclusive "old boys' club" that displays marked resistance to the admission of women. Despite the increasing presence of women in scientific research in the second half of the twentieth century, the percentage of women in the Academy, generally no more than 5–10% of the total membership, hardly changed from 1970 to 1992.

In 1945, in his landmark report *Science, the Endless Frontier*, Vannevar Bush, the director of the Office of Scientific Research and Development (OSRD) during World War II, urged the government of the United States to continue its tradition of opening new frontiers by recognizing that although the frontiers represented by new lands for settlement had essentially disappeared, the frontiers of science remained. Like Bacon and Descartes, Bush argued that scientific progress would lead to improvements in human health, well-being, and security. *Science, the Endless Frontier* had been written in response to a letter from President Franklin D. Roosevelt designed to elicit a study of how OSRD's wartime experience might be used to improve the national health and create new enterprises, new jobs, and a better national standard of living after the war.

The government agencies of special interest to twentieth century researchers in the life sciences are the National Institutes of Health (NIH) and the National Science Foundation (NSF). The National Institute of Health was officially established in 1930 to promote research vital to the health and welfare of the nation. The roots of this agency can be traced back to 1887 when Dr. Joseph J. Kinyoun was awarded $300 to establish a bacteriological laboratory at the Marine Hospital on Staten Island as part of the United States Public Health Service. After World War II the National Institute of Health began to proliferate into a series of institutes designed to stimulate disease-oriented research and the training of future scientists. The National Science Foundation, an institution originally dedicated to the support of pure research, was not created until 1950. Obviously quite different from the scientific societies generated by the first scientific revolution, these organizations illustrate the continuing interaction between science and society as negotiated in terms of institutional sponsorship and support. During the years between World War I and World War II, private foundations, such as those established by Andrew Carnegie and John D. Rockefeller, orchestrated and dominated the system of extramural grants that supported much of the scientific research carried out in the United States. It was only in the wake of World War II that "big science" first emerged and the federal government became the great patron of American science.

MICROSCOPES AND THE SMALL NEW WORLD

During the seventeenth century many instruments fundamental to modern experimental science were either invented or first put into use on a significant scale. The projects and goals of academicians encouraged them to adopt new instruments, such as the telescope, microscope, air pump, and barometer, and improve the precision and accuracy of older measuring devices, including clocks, rulers, scales, and thermometers.

Instruments that apparently extended the range of the human senses also created uncertainty about the relationship between the observer and the object under investigation, changed the nature of the questions put to nature, and often intensified old debates. The air pump, for example, was used to answer old questions about the nature of air and the existence of the vacuum. All observers could agree that pumping air out of various kinds of vessels allowed the experimentalist to suffocate mice, put out candles, and so forth, but those who continued to deny the existence of the vacuum argued that removing useful air simply allowed some unknown and subtle entity to enter the vessel and take its place.

Galileo might well assert that "anyone can see through my telescope," but that did not ensure that all observers would interpret what they saw in the same way or that skeptics would not continue to argue that the apparent images were generated by the device itself and bore no relationship to objects in the distant heavens. In other words, instruments can be regarded as untrustworthy as well as theory-laden entities. The telescope allowed scientists to see the immense and distant heavens more clearly than ever, but the stars and planets had been the object of study and speculation for thousands of years. The microscope gave observers access to a small and secret new world on earth. Those who did accept the value and reliability of information gathered with the new instruments had to come to terms with the end of the Baconian ideal of assembling all observable facts about the world because it was possible that further advances in instrumentation would engender infinite numbers of observations about entirely unknown worlds. Even the most enthusiastic investigators had to admit that using the new instruments correctly was quite difficult and that sometimes the instruments exaggerated the faults and foibles of individual investigators.

The art of cutting and polishing stones is quite ancient, as indicated by the discovery of lens-shaped gems in the ruins of Nineveh, Pompeii, and Herculaneum. Such gems may be quite beautiful, and looking through them might be amusing, but they would probably not be of much use to people with defective vision. The Emperor Nero is said to have watched performances in the arena with the aid of a jewel having curved facets which he held up to one eye. Although it is possible that the jewel contained concave facets to correct myopia (nearsightedness), there are few other reports of such uses of lenses until the thirteenth century. There is good evidence that the ancients knew that a convex lens could be used to converge the rays of the sun and act as a burning glass. According to Pliny, doctors produced therapeutic burns on their patients by placing crystalline spheres in the path of the rays of the sun. Pliny did not, however, mention the use of lenses for aiding human vision or for magnifying images. The problem of shortsightedness was considered incurable, but Pliny noted that tired eyes might be strengthened or rested by gazing at the world through green stones or emeralds. The Roman philosopher Seneca noted that small, indistinct writings appeared larger and clearer when seen through a globe filled with water, but he may have thought that water itself caused magnification.

An ingenious effort to analyze the way convex lenses produced a magnified image was made by Ibn al-Haitham (962–1038), known to Europeans as Alhazen. His treatise was translated into Latin in the twelfth century as the *Opticae Thesaurus Alhazeni Arabis*. Alhazen dealt with experimental means of proving various propositions concerning the action of lenses and attempted to relate anatomical features of the eye to the physics of vision. Although Alhazen was interested in philosophical, mathematical, and medical aspects of optics and knew that a sphere could be used to create an enlarged image, he did not suggest practical applications of this phenomenon. In his discussion of the work of Alhazen and his own studies of the properties of light, Roger Bacon (1214–1292) drew attention to the magnifying properties of lenses and the design of burning glasses, mirrors, and magnifiers. Some passages in his *Perspectiva* seem to describe telescopes, but modern scholars suggest Bacon was actually referring to objects that were inside well-known magnifiers, such as a glass dome, and illusions that could be created with sets of mirrors. Bacon did note that old men with weak eyes could use lenses curved on one side as an aid to reading.

Medieval scholars could have used lenses held in the hand just above a book or set directly on the page like some modern magnifiers, but this would make it very difficult to turn pages or write. This problem was solved towards the end of the thirteenth century with the invention of spectacles, a device that secured appropriate glass lenses in front of the eyes. The inventor of the first spectacles might have been a glazier, that is, an artisan who created windows, glass ornaments, and disks, but the grave marker of Salvano d'Aramento degli Amati, a nobleman of Florence (d. 1317), states that he had invented spectacles and kept his methods secret. Alessandro della Spina of Pisa (d. 1313) claimed that he had learned the secret of making spectacles and shared the discovery with others. Whatever the truth of such claims, by the beginning of the fourteenth century, spectacles were mentioned quite frequently as a well-known but still much admired invention. Spectacles were rare and expensive items that were proudly mentioned in account books and wills as valuable property. Like literacy itself, eyeglasses became a mark of distinction and wisdom. Ancient sages and saints were depicted with quite anachronistic spectacles in their hands or on their noses.

The use of spectacles must have occasioned a profound effect on attitudes towards human limitations and disabilities. Certainly spectacles made it possible for scholars and copyists to continue their work, but more fundamental, they accustomed people to the idea that certain physical limitations could be transcended by the use of new technologies. These early corrective lenses were all of the convex or converging type and could not help victims of myopia. It was not until the sixteenth century that the use of concave lenses to correct nearsightedness was first mentioned, but from the sixteenth century on spectacles were generally mentioned in every book dealing with eye disorders. Giambattista della Porta's *Natural Magic* (1589) contains a section called "Of Strange Glasses," which describes the use of lenses to correct defects in human vision. Della Porta explained that concave lenses were used to see things that were far away, while convex lenses were used to see things close to the eye. He claimed that he had been able to combine convex and concave lenses to help friends with vision problems so that they could see things both far away and nearby more clearly.

By the sixteenth century the making of lenses was an established industry and spectacle makers had accumulated a good deal of empirical knowledge about optics. The invention of the telescope and the microscope was probably a somewhat accidental outcome of the art of making spectacles. Several texts and letters from the late sixteenth century seem to refer to the use of primitive telescopes. The first microscopes appeared sometime during the period from 1590 to 1609; the idea of the instrument spread very quickly, and many ingenious modifications and variations were soon introduced. Credit for the invention has been attributed to three Dutch spectacle makers—Hans Janssen, his son Zacharias, and Hans Lippershey—but some writers cited Galileo as the inventor of both the microscope and the telescope. Although the telescope designed by Galileo could be used as a microscope by increasing the distance between its lenses, it provided only a very restricted field of view. A description of such instruments written in 1614 notes that the stars could be studied by using a lens system in a tube about 2 feet long; but to study tiny objects, the tube had to be two or three arms' length in size. While Galileo did not actually invent the telescope, his astronomical discoveries revolutionized ideas about the heavens. With the telescope he discovered the moons of Jupiter and the phases of Venus; under the microscope he saw flies that seemed to be as big as lambs.

Perhaps as early as 1590 Zacharias or Hans Janssen created a crude compound microscope by combining a concave and a convex lens at the ends of a brass tube which was about an inch in diameter and 18 inches long. There is no evidence that the Janssens ever made any significant observations with their magnifying device. Cornelius Drebbel seems to have obtained one of the early instruments constructed by the Janssens. After adding his own improvements, he attempted to bring the device to the attention of scientists and scholars. Improvements in the construction of lenses have been attributed to Lippershey, who constructed an unusual binocular telescope.

Even with magnification no greater than 10-fold, the first microscopes provided exciting new images; creatures just barely visible to the naked eye became complex and bizarre monsters. The ubiquitous flea was such a popular object of study that magnifying lenses were often referred to as flea glasses. Some philosophers warned against believing in the images created by unnatural devices that misled the senses; less sophisticated skeptics considered use of such instruments a form of witchcraft. For example, George Stiernhielm, a Swedish poet who amused himself with scientific experiments, was denounced as a sorcerer and an atheist by a clergyman who had been persuaded to look at a flea through a magnifying glass.

With the new magnifying glasses, seventeenth-century scientists assembled a remarkable body of observations and experiments, beginning with the work of Francesco Stelluti in 1625. His strikingly detailed illustrations of the parts of the bee, obtained at magnifications of 5 and 10 diameters, were published by the Academy of the Lynx Eye. Giovanni Borelli (1608–1679), better known for his studies of muscles, investigated objects as varied as the blood circulation, nematodes, textile fibers, and spider eggs. Athanasius Kircher (1602–1680) used the microscope in his studies of putrefaction and disease. Although many seventeenth-century scientists made some use of the microscope, Antoni van Leeuwenhoek, Marcello Malpighi, Robert Hooke, Jan Swammerdam, and Nehemiah Grew raised microscopy to a level not superseded until the middle of the nineteenth century.

Antoni van Leeuwenhoek (1632–1723), a man of immense energy and curiosity, was the most ingenious microscopist of the seventeenth century. The scope of his investigations was so extensive that well into the nineteenth century scientists were warned that they should take care to examine Leeuwenhoek's writings before claiming to have made an original microscopic observation. Leeuwenhoek's father, who died when Antoni was only 6, was a basket maker, and his mother was the daughter of a brewer. The family lived in Delft, Holland, a city famous for the production of beer, textiles, and chinaware. When his mother remarried, Leeuwenhoek was sent to grammar school at Warmond, a village near Leiden. At the age of 16, Leeuwenhoek was employed as bookkeeper and cashier by a linen draper (cloth merchant) in Amsterdam. He returned to Delft 6 years later and bought a house and shop. In 1660 he began to add a variety of civil service positions, including usher, chief warden, and wine gauger, to his own business. Despite the demands of his business and municipal duties, he was able to devote so much of his time to his hobbies that he was accused of neglecting his family. Perhaps there was some merit in this charge; he survived two wives and all but one of his six children. Yet Leeuwenhoek himself lived, as recorded on his tomb, for 90 years, 10 months, and 2 days.

Leeuwenhoek's career as a scientist began when he was about 39 years old, but his inspiration for making microscopes is uncertain. One possibility is that he got the idea while working with special lenses that linen drapers used at the time to inspect the quality of cloth. Another possibility is that Leeuwenhoek learned about microscopes from someone with a real scientific interest in the device. Many possible mentors have been suggested, but when the ages and career paths of the various individuals are taken into account, it appears likely that Robert Hooke's *Micrographia* (1665) could have been the model for Leeuwenhoek's early microscopic work. The preface of the *Micrographia* explains how to make and use simple magnifiers, also known as single microscopes, although Hooke found these instruments very difficult to use and preferred to use the compound or multilens microscope. Some apparently random specimens that Leeuwenhoek sent to the Royal Society actually reproduce items described by Hooke. By 1674, however,

A portrait of Antoni van Leeuwenhoek by Jan Verkolje in 1686

Leeuwenhoek was publishing accounts of his work that were superior to Hooke's. Leeuwenhoek was interested in searching for truly novel entities, while Hooke was primarily interested in magnifying conventional objects.

Leeuwenhoek's first simple microscope or magnifier was a tiny lens, ground by hand from a globule of glass, clamped between perforated metal plates; the device had an attached specimen holder. As his interest and skills increased, Leeuwenhoek learned to make very small biconvex lenses of short focal length. With these tiny, high-powered lenses, Leeuwenhoek's simple microscopes were superior to most seventeenth-century optical systems, whose multiple lenses invariably produced multiple aberrations and artifacts. During the 50 years he conducted his microscopical researches, Leeuwenhoek probably produced more than 500 lenses. When he died, more than 400 microscopes and magnifying glasses were found. A cabinet with 26 instruments and extra lenses was bequeathed to the Royal Society. Members of the Society tested some of these lenses. Magnifying powers of from 50- to 200-fold were found, but unfortunately these instruments were lost within 100 years of Leeuwenhoek's death. The magnification of Hooke's compound microscope was generally comparable to the low power range of a conventional modern microscope; he could obtain a direct magnification of about 20–50 times. Using good single lenses about 10 times as powerful as Hooke's compound system, Leeuwenhoek generally worked at a magnification of about 100- to 200-fold. Most of his lenses were made by grinding and polishing selected fragments of glass or even grains of sand, but some were probably made from molten glass.

Scientists and historians who were unsuccessful in imitating Leeuwenhoek's work generally concluded that he used special secret methods or that he might have used compound lenses. Others have demonstrated that simple microscopes and appropriate methods of sample preparation and illumination can produce the results reported by Leeuwenhoek, that is, images not significantly inferior to the magnifying power of modern optical instruments. Leeuwenhoek's methods made it possible to see cells, nuclei, molds, and microbes. Various bacteria would be just visible and pond infusoria would be clearly visible with a good Leeuwenhoek microscope. While Leeuwenhoek was making his simple microscopes, other instrument makers were turning out instruments that look much like the modern compound microscope. Nevertheless, many distinguished nineteenth-century scientists, such as Robert Brown and Charles Darwin, were still using simple microscopes in the mid-nineteenth century. Victorian compound microscopes were more ornate, complicated, and fashionable, but they were not really better than the best of the simple microscopes.

In one of his early letters to the Royal Society, Leeuwenhoek described himself as lacking in style for written expression because he was brought up only "to business," not to languages or learning. Moreover, he frankly informed the learned gentlemen that he preferred to keep his methods to himself because he did not gladly suffer "contradiction or censure from others." Perhaps Leeuwenhoek's inability to read the learned Latin treatises of his contemporaries was an asset that kept him remarkably free of prevailing philosophical dogmas and preconceptions. Nevertheless, he did not work in total intellectual isolation. He read the work of Dutch authors, Dutch translations of standard works, studied the illustrations in books he could not read, and, when necessary, he invoked the help of friends and professional translators. Perhaps he missed some useful information, but he maintained his independence of mind. Even the knowledge that ignorant people considered him a magician who showed people things that did not exist and that some learned men did not accept the truth of his writings did nothing to diminish his confidence in his own observations. Leeuwenhoek regarded speculation as something that belonged to an academic world that did not concern him. Generally, he presented his observations as facts, but he was certainly willing to suggest possible interpretations.

Leeuwenhoek's work became known to the Royal Society through his friend the physiologist Regnier de Graaf. Letters from Leeuwenhoek to the Royal Society were edited, translated into English or Latin, and published in the *Philosophical Transactions*. In 1680 Leeuwenhoek was unanimously elected a Fellow of the Royal Society. For his benefit, the certificate was written in Dutch. Although he treasured this honor, he often saw fame as a nuisance that brought many visitors in its wake, including heads of state and curiosity seekers. To cope with interruptions, Leeuwenhoek seems to have kept certain simple demonstrations on hand, but he kept his best microscopes to himself. Despite his reluctance to suffer interruptions, some visiting scholars and scientists presumably brought him useful information about the work of other microscopists.

The majority of Leeuwenhoek's 400 letters to the Royal Society have been preserved, but it was not until 1981 that Brian Ford discovered several packets of specimens still attached to a letter dated June 1, 1674. Among the specimens that have been identified were fine sections of cork, sections of elder-pith, bits of dust, spores, house mites, pollen grains, cotton seeds, sections of the optic nerve of a cow, and dried specimens of pond microbes. The finest parts of Leeuwenhoek's hand-cut section of cork were remarkably thin and the cell walls were clearly visible when examined by scanning electron microscopy. Some specimens contained a few blood cells, which presumably came from Leeuwenhoek's razor. Another packet contained a sample of dried algal mats, which were called "paper from heaven." After looking at this material under a microscope, Leeuwenhoek concluded it was a sheet of dried plant material only superficially resembling paper. When these samples were reconstituted, some of the microscopic pond organisms discovered by Leeuwenhoek could be seen.

Unencumbered by a rigid philosophical system or a fashionable theoretical program, Leeuwenhoek allowed his insatiable curiosity to determine the course of his microscopical researches. His work reflects common sense, pragmatism, originality, and lack of scholastic prejudices. Crystals, minerals, plants, animals, water from different sources, scrapings from his teeth, saliva, seminal fluid, even gunpowder were examined under his lenses. Unfortunately, Leeuwenhoek was a secretive man who could be quite contemptuous of those who lacked his own curiosity, dexterity, and persistence. Thus, when asked why he did not teach his methods to others, he responded that he saw no reason to train young people to grind lenses because most students lacked curiosity and sought instruction only as a means of making money or to earn a reputation. Such motives, he contended, did not inspire people to discover the world hidden from our sight.

While Leeuwenhoek obviously took great joy in his work, he also took pains to make it as accurate and quantitative as possible. Having no standard units of measurement available for microscopic work, he used simple objects for making comparisons, such as strands of hair or grains of sand, and stated his measurements in terms of fractions or multiples. For some objects he offered size estimates in terms of more complex "biological units" such as the eye of a louse. His eminently practical scale allowed him to communicate with his contemporaries and provided good clues for historians and scientists attempting to recreate his experiments.

In the course of his long and vigorous career, Leeuwenhoek made so many discoveries that it is impossible to cover them in any detail. His studies of microorganisms, sperm cells, and the capillaries were especially important, but he also contributed to pioneering studies of the life history of insects, structural studies of plant and animal tissues, parasitology, microbiology, and the battle against the doctrine of spontaneous generation. Leeuwenhoek studied the mating behavior and life cycle of eels, intestinal parasites, insects, and other creatures said to be spontaneously generated. Under the microscope, even mites, bees, and fleas were shown to be very complex creatures, and Leeuwenhoek was convinced that even the smallest microscopic creatures were produced by parents like themselves. Although Leeuwenhoek also studied reproduction in plants, he apparently thought that the flowers were essentially decorative, and he did not seriously study their reproductive parts.

In 1661 Marcello Malpighi had observed the capillaries and blood corpuscles within them, but he apparently thought that the blood cells were fat globules. Leeuwenhoek was probably unaware of Malpighi's work when he rediscovered the blood corpuscles in 1674 and the capillaries in 1683. He attempted to measure the erythrocytes and their nuclei in fishes, frogs, and birds, and he studied the red blood cells of humans and other mammals. In demonstrating that the capillaries link the arteries and veins, Leeuwenhoek helped to complete Harvey's work on the circulation of the blood. Leeuwenhoek called attention to the fact that the passage of blood from the arteries into the veins took place in vessels so thin that only one corpuscle could be driven through at a time. From numerous observations of the capillary network in the tails of tadpoles and eels, he ventured the opinion that the same thing must occur in humans but could not be seen because of the thickness of the skin.

In 1677 Johan Ham, a medical student, seems to have made the first observation of sperm in the seminal fluid of a man with gonorrhea. Ham, who told Leeuwenhoek about this discovery, assumed that the "animalcules" were products of putrefaction related to venereal disease. The idea that sperm were a sign of disease proved very persistent despite Leeuwenhoek's demonstration that these entities were found in the semen of healthy men and animals, including dogs, rabbits, birds, amphibians, and fishes. Leeuwenhoek concluded that the animalcules were normal constituents of semen throughout the animal kingdom. In a letter to the Royal Society he acknowledged that whole universities full of scholars refused to believe that there were living creatures in the male seed, but their opinions did not concern him because he knew that he was right. His studies of sperm and eggs from fish and frogs led him to reject the prevailing view that vapors from seminal fluid were responsible for fertilizations. Although he was unable to demonstrate the actual union of sperm and egg, he believed that spermatozoa penetrated the egg. Assuming that motion was essentially equivalent to life, he was convinced that the sperm provided the origin of new animal life; the egg, in contrast, appeared to have no independent capacity for motion.

In 1674 Leeuwenhoek recorded his discovery of the true nature of microorganisms. Since he assumed that motility was characteristic of living beings, he concluded that the tiny moving things that he had observed under the microscope must be minute animals. About 30 of his letters to the Royal Society deal with microscopic organisms, including rotifers, protozoa, and bacteria. His reports generated a great deal of excitement and also some skepticism, to which Leeuwenhoek responded by sending testimonials from reliable witnesses such as doctors, jurists, and ministers.

Like Leeuwenhoek, Jan Swammerdam (1637–1680) was one of the great Dutch scientists produced in that country's golden age. In outlook, education, and methodology, Leeuwenhoek and Swammerdam differed in everything but their devotion to testing the limits of microscopy. Almost obsessively methodical in his approach to his work, Swammerdam lived a disorderly, restless, and tormented existence. His father was a prosperous apothecary who was interested in the natural sciences, but he insisted that his son should take religious orders rather than study natural history. As a compromise, Swammerdam was sent to the medical school at Leiden. Interested in research rather than medical practice, Swammerdam was constantly at odds with his father, who seems to have been a stingy, cantankerous man obsessed with the idea that his adult son should support himself rather than waste his time studying insects.

Using the microscope as a tool rather than an end in itself, Swammerdam undertook systematic studies of entomology and comparative anatomy. He investigated the fine structure of insects, plants, animals, and humans and discovered the minute "seeds" of ferns. Fascinated by the activities of insects, Swammerdam willingly let lice bite him so that he could observe the operation of their mouth parts. His dissections of human cadavers led to important discoveries about the uterus, lymphatic system, spinal medulla, and the organs of respiration. His proof that the lungs of a newborn mammal would float in water if the animal had begun to breathe on its own but would sink if respiration had not begun had important implications for forensic science.

After publishing a general history of insects in 1669, the impoverished and sickly entomologist contracted malaria, a mosquito-borne disease that was to trouble him for the rest of his life. When his father sent him to the countryside to recuperate, Swammerdam used the time for his classic research on the mayfly. For Swammerdam, the study of the mayfly was not just a scientific account of a rather insignificant insect, but a deeply felt reflection of his religious commitment. Knowledge of the brief life of the mayfly might, he hoped, give human beings a vivid image of the shortness of earthly existence and so inspire them to a better life. Chronically ill and desperate for money, he tried to sell his collections of specimens and even his microscopes. Eventually he became involved in religious mysticism and abandoned science as a vain and unworthy enterprise.

The bulk of Swammerdam's writings were not published until 1737. After a precarious existence in various private collections, his manuscripts were purchased by Herman Boerhaave, who published them in Dutch and Latin with plates engraved from Swammerdam's own drawings. Swammerdam's *Bible of Nature* was the first major study of insect microanatomy, transformations, and classification. The work also contains observations of marine invertebrates and amphibian metamorphosis. Working with minute insects, Swammerdam had constructed special instruments that allowed him to refine the art of dissection. Like Leeuwenhoek, Swammerdam had many different microscopes for specialized purposes. His dissecting tools included scalpels carefully ground under a magnifying glass and fine glass tubes that could be used to inject air, colored liquids, ink, or wax into his specimens.

During the eighteenth century, Swammerdam's work on insects was used in support of the theory of embryological development known as preformationism. His studies of the stages of insect metamorphosis indicated that it was a precisely predetermined sequence of events during which later stages grew from tiny parts already present at earlier stages. Comparing the development of humans, insects, and amphibians, Swammerdam suggested that a process like molting occurred in each case; that is, the new parts developed under the old skin, which was then shed.

Leeuwenhoek and Swammerdam proved that the so-called lower creatures were complex and interesting and that careful studies of their life cycle could help dispel the myth of spontaneous generation. Both argued that the microscope revealed that even apparently simple creatures consisted of complex parts that were unlikely to arise from mud and slime. The microscope could also be used as a means of extending and refining classical approaches to the study of the structure and function of the human body, as in the work of Marcello Malpighi (1628–1694). Second only to Leeuwenhoek for the breadth of his interests and his enthusiasm for microscopic studies, Malpighi was a pioneer in physiology, embryology, plant anatomy, comparative anatomy, and histology. While studying philosophy at Bologna, Malpighi became interested in the new experimental approach to natural philosophy. Bartolomo Massaari, professor of anatomy, invited Malpighi to use his private library and become a member of his scientific academy. After completing his studies of medicine and philosophy, Malpighi established a private practice and obtained a position as lecturer in logic at the University of Bologna, where he became notorious for his anti-Galenist views. In 1656 a chair of theoretical medicine was created for him at the University of Pisa. Malpighi found the intellectual climate of Pisa a decided improvement, especially after he became a member of the *Accademia del Cimento* (Academy of Experiments). The international nature of the seventeenth-century scientific community is reflected in Malpighi's correspondence with Henry Oldenburg,

Secretary of the Royal Society, and his election as a Fellow of the Royal Society.

Especially interested in the physiology of respiration, Malpighi conducted extensive microscopic examinations of the capillaries, the fine structure of the lungs, kidneys, cerebral cortex, plant microanatomy, and the natural history of invertebrates. Historians have often called attention to the fact that Malpighi was born in 1628, the year in which William Harvey published *De motu cordis,* and that it was Malpighi who demonstrated the physical existence of Harvey's hypothetical anastomoses. When Malpighi began his studies in 1660, the lung was thought of as a fleshy organ consisting of a porous parenchyma. Without the aid of the microscope, the smallest divisions of the blood vessels and the ramifications of the windpipe were, of course, invisible. Anatomists assumed that blood and air simply mixed within the parenchyma of the lung, but Malpighi demonstrated that the lung was composed of membranous vesicles filled with air. Further investigation revealed that blood did not leak out of its proper vessels into the air spaces, but made its way from artery to vein through the minute structures known as capillaries. By washing out the blood, inflating the lungs through the windpipe, and allowing the tissues to dry, Malpighi was able to trace the ramifications of the trachea or windpipe to a network of thin-walled bladders.

Despite the crudity of his instruments and what he called his own clumsiness, through reason, observation, and experiment, Malpighi linked his structural studies to new theories of function for the lungs and kidneys. Urging others to follow his example, he explained that he did not come to his conclusions about the structure of the body by means of books, but through long, patient, and varied use of the microscope. While his instruments and techniques were limited, Malpighi carefully described his use of mercury and wax injections, methods for staining microscopic preparations with ink and other readily available liquids, and the use of various kinds of lenses and means of illuminations. Malpighi explored the fine structure of many organs, always trying to trace the smallest elements of the tissue under study as a means of understanding its physiological significance. The single best known microscopic observation of the seventeenth century, however, is probably the description of the cellular structures that Robert Hooke found in cork.

Overshadowed by his great rival Sir Isaac Newton, Hooke was one of the seventeenth-century's greatest inventors and designers of scientific instruments. His contributions to astronomical instrumentation were particularly valuable, but he also made many suggestions for the improvement of the microscope and methods of illumination. Indeed, the study of light was one of his main theoretical preoccupations. As Curator of the Royal Society

Robert Hooke was responsible for the demonstrations performed at the weekly meetings. In 1663 he was assigned the task of undertaking an extensive program of microscopical investigations so that he could present at least one novel observation at each meeting. Hooke presented studies of various insects, such as fleas, lice, and gnats, and demonstrated different types of hair, various textiles, the point of a needle, the edge of a razor, molds, moss, and the cells seen in sections of freshly cut cork. Microscopic studies were quite difficult and even painful for Hooke who suffered from an eye disorder.

Hooke's *Micrographia* (1665), which greatly stimulated interest in microscopy, was a popular treasury of information about microscopic technique and illustrations of the wonders that appeared when familiar objects were magnified. Although writing for a general audience, Hooke used the *Micrographia* as a vehicle for his theoretical work and speculations about the nature of light, the relationship between respiration and combustion, and the origin of fossils. Hooke described the streaming juices in live plant cells, but his colleague Nehemiah Grew (1628–1712) carried out more extensive microscopic studies of plant and animal anatomy.

Nehemiah Grew was the only son of the Reverend Obadiah Grew, an anti-Royalist clergyman who lost his position when Charles II was restored to the throne. After graduating from Cambridge, Grew left England to study medicine at the University of Leiden. Having established a successful private practice, Grew devoted much of his energy to studies of the anatomy of plants and animals. As he explained in an essay presented to the Fellows of the Royal Society in 1671, his research was guided by the idea that animals and plants must possess fundamentally similar structures because both were designed by "the same Wisdom." When Henry Oldenburg died in 1677, Grew became Joint Secretary of the Royal Society with Robert Hooke. Grew's treatise, *The Comparative Anatomy of Stomachs and Guts begun*, appeared as an appendix to a catalog of specimens in the museum of the Royal Society. Modest about his work, Grew similarly entitled part of his great work on plants *The Anatomy of Vegetables begun* (1682). This study was the first part of *The Anatomy of Plants,* which included *The Anatomy of Roots, The Comparative Anatomy of Trunks*, and finally *The Anatomy of Leaves, Flowers, Fruits and Seeds*.

The Anatomy of Plants helped to revive the science of botany, which had stagnated, except for the production of herbals of dubious value, since the time of Theophrastus. Grew studied the vascular tissue of plants, noted its tubular nature, and discussed its different distribution in roots and stems. Although his discussion was more detailed than that of Malpighi, Grew acknowledged Malpighi's priority in the discovery of the vascular system of plants. Grew speculated about the possibility that flowering plants might undergo sexual reproduction and suggested that the flowers carried the sexual organs of the plant. He recognized the pistil as the female part but was uncertain about the purpose of the stamens. Calling attention to the fact that the pistil and stamens were sometimes present in different flowers and sometimes appeared in the same flower, he suggested that some flowers were hermaphroditic like snails. In 1694, 12 years after the publication of Grew's *Anatomy of Plants*, Rudolph Camerarius described plant sexuality based on anatomical and experimental studies.

Comparative Anatomy of Stomachs and Guts begun was the first zoological book with the phrase "comparative anatomy" in the title and the first book to deal with one system of organs by the comparative method. Viewing human beings as the highest type, Grew compared all other animals to the human standard. Approaching more physiological ideas, Grew argued that the morphological significance of any organ could not be grasped until its function had been considered. This point was illustrated by a discussion of the foods eaten by each animal and the type of gut required to process different foods. By studying freshly killed animals, Grew was able to see the peristaltic action of the gut and study the villi of the mucosa. Patient and thorough in his research, Grew dissected the guts of 40 different species of birds alone. Many species of fish were also examined, though clearly Grew found the mammals most worthy of study. The strong religious influence that pervaded Grew's life appeared most clearly in his last work, *Cosmologica Sacra* (1701), in which he attempted to demonstrate the wisdom of God in the excellence of His creations.

Unfortunately, none of the great seventeenth-century microscopists seems to have been willing or able to train pupils or attract disciples capable of continuing their work. There were, on the other hand, critics who regarded the instrument and its use as somewhat sacrilegious, that is, an attempt to pry into matters God did not want human eyes to see. Some of the most respected physicians and philosophers of the late seventeenth century, such as Thomas Sydenham and John Locke, condemned microscopy as a distraction irrelevant to understanding the nature of life, disease, and human beings.

During the eighteenth century, the microscope became for the most part a toy to entertain wealthy amateurs and self-styled virtuosi. The instrument that can be considered a system of "brass and glass" became one in which the "brass" aspects were the subject of much elaboration, sometimes in the form of silver and gold ornamentation, while the lenses, which determine

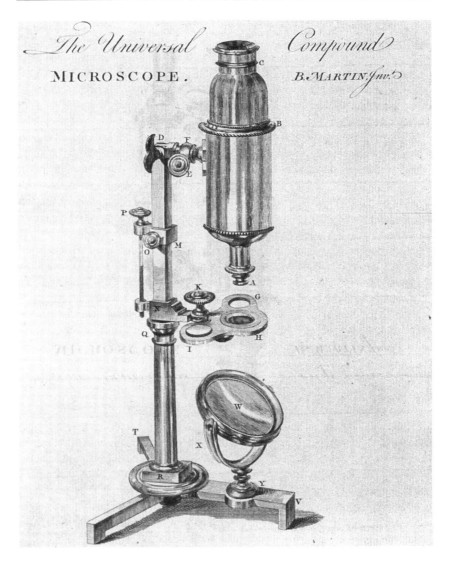

An eighteenth-century universal compound microscope

resolution and magnification, were often neglected. Still, the microscope became immensely popular in the eighteenth century among wealthy amateurs who developed considerable skill, if not critical judgment, in their attempts to amuse themselves and impress their friends. Their willingness to pay for quality instruments stimulated instrument makers to improve the mechanical design and convenience of the instrument and to provide better forms of sample preparation and manipulation.

The remarkable accomplishments of the seventeenth century microscopists were not significantly advanced until the middle of the nineteenth century, when serious problems with the lens systems were finally solved. Chromatic and spherical aberrations were the most serious problems, although poor quality glass, which produced lenses that were cloudy and contained distracting bubbles, was also troublesome. As more convex lenses were fashioned to increase magnification, lenses became more like prisms, separating light into different wavelengths. Although some physicists thought that the problem of chromatic aberration was insoluble, not all lens makers were willing to accept this declaration as the final word on the subject. Chromatic and spherical aberration were less of a problem with single lenses than for compound microscopes, but most microscopists preferred the compound microscope because of its size and shape. The length of the body of the instrument allowed it to stand on a table at a convenient height; thus, the microscopist could easily take notes while using the instrument. The simple microscope had to be held in the hand; note taking involved frequent changes of position.

While it is true that chromatic aberration is inevitable when single lenses are used with ordinary light, the problem could be avoided by using monochromatic light or by constructing doublet or triplet lenses with different indices of refraction. The latter solution was primarily inspired by considerations of the lens system of the human eye. The construction of achromatic lenses is generally attributed to John Dollond (1706–1761), but he may have learned about previous attempts to create such lenses by combining a convex lens of crown glass with a concave lens of flint glass. John Dollond was a silk weaver and an amateur scientist with a special interest in theoretical optics, including the work of Isaac Newton and the mathematician Leonhard Euler (1707–1783). After joining his son's optical business in 1752, Dollond began investigating the problem of making achromatic lenses. Using a combination of crown glass and flint glass, Dollond created a successful lens system for a telescope and took out a patent. Similar solutions were reached by others at about the same time, but it was Dollond who won the Copley Medal of the Royal Society for a telescope using his achromatic lenses. In

1761 the former silk weaver was elected a Fellow of the Royal Society and appointed optician to the king.

In the 1820s Giovanni Battista Amici (1786–1863), astronomer and mathematician, initiated a series of improvements in the construction of the microscope. After serving as professor of mathematics at the University of Modena from 1815 to 1825, Amici became astronomer to the grand Duke of Tuscany and director of the observatory at the Royal Museum in Florence. Using his improved microscope, Amici made some of the earliest observations on the growth of the pollen tube. The nature of fertilization in flowers was not, however, completely clarified until the 1880s. A solution to the problem of the accumulation of spherical errors in lens combinations was provided in 1830 by Joseph Jackson Lister (1786–1869), father of Joseph Lister, the surgeon who developed the antiseptic system. Lister's design was considered especially significant because it was based on optical theory rather than trial and error. The Lister-type microscope, with its modern system of optics and familiar triangular configuration, however, was named for John Joseph Lister, who was not related to Joseph Lister.

Microscopists are, of course, concerned with both resolution and magnification. Resolution is the ability to distinguish two separate points or details on an object as separate points. Since in practice it is resolution that determines the limits of the details that can be studied, resolution is more significant than magnification. Several levels of magnification are possible with simple magnifiers and ordinary light microscopes. At about 10-fold magnification familiar objects can be seen more clearly. Low-power microscopy, which involves magnifying objects about 50 times, reveals interesting anatomical details in plant and animal sections and is useful in visualizing larger pond organisms; Hooke usually worked at this level. A magnification of about 300-fold reveals the interior components of cells, including the nucleus and mitochondria, and various microorganisms, including bacteria. Even higher magnification and maximal resolution can be achieved with oil-immersion microscopy.

At least in principle, immersion microscopy had been anticipated by Robert Hooke for use with the simple microscope. Amici is often credited with priority in the introduction of immersion lenses, but Sir David Brewster (1781–1868) had adopted immersion microscopy as early as 1812. Attempting to develop an achromatic lens, Brewster suggested that the front element of a microscope's objective lens could be immersed in the liquid in which the object of study was mounted. Brewster may have been the first microscopist to use filters to produce monochromatic light. These ideas were elaborated in his work *A New Treatise on Philosophical Instruments* (1813).

Working independently of Brewster, Amici had developed immersion microscopes as a means of diminishing the loss of light in high-power lens systems in order to improve definition. Oil-immersion microscopes can provide magnification up to 600-fold and resolution near the limits imposed by the nature of light.

Further advances in microscopy resulted from the collaboration of the physicist Ernst Abbe (1840–1905) and the instrument maker Carl Zeiss (1816–1888). One of Abbe's goals was to remind microscopists that increased magnification without improved resolution was futile. In their skills and education, Abbe and Zeiss nicely complemented each other; one had expertise in theoretical science and mathematics, and the other had essential practical experience. Their attempts to develop achromatic compound lenses were frustrated by the limited range of optical glasses then available, but their work stimulated the search for new kinds of glass. Abbe developed the apochromatic objective in 1886, improved the oil-immersion microscope, and developed oil-immersion objectives with at least eight component lenses. The substage condenser was developed in the 1870s to complement his high-power objectives.

Phase contrast microscopy, a modification of the light microscope, was developed by Fritz Zernike (1888–1966), professor of physics at the University of Groningen and winner of the 1953 Nobel Prize. Structures that differ in refractive index from the surrounding matrix appear as differences in intensity under phase contrast microscopy. This approach was particularly suited to the study of biological specimens, including living cells, which are often colorless and transparent. In favorable cases the phase contrast microscope obviated the need for dehydration, fixation, staining, and other aspects of sample preparation that are likely to introduce artifacts. Although the production of phase contrast microscopes began during the early years of World War II, such instruments were not generally available until after the war.

Improvements in the light microscope, as well as new methods of sample preparation, made it possible to explore the fine structure of complex organisms and the nature of microorganisms. After many of the technical difficulties of optical microscopy had been solved, however, one fundamental problem—the finite size of light rays—put a limit on the usefulness of the instrument. To obtain higher magnification and resolving power, microscopists had to find ways of working with rays having a shorter wavelength than ordinary light. One approach to extending the limits of resolving power was to use ultraviolet microscopes. Since glass filters out short ultraviolet rays, standard lenses had to be replaced with quartz-transmitting or glass-reflecting lenses. In theory, this should have doubled resolution; however, this approach creates many practical problems, foremost being the fact that the human eye is not sensitive to light in the ultraviolet range. Originally, in ultraviolet microscopy the final image was recorded on a photographic plate and the final prints were studied, but eventually such microscopes were combined with television cameras. Fused quartz lenses were used in the original ultraviolet microscopes, but synthetic fluorite was also found to be appropriate. Coverslips and slides also had to be made of fused quartz, and substitutes for conventional mounting media and immersion oils were required. All these factors made the use of ultraviolet microscopy complicated and expensive. During the 1930s such microscopes were used to locate nucleic acids within the cell and to study their behavior during cell division.

Biological studies have been profoundly affected by the development of the electron microscope, an instrument some 100 times as acute as the light microscope. It has been said that in retrospect the electron microscope was an obvious invention since the basic discoveries that made the instrument possible occurred in the 1850s, but the first primitive instruments were not produced until the 1930s. On the other hand, while the theoretical possibility of the instrument might have been obvious in the early twentieth century, most physicists thought that the device could not be practical because almost everything put under the electron beam would burn to a cinder. Given the many fundamental problems that had to be resolved before the electron microscope became a useful tool, it would seem impossible for any one individual to claim credit for having invented it. However, this did not prevent several patent battles involving industrial firms, independent scientists, research laboratories, and universities in Germany, Brussels, and the United States. By 1932 the German physicists Max Knoll and Ernst Ruska had built a model they called a super-microscope. They were able to obtain magnifications of about 400-fold, but most of their images were so out of focus that they were of no practical use. By 1933 Ruska had built the instrument that might be regarded as the ancestor of the first generation of successful electron microscopes. Pictures of a piece of aluminum foil and bits of cotton fiber were obtained at a resolution of about 500 angstroms. Within 2 years an improved instrument produced photographs of the legs and wings of a housefly at a resolution of about 400 angstroms. About the same time, Ladislaus Laszlo Marton at the University of Brussels built a microscope, which he planned to use in the study of bacteria. Because of the fuzzy details and the fading of the image, his final results were no better than those from ordinary microscopes.

In 1934 Marton published the first successful electron micrograph of a biological specimen: a thin section of a sundew leaf impregnated with osmium salts. It was soon apparent that the major problem for microscopists would be preparing specimens in a way that would exploit the possibilities of the new instrument. As Marton discovered after publishing his photographs in the British journal *Nature* in 1934, not everyone was excited about the prospects of seeing objects previously invisible. When Marton gave a talk about the electron microscope to an audience of medical and biological scientists, immunologist and Nobel Laureate Jules Bordet protested, more or less facetiously, that biologists were having enough trouble interpreting the images produced by the light microscope. Marton published what he considered his first successful bacterial micrographs in 1937. As scientists and instrument makers became convinced of the potential of the electron microscope, several teams entered the competition to produce a practical scientific instrument and manufacturers of electrical instruments decided to explore its commercial possibilities. Research on the electron microscope was somewhat slowed by World War II, but early versions of the instrument were used for the study and development of war materials such as metal alloys, synthetic rubber, catalysts, and cements. After the war, a new wave of research, theoretical insights, technical improvements, and commercially successful instruments appeared. In Germany during the war years, rivals conducted bitter priority battles over patent rights to the electron microscope. The winner of the patent wars might be considered the inventor of the electron microscope as a matter of law, but the instrument was clearly the result of the work of many people.

After high-resolution electron microscopy had become routine, new forms of microscopy, like scanning tunneling electron microscopy, atomic force microscopy, near-field optical microscopy, and acoustic microscopy, made it possible to see in an entirely new way. But even such innovations are in a sense a continuation of the basic wonder of the microscope. As noted by microscopists and historians of science, the microscope is different from the telescope in that the objects to be studied by the telescope are obvious and the desire to study the objects in the heavens was very ancient, but once microscopists got beyond the stage of admiring magnified fleas, they were able to cross a threshold into an entirely new world in which the nature of the once invisible objects was even more mysterious than that of the heavenly bodies. Potentially, the microscope offered not just a new way of seeing, but a revolutionary change in the world of thought and imagination.

SUGGESTED READINGS

Abbot, E. A. (1985). *The Life and Works of Francis Bacon*. London: MacMillan.
Andrande, E. N. da C. (1960). *A Brief History of the Royal Society*. London: The Royal Society.

Ben-David, J. (1991). *Scientific Growth: Essays on the Social Organization and Ethos of Science*. Berkeley, CA: University of California Press.

Berman, M. (1978). *Social Change and Scientific Organization: The Royal Institution, 1799–1844*. Ithaca, NY: Cornell University Press.

Birch, T. (1968). *The History of the Royal Society of London for Improving of Natural Knowledge from Its First Rise*. Introduction by A. Rupert Hall. 3 vols. 1756–57. New York: Johnson Reprint Corp.

Boffey, P. (1975). *The Brain Bank of America*. New York: McGraw-Hill.

Bradbury, S., and Turner, G. L. E., eds. (1967). *Historical Aspects of Microscopy*. Cambridge: W. Heffer & Sons Ltd.

Casida, L. E. (1976). Leeuwenhoek's observations of bacteria. *Science 192:* 1348–1349.

Clark, G., and Kasten, F. H. (1983). *History of Staining*. 3rd ed. Baltimore, MD: Williams & Wilkins.

Clay, R. S., and Court, T. H. (1932). *The History of the Microscope*. London: Charles Griffin.

Cozzens, S. E., and Gieryn, T. F., eds. (1990). *Theories of Science in Society*. Bloomington, IN: Indiana University Press.

Crane, D. (1972). *Invisible Colleges: Diffusion of Knowledge in Scientific Communities*. Chicago, IL: University of Chicago Press.

Crowther, J. G. (1960). *Francis Bacon: The First Statesman of Science*. London: Cresset.

Descartes, R. (1965). *Discourse on Method: Optics, Geometry, and Meteorology*. Trans. by P. J. Olscamp. Indianapolis, IN: Bobbs-Merrill.

Descartes, R. (1972). *Treatise of Man*. Trans. by Thomas Steele Hall. Cambridge, MA: Harvard University Press.

Dobell, D. (1932). *Antony van Leeuwenhoek and His "Little Animals."* New York: Harcourt, Brace. (Reprinted, Dover Publications, 1960)

Dupree, A. H. (1986). *Science in the Federal Government: A History of Policies and Activities*. Cambridge, MA: Belknap Press of Harvard University.

Flam, F. (1992). National Academy of Sciences. What should it take to join science's most exclusive club? *Science 256:* 960–961.

Ford, B. J. (1985). *Single Lens. The Story of the Simple Microscope*. New York: Harper & Row.

Frängsmyr, T., ed. (1990). *Solomon's House Revisted. The Organization and Institutionalization of Science*. Canton, MA: Science History Publications.

Fujita, H., ed. (1981). *History of Electron Microscopes*. Tokyo: Japan Scientific Societies Press.

Ganelius, T., ed. (1986). *Progress in Science and Its Social Conditions*. New York: Pergamon.

Gillespie, C. C. (1980). *Science and Polity in France at the End of the Old Regime*. Princeton, NJ: Princeton University Press.

Grew, N. (1965). *The Anatomy of Plants*. New York: Johnson Reprint.

Hahn, R. (1971). *The Anatomy of a Scientific Institution: The Paris Academy of Sciences, 1666–1803*. Berkeley, CA: University of California Press.

Hall, M. B. (1984). *All Scientists Now: The Royal Society in the Nineteenth Century*. New York: Cambridge University Press.

Harré, R. (1981). *Great Scientific Experiments. 20 Experiments that Changed our View of the World*. Oxford: Paeidon Press.

Hawkes, P. W., ed. (1985). *The Beginnings of Electron Microscopy*. Orlando, FL: Academic Press.

Hawley, G. G. (1946). *Seeing the Invisible: The Story of the Electron Microscope*. New York: Knopf.

Hindle, B. (1956). *The Pursuit of Science in Revolutionary America, 1735–1789*. Chapel Hill, NC: University of North Carolina Press.

Hooke, R. (1961). *Micrographia*. New York: Dover Publications.

Hull, D. L. (1988). *Science as a Process. An Evolutionary Account of the Social and Conceptual Development of Science*. Chicago, IL: University Chicago Press.

Hunter, M. (1981). *Science and Society in Restoration England*. New York: Cambridge University Press.

Kearney, H. F. (1971). *Science and Change, 1500–1700*. New York: McGraw-Hill.

Kohler, R. E. (1991). *Partners in Science: Foundations and Natural Scientists, 1900–1945*. Chicago, IL: University of Chicago Press.

Kohlstedt, S. (1976). The Formation of the American Scientific Community: The AAAS, 1848–60. Urbana, IL: University of Illinois Press.

Kuhn, T. S. (1970). *The Structure of Scientific Revolutions*. 2nd ed. Chicago, IL: University Chicago Press.

Leeuwenhoek, A. van (1941). *The Collected Letters of Antoni van Leeuwenhoek*. Amsterdam: Swets and Zeitlinger.

Leeuwenhoek, A. van (1977). *The Select Works of Antony van Leeuwenhoek (1632–1723)*. Translated by S. Hoole. Facsimile edition. New York: Arno Press.

Lindberg, D. C. (1976). *Theories of Vision from al-Kindi to Kepler*. Chicago: University of Chicago Press.

Marton, L. L. (1968). *Early History of the Electron Microscope*. San Francisco, CA: San Francisco Press.

McClellan, J. E. (1985). *Science Reorganized: Scientific Societies in the Eighteenth Century*. New York: Columbia University Press.

Merton, R. K. and Gaston, J., eds. (1977). *The Sociology of Science in Europe*. Carbondale, IL: Southern Illinois University Press.

Morrell, J., and Thackery, A. (1981). *Gentlemen of Science. Early Years of the British Association for the Advancement of Science*. New York: Oxford University Press.

Nicholson, M. (1956). The microscope and the English imagination. In *Science and Imagination*. Ithaca, NY: Cornell University Press.

Nye, M. J. (1986). *Science in the Provinces. Scientific Communities and Provincial Leadership in France, 1860–1930*. Berkeley, CA: University of California Press.

Oleson, A., and Brown, S. C., eds. (1986). *The Pursuit of Knowledge in the Early American Republic: American Scientific and Learned Society from Colonial Times to the Civil War*. Baltimore, MD: Johns Hopkins University Press.

Oleson, A., and Voss, J., eds. (1979). *The Organization of Knowledge in Modern America, 1860–1920*. Baltimore, MD: Johns Hopkins University Press.

Olson, R. (1991). *Science Deified and Science Defied. The Historical Significance of Science in Western Culture*. Berkeley, CA: University of California Press.

Ornstein, M. (1936). *The Role of Scientific Societies in the Seventeenth Century*. Chicago, IL: University of Chicago Press.

Paul, H. W. (1985). *From Knowledge to Power. The Rise of the Science Empire in France, 1860–1939*. New York: Cambridge University Press.

Perez-Ramos, A. (1989). *Francis Bacon's Idea of Science and the Maker's Knowledge Tradition*. New York: Oxford University Press.

Porta, della G. (1957). *Natural Magick*. New York: Basic Books, reprint.

Rogers, J. T. (1982). What Leeuwenhoek saw. *Scientific American 246:* 79–80.

Roszak, T. (1973). *Where the Wasteland Ends*. New York: Doubleday.

Schiebinger, L. (1989). *The Mind Has No Sex? Women in the Origins of Modern Science*. Cambridge, MA: Harvard University Press.

Schierbeek, A. (1967). *Jan Swammerdam 1637–1680, His Life and Works*. Amsterdam: Swets and Zeitlinger.

Shapiro, B., and Frank, R. G., Jr. (1979). *English Scientific Virtuosi in the Sixteenth and Seventeenth Centuries*. Los Angeles, CA: University of California Press.

Shea, W. R. (1991). *The Magic of Numbers and Motion: René Descartes' Scientific Career*. Canton, MA: Science History Publications.

Smith, A. G. R. (1972). *Science and Society in the Sixteenth and Seventeenth Centuries*. New York: Harcourt, Brace, Jovanovich.

Sprat, T. (1958). *History of the Royal Society*. Facsimile of 1667 edition, edited with critical apparatus by J. I. Cope and H. W. Jones. St. Louis, MO: Washington University Press.

Stimson, D. (1968). *Scientists and Amateurs: A History of the Royal Society*. New York: Greenwood.

Stroup, A. (1990). *A Company of Scientists. Botany, Patronage, and Community at the Seventeenth-Century Parisian Royal Academy of Sciences*. Berkeley, CA: University of California Press.

Taylor, D. L., Nederlof, M., Lanni, F., and Waggoner, A. S. (1992). The new vision of light microscopy. *American Scientist 80:* 322–335.

Turner, G. L'Estrange (1980). *Antique Scientific Instruments*. Poole, Dorset: Blandford Press.

Whitley, R. (1985). *The Intellectual and Social Organization of the Sciences*. New York: Oxford University Press.

Ziman, J. (1987). *An Introduction to Science Studies*. New York: Cambridge University Press.

Ziman, J. (1987). *Knowing Everything About Nothing. Specialization and Change in Scientific Careers*. New York: Cambridge University Press.

5

PROBLEMS IN GENERATION:
ORGANISMS, EMBRYOS, AND CELLS

Long before philosophers began to debate the nature of generation, human beings had been tending crops, pollinating plants, selecting desirable traits in domesticated animals, and castrating men and other animals. Thus, when philosophers and naturalists began to explore fundamental questions about generation, they could build their theories on a rich but confusing accumulation of ideas and observations that had been transmitted in stories, myths, and folklore. *Generation*, a term that once encompassed a broad domain that now includes highly specialized areas in genetics, reproduction, embryology, and regeneration, has become so obsolete and archaic that many biologists are unfamiliar with its original meaning. Generation originally encompassed some of the most exciting aspects of modern biology—reproduction, embryology, development and differentiation, regeneration of parts, and genetics. In its classical usage, the term generation referred to the coming into existence of new individual organisms, animals or plants, regardless of the methods involved: sexual, asexual, or spontaneous generation.

Fundamental questions about generation were posed by early Greek philosophers and physicians. Why were offspring similar to but usually not identical to their parents? What caused births that deviated from the normal range, and what produced strange and monstrous births? In sexual reproduction, what contributions were made by the two sexes? What caused organisms to grow from simple, even unrecognizable forms into adults like their parents? How was it possible for some species to regenerate lost parts while others were permanently mutilated by even trivial accidents? Some fragments from the writings of Alcmaeon show that the chick egg had already been chosen by natural philosophers in the sixth century B.C. as a model system for investigating embryology. A proposal for a systematic embryological study appeared in the Hippocratic *De natura pueri*. Given the limited opportunities for obstetrical investigation, the physician could attempt to broaden his understanding of embryological development by selecting and incubating about 20 fertile chicken eggs; one egg could be taken each day and opened for examination. There is little evidence, however, that this systematic program was actually carried out during the time of Hippocrates or even by Aristotle.

One of the great debates that runs through the study of generation was initiated by Aristotle in his discussion of the alternative models of development known as *preformation* and *epigenesis*. Preformationist theories hold that an embryo or miniature individual preexists in either the egg of the female or the semen of the male and begins to grow into its adult form as the result of the proper stimulus. Many preformationists believed that all the embryos that would in due course develop into adults had been formed by God at the Creation and were somehow encased within each other. The theory of epigenesis, which Aristotle favored, is based on the belief that each embryo or organism is gradually produced from some initial undifferentiated mass by means of a series of steps and stages during which new parts are added. According to Aristotle, the female parent contributed only unorganized *matter* to the new individual, while the male provided the all important principle of *form*. Despite the ambiguity of Aristotle's use of the terms form and essence in *Generation of Animals*, most philosophers believe that he did not mean to include accidental, material features of the organism in question or specific individual forms. Aristotle apparently believed that the embryo developed towards what could be called "parental likeness." All the parts were not formed simultaneously, but the *potential* for the whole organism existed in the semen. The parts did not form simultaneously because the first part to be formed, the heart, was of necessity the one that possessed the principle of growth. When the *movements* proper to form could not gain control over matter, a deformity or monstrosity developed. Given the fact that Aristotle regarded the female as a "damaged male," characterized by a deficiency of heat, the semen presumably failed to live up to its potential in about half of all births. Like his predecessors, Aristotle called attention to the developing chick egg, but William Harvey thought that Aristotle's discussion of this system reflected only a few observations. Nevertheless, Aristotle's theory of causes and his ideas about the soul, the nature of semen, the conceptual mass, the primacy of the heart, the roles of the two sexes, the nature of epigenetic development, and fetal nutrition formed the core of embryological thought for some 2000 years.

While Galen often harangued his readers about the importance of direct observation, like Aristotle, he did not carry out a systematic program of observations on the developing chick egg. Galen described the fetal membranes, the blood vessels of the uterus, the fetal organs, and the foramen ovale, including the way it closes after birth. In *On the Semen*, Galen disputed Aristotle's idea that the female does not form any counterpart to the male semen. The female testes (ovaries), Galen argued, secrete semen into the horns of the uterus where the male and female sexual products mix with each other and with blood furnished by the mother to form the fetus. As the fetus developed, the liver, heart, and brain appeared. After the formation of the liver, fetal life was rather like that of a plant, because it was nourished under the influence of the vegetative soul. With the formation of the heart, fetal life was like that of an animal because the heart served as the source of heat. The brain was the last of the important organs to be formed because the fetus did not hear, taste, smell, or engage in any voluntary action, nor did it have any intellectual faculties. In accordance with Galen's teleological principle, therefore, it did not need a brain until it was ready to be born.

In *On the Usefulness of the Parts,* Galen argued that human beings are the most perfect of all animals, but man is more perfect that woman because the male has more innate heat, which is Nature's primary instrument. This can be seen in the anatomy of the reproductive organs. Because of the deficiency of heat that caused a fetus to become a female, after the "generative parts" had been formed they were unable to project themselves to the outside, in contrast to the masculine pattern of development. The female reproductive organs were, therefore, essentially defective or inside-out versions of the perfect male parts; for example, the uterus was an inversion of the scrotum. The effects of castration, Galen asserted, proved that the testes played an important part in the formation of semen and exerted a profound effect on development. Castration removed heat, virility, and manliness, which caused the body of a eunuch to become more like that of a woman. Lack of heat in the female resulted in the accumulation of fat. As proof of this concept of women as imperfect men, Galen referred to anecdotes concerning women who had been transformed into men. The reverse transformation was impossible, Galen asserted, because nature always strives towards perfection from the lower to the higher form.

In general, ideas about generation underwent little fundamental change from the time of Galen to that of William Harvey. Although some scholars of the Middle Ages had begun to suspect that Aristotle's writings had not exhausted the potential limits of wisdom, even the reform-minded anatomists of the sixteenth century hardly emerged from his shadow, at least in so far as the study of generation was concerned, nor did they consider embryology a separate and distinct field worthy of systematic study. Given the lack of opportunities for work on human development, it is not surprising that Vesalius and his contemporaries generally based their discussions of embryology on dissections of other animals and perpetuated many Aristotelian and Galenic ideas about the human fetus. For example, Realdo Colombo described the dissection of a human fetus he believed to be about one month old, but he noted that he could not study other phases of human development because of the insufficient supply of aborted fetuses. Nevertheless, he criticized other anatomists who had brashly applied what they had imperfectly observed in the lower animals to the human fetus. Many of his observations, however, were also problematic. At the University of Padua, students of Girolamo Fabrici were introduced to embryology as well as comparative anatomy. During his anatomical lectures, Fabrici dissected a woman who had died in labor, demonstrated the foramen ovale and the anatomy of the fetal horse and sheep, and closed a long discussion of development with the vivisection of a pregnant ewe. Fabrici published two important embryological works: *On the Formed Fetus* and *On the Development of the Egg and the Chick*. Despite the fact that Fabrici worked without magnifying lenses, the illustrations in these texts are remarkably detailed. The text, however, reflects the tension between original observation and deference to authority that haunted sixteenth-century scientists. While trying to preserve the framework provided by Aristotle and Galen, Fabrici struggled for accuracy and originality. Nevertheless, nineteenth-century naturalists rejected much of Fabrici's embryological work as "sheer Galenism," and Karl Ernst von Baer complained that the theoretical aspects of Fabrici's embryology were so annoying that it was almost impossible to read more than two pages at a time. In discussing his own embryological investigations, William Harvey later said that he had been guided primarily by Aristotle, among the ancients, and Fabrici, among the moderns.

Harvey's *On the Generation of Animals* did not appear in print until 1651, some 23 years after the publication of his treatise on the circulation of the blood. Nevertheless, the treatise on generation was not simply a lesser work of Harvey's old age; it was the outcome of many years of research concerning a problem of profound importance to Harvey as an experimental scientist and natural philosopher. More than any other aspect of his work, the researches on generation demonstrate the ingenuity of Harvey's experimental approach and the difficulties posed by the lack of reliable instrumental and theoretical guides. Having set out to provide experimental proof for the venerable Aristotelian theory of epigenetic development, Harvey came close to demonstrating that the central dogma on which all prevailing theories of generation rested was absolutely wrong. From the time of Aristotle, the embryo had been regarded as a combination of matter from the female and the form-building principle from the male; the embryo supposedly formed by coagulation or precipitation in the uterus immediately after mating. With pregnancy established, blood from the female was supposed to nourish the embryo and provide additional matter for growth.

For his experiments on generation Harvey generally used the developing chick egg and deer from the king's Royal Park. Searching for the seminal mass from which the embryo was supposed to originate, Harvey examined uteri from numerous does during and after the mating season. A careful series of dissections revealed that until several weeks after mating nothing substantial could be found in the uterus. Like Aristotle and Fabrici, Harvey concluded that the male made no material contribution to the developing embryo; nevertheless, the semen was essential because its fertilizing power affected not just the egg, but somehow made the uterus and the entire female body fertile. The semen, according to Fabrici, exerted its fertilizing power over the uterus by emitting some "spiritous substance." Harvey rejected this idea as being unnecessary, unproved, and excessively obscure. His own explanation likened the process of fertilization to the spread of a *contagion*, a vague term that denoted that which is transferred by contact. The seed of the male, Harvey argued, affected the female with a kind of contagion that transferred to her its prolific quality, which then made the eggs, the uterus, and the hen herself fruitful. This idea could be applied to the changes that occur in pregnant mammals. When human females become pregnant, Harvey noted, it was not only the embryo that developed and changed; the whole system of the female was affected as indicated by the enlargement of the breasts, the swelling of the uterus, and other well-known symptoms of pregnancy. While Harvey admitted that he could not explain the way in which this putative contagion accomplished its mission he was sure that the concept was useful. It has been said that a good theory need not be true, but it must be fruitful. In this sense Harvey's theories, like his hypothetical fertilizing contagion, were ultimately fruitful. As in the study of the heart and blood, beginning with respect for Aristotle and ancient philosophy, Harvey's investigations resulted in the wreckage of the ancient foundations of the prevailing paradigm. In the case of his studies of generation, however, Harvey was unable to construct a new theoretical framework or provide a satisfactory solution to the problem. Careful studies of the developing chick egg convinced Harvey that generation proceeded by epigenesis, that is, the gradual addition of parts.

Whereas Fabrici had concluded that most animals arise from eggs, Harvey extended this concept to all animals as epitomized in his dictum: *Omne vivum ex ovo* (all living beings arise from eggs). All animals arose from "a certain corporeal something having life in potentia, or a certain something existing per se," which could be called an egg. The egg in Harvey's *Omne vivum ex ovo*, however, was actually the fetus at a stage large enough to be seen without a magnifying lens. While Harvey's "egg" was as invisible as his anastomoses between the arteries and veins, it was just as logically necessary. Harvey was one of the last major advocates of epigenesis during the late seventeenth century and for most of the eighteenth century. The mechanistic view of the living world, which his own work on the movement of the heart and the blood had done so much to support, led to a preoccupation with preformationist models that seemed to fit the mechanistic mode of explanation better than epigenetic theories.

The preformationist theories that enjoyed a prominent place in eighteenth-century science had philosophical roots extending back to Democritus, Empedocles, Lucretius, Seneca, and the Church Fathers. During the seventeenth century several naturalists claimed to have seen the rudiment of the chick embryo in the egg before incubation and the plant in the ungerminated seed. In overcoming the authority of Aristotle and Harvey, preformationists drew on the increasing acceptance of the mechanical philosophy, based on the work of Galileo, Newton, and Harvey, and microscopic techniques that allowed naturalists to see details of embryological development at earlier stages. Newtonian or mechanical philosophy was a primary characteristic of Enlightenment thought; that is, an emphasis on the rational and objective mind engaged in the analysis of a universe explicable in terms of matter and motion and the conviction that explanations derived from the physical sciences applied equally to heaven and earth. Frustrated in their attempts to establish a mechanism by which parents could fabricate the unseen entity that ultimately became a complete organism, some naturalists denied the idea of the occurrence of individual development altogether. Instead, they adopted the idea that new individuals grew from preformed miniatures created by God.

Marcello Malpighi's descriptions of the developing chick egg reflect the paradoxical problem of embryological models subjected to microscopic observation; proponents of both epigenesis and preformation were able to study the same systems and come to very different conclusions. Thus, although in general development seemed to occur gradually, Malpighi believed that some parts were present in the chick egg at the beginning of the incubation period. The heart, for example, seemed to exist before it started beating. On many occasions Malpighi complained about the poor quality of his microscopes, but he was able to reveal details of earlier stages of the embryo than had ever been seen before. Malpighi generally avoided making dogmatic statements about the mechanism of generation, but his contemporary Jan Swammerdam argued that the four stages in the life of an insect— egg, larva, pupa, and adult—were packaged within each other as in a series of boxes. Because he thought that the butterfly could be extracted from the caterpillar, Swammerdam concluded that no new parts were formed during insect metamorphosis. Similarly, the limbs of the frog could be dissected out from under the skin of the tadpole some time before they would normally emerge.

In essence, Swammerdam believed that what appeared to be the generation of new individuals involved the sequential emergence of an embedded series of individuals that had been brought into existence at the creation of the world. Nicolas de Malebranche (1638–1715), Cartesian philosopher and priest, reformulated Swammerdam's rather vague preformationist ideas into a philosophical scheme involving an endless series of embryos preexisting within each other like a nest of boxes. The rapid and enthusiastic emergence of support for the doctrine of preformation by preexistence has led many scholars to conclude that it must have resonated with the deepest philosophical needs of many eighteenth century naturalists as they struggled to deal with the failure of mechanical epigenesis, as well as the disturbing materialist and atheistic implications of mechanical epigenesis. Such a theory implied that unorganized matter could produce living beings without the intervention of the Divine Creator. Preformationist theories, in contrast, could be allied with the theological doctrine that all human beings were quite literally the offspring of Adam and Eve. But the boxes-within-boxes theory allows only one parent to serve as the source of the preformed individual; the egg was well known for many species and seemed the reasonable carrier for the new organisms. Initially, the role of the male seemed limited to initiating development, but when the microscope revealed great numbers of active little animals in the semen, it seemed possible that the male parent might carry the preformed individual.

While some microscopists suggested that sperm were either associated with disease or were themselves permanent parasites, Antoni van

Leeuwenhoek believed that they were normal constituents of semen. In 1677 Johan Ham, a medical student at Leiden, discovered little animals in semen from a man with gonorrhea. Assuming that the animalcules were associated with the disease, Ham noted that they appeared to die when the patient took turpentine as a remedy. On one of his visits to Leeuwenhoek, Ham brought with him a small glass phial containing the spontaneously discharged semen of a man suffering from gonorrhea. Confirming Ham's observations, and extending his investigations to the semen of healthy men and other animals, Leeuwenhoek studied the appearance, motility, survival, and other characteristics of these animalcules. He explained that the observations of human semen were not obtained by any "sinful contrivance," but were made with "the excess with which Nature provided me in my conjugal relations." In fresh human semen, examined before "six beats of the pulse had intervened" between ejaculation and microscopic analysis, the admirable Leeuwenhoek found a great number of living animals. Noting that these animalcules were smaller than the red corpuscles of the blood, he concluded that a million of them would not equal the size of a large grain of sand. The little animals had a round body and a long thin tail, which they used to swim about with

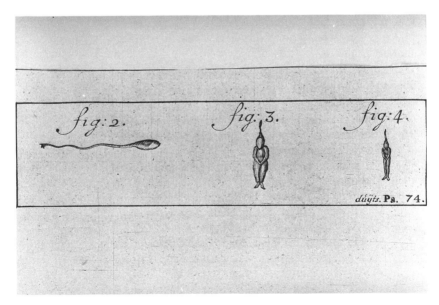

An engraving of three sperm accompanied a letter sent by Antoni van Leeuwenhoek to Nehemiah Grew in March 1678. Two of the sperm show a small man, or homunculus, inside.

the snake-like motion of eels. Convinced that he had detected two sorts of animalcules in semen from humans and dogs, Leeuwenhoek suggested that they might give rise to male and female offspring. He did not, however, believe that it would be possible to see the future adult in the sperm any more than it was possible to see a tree in an apple seed.

When he first described these observations in a letter to the Royal Society, Leeuwenhoek suggested that the Secretary should feel free to destroy them if he found them disgusting or scandalous. Fortunately, the Fellows of the Royal Society were intrigued rather than scandalized, and a Latin translation was published in the *Philosophical Transactions*. In all, Leeuwenhoek studied the spermatozoa of about 30 different kinds of animals, including mammals, birds, amphibians, fish, molluscs, and arthropods. He also investigated the ovaries of cows and sheep to study the Graafian follicles. In opposition to Regnier de Graaf, Leeuwenhoek argued that these putative eggs could not possibly get to the uterus because they would have to pass through the fallopian tubes, which were no larger in diameter than a pin, whereas the alleged eggs were as large as peas. Although Leeuwenhoek had considerable respect for learned authorities, he had no hesitation in disagreeing with them on the basis of his own observations. It was obvious, Leeuwenhoek noted in his letter to the Royal Society, that all of the learned authorities who disagreed with him were mistaken. On the other hand, despite Leeuwenhoek's observations concerning parthenogenesis in aphids, he remained convinced that new organisms must arise from the sperm. When Leeuwenhoek discovered that female aphids could reproduce without mating, he attempted to resolve the paradox of parthenogenesis by considering all the tiny new individuals functionally equivalent to spermatozoa. In his last letter concerning aphids, however, he simply said that all aphids belonged to the female sex and were viviparous. Parthenogenesis in aphids, the budding of hydra, the formation of daughter colonies within *Volvox*, and other observations first made by Leeuwenhoek were used by later scientists to support alternative theories of preformation.

Preformationist doctrines won the support of many competent and cautious scientists and philosophers, but they also attracted extremists who argued that the whole human race and all its parasites had preexisted in the ovaries of Eve, and others who saw homunculi (tiny people) stuffed into spermatozoa. Some animalculists were quite convinced that their microscopes allowed them to see sperm displaying behaviors characteristic of the types of animals to which they would give rise. Other eighteenth-century naturalists rejected both ovist and spermist preformationist views. For example, in his monumental *Natural History* Georges Buffon attempted to find a mechanistic explanation for how the world came to its present state. He suggested that the embryo was formed from organic particles in the male and female semen with the aid of an internal mold that directed the particles to their proper places. Pierre Louis Moreau de Maupertuis argued in *Venus physique* (1745) that particles from each part of both parents contributed to the seminal fluid from which the embryo was formed in the womb. Some obscure force, vaguely reminiscent of gravity, drew the particles to the embryo, although Maupertuis could not say how they could recognize the proper place.

Some of the most respected scientists of the eighteenth century, namely, Albrecht von Haller (1708–1777), Charles Bonnet (1720–1793), Lazzaro Spallanzani (1729–1799), and René Antoine Ferchault de Réaumur (1683–1757), were proponents of ovist preformation. In terms of the range and significance of the philosophical questions posed, the most important of the eighteenth century debates between an epigenesist and a preformationist was that carried out by Caspar Friedrich Wolff (1733–1794) and Albrecht von Haller, who was best known for his work as a physiologist. The debate between Haller and Wolff concerning the nature of embryological development provides a valuable case study in scientific controversy and the role of observation, explanation, and metaphysical assumptions in establishing the position taken by scientists with opposing viewpoints. Haller is also especially interesting for having accepted and then rejected competing theories of embryological development. At first he accepted the spermist preformation explicated by his mentor Hermann Boerhaave. During the 1740s, influenced by Abraham Trembley's (1710–1784) discovery of the freshwater polyp and its ability to produce separate and complete new animals after it had been cut in half, Haller converted to epigenesis. If half of an animal like the hydra could form two new individuals when cut in half, he reasoned, then the preformed germ of the new individual could not preexist in either the egg or the sperm and the new parts must gradually form, perhaps out of some unorganized fluid. Finally, in the 1750s Haller adopted ovist preformationism based on observation, experience, and religious considerations arising from his belief that it was necessary to see God rather than chance at work in embryological development. Spontaneous generation was rejected along with epigenesis on the grounds that this would mean that matter, without the action of God, could produce living beings.

As a preformationist, Haller's own studies of the developing chick egg convinced him that the embryo could be found in the egg, along with all the essentials for maturation, before the egg emerged from the chicken. As early as the 4-day stage of the chick egg, Haller believed that he could see continuity between the membranes in the yolk sac and the intestines of the embryo. Therefore, he concluded, the embryo is already present in the unfertilized egg. Conception, he assumed, occurred when the male semen stimulated the egg; the inherent irritability of the heart then caused it to beat. The beating of the heart led to the movement of fluids through the transparent embryo, inflating and solidifying its parts and causing a transition from transparency to opacity, thus making the fetus visible. However, the evidence from these experiments was not the only source of his conversion; Haller had long been deeply troubled by the implications of the mechanistic theories of epigenesis put forth by Georges Buffon and other naturalists. In his review of the writings of the "shrewd and learned" French naturalist, Haller expressed many reservations and objections. Above all Haller objected to Buffon's notion of an interior mold in which organic particles from the semen were somehow drawn together and put into their proper places without the aid of a "Building Master." No Buffonian "penetrating force" seemed capable of providing the organization needed to create the "wonderful plan of the human body" even if such forces could create crystals and snowflakes. Buffon's model seemed to imply that matter and innate forces could create life without the Creator. This was intolerable to Haller, convinced as he was that God was the prime mover in all of nature, and the source of gravity, irritability, magnetism, and any other forces allegedly possessed by matter. Given such strong motives for allegiance to preformation, Haller was ready to interpret Charles Bonnet's studies of parthenogenesis in aphids as compelling evidence of ovist preformationism. Extremely pious, like his colleague Albrecht von Haller, Bonnet similarly sought to find a comforting harmony between natural philosophy and religious doctrine. After an initial loose attachment to epigenetic views of development, Bonnet became convinced of the merits of preformationist theory and worked out the details of a more sophisticated encapsulation theory. In this he was influenced by his own research and his correspondence with Haller.

While studying the reproduction of aphids, Bonnet found that females that hatched during the summer gave birth to live offspring without fertilization. In the autumn the new generation of males and females mated and the females laid eggs. By carefully segregating young females Bonnet was able to raise 30 generations of virgin aphids by parthenogenesis. According to Bonnet, the first female of every species contained within her ovaries the miniature precursors of all future members of that species. Male semen was needed to initiate growth, but the female parent provided the nutrition needed for development. In order to cope with troublesome cases and apparent exceptions, preformationist theory had to become ever more complex and convoluted. For example, while the germs in higher animals might be confined to the reproductive organs, in lower animals the germs had to be scattered throughout the body. Such reasoning was essential in explaining the behavior of hydra and similar polyps, which reproduced by budding, and the ancient observation that certain creatures could regenerate lost parts or form complete new individuals from segments.

There was some doubt in Bonnet's mind as to whether the germ was actually the particular individual or merely a generalized member of the species. Eventually he adopted the idea that the germ must carry the imprint of the species rather than that of the individual. That is, the original germ determined only the species characteristics, while natural processes and factors encountered during gestation, such as the mother's health, size, and nutritional status, determined its particular fate. Once initiated, the unfolding and growth of the germ could be regarded as a purely mechanical process in which nutriments were absorbed into the appropriate parts of the expanding embryo. By allowing for the effects of accidents, disease, and poor nutrition, this theory exonerated God in the matter of "monstrosities," even those that appeared over several generations. Recognizing the many difficulties posed by this theory, Bonnet acknowledged that preformation theory had to be understood as a major triumph of rational thought over direct observational experience.

Further support for the ovist preformationist model was provided by Lazzaro Spallanzani, one of the great figures in experimental physiology and the eighteenth century's foremost authority on fertilization phenomena. Originally a student of law, Spallanzani was drawn to science by his cousin Laura Bassi (1711–1778), a remarkable woman who had earned a doctorate in philosophy at Bologna and was the first woman to hold a chair of physics at any European university. Famous for her work in mechanics, Bassi was appointed to an extra, essentially temporary chair of physics, but it has been suggested that she did not actually teach in a public capacity. Spallanzani took religious orders, but he also held various university appointments and traveled widely. Versatile and energetic, Spallanzani taught mathematics, physics, philosophy, and Greek. His research interests included digestion, blood circulation, respiration, regeneration, spontaneous generation, the reproduction of plants and animals, the senses of bats, the nature of sponges, and the electrical activity of the *Torpedo*. Spallanzani's battle against the doctrine of spontaneous generation led to his study of the animalcules in semen. Having conducted studies on the nature of regeneration, Spallanzani chose to use amphibians, such as frogs, toads, and newts, as a model system. Through his work on regeneration, Spallanzani became interested in the life history of amphibians and their point of origin, which he concluded was undoubtedly the egg. In his *Prospectus Concerning Animal Reproduction*

(1768), he claimed to have discovered the preformed germ of the tadpole in the eggs of the frog. Later, in his *Dissertations Relative to the Natural History of Animals and Vegetables* (1789), he assured his readers that further studies of more animal species had substantiated his previous claim that the germ preexisted in the eggs of the female before fertilization occurred. Moreover, he had convinced himself that the preexistence of the germ was a general law of nature. Studying the development of amphibian eggs he found that frog eggs while still inside the female increased in size; such growth, he assumed, could only be explained if the organism already existed. The tadpole, in some condensed or miniature form, was, therefore, already present in the egg, ready to unfold itself in the presence of the fecundating semen of the male. Although the future organism was already present in the egg, it had to undergo many changes before assuming its true shape and identity. Observations of the development of the eggs of birds, reptiles, and insects were also integrated into Spallanzani's arguments as evidence of preformation.

Since the mammalian ovum was not discovered until 1827 and the union of sperm and egg nuclei was not seen until the end of the nineteenth-century, eighteenth century discourse about the nature of the process known as fecundation or fertilization was necessarily cast in very different terms. Many naturalists argued that fertilization never occurred outside the body of the mother. Some believed that the egg had nothing to do with fertilization, but existed only as a means of nourishing the new germ of life. Others denied any significant role in generation to the sperm, or explained them away as a kind of parasite, perhaps spontaneously generated like other infusoria. Given the antiquity of evidence that male semen was essential for development and the excitement generated by microscopic studies of the nature of semen, arguments were raised about which part represented the fertilizing power: spermatic animals, liquid, or some vaporous entity called the *aura seminalis*, an invisible power vaguely analogous to gravity or the action of a magnet.

An ingenious experimentalist, Spallanzani provided rigorous proof that fertilization in frogs was external and that it was not caused by some mysterious vapor. Having observed the normal mating behavior of frogs, he performed experiments in which female frogs were killed while the eggs were being emitted. Eggs that had been discharged and had come in contact with the semen developed normally, but eggs dissected from within the body of the female did not develop. To confirm his hypothesis that fertilization was external and that contact with semen was necessary for development, Spallanzani designed special tight-fitting taffeta overalls for his frogs. Despite these unique costumes, the frogs still attempted to mate; eggs were discharged by the females, but none of them developed. When some of the eggs were mixed with semen that had been retained by the trousers, normal development took place; eggs that were not mixed with semen disintegrated. Bonnet was very excited about Spallanzani's invention of a method of artificial fertilization and suggested various exotic hybrids that might be created. In a letter to Spallanzani, Bonnet suggested that his discovery of artificial fertilization might one day be applied to the human species. Indeed, not long afterwards John Hunter (1728–1793), physician and anatomist, very discreetly supervised artificial insemination on behalf of a patient with a rather peculiar problem, and thus allowed him to father a child.

A remarkable series of experiments was triggered by Bonnet's suggestion that Spallanzani use his artificial fertilization system to answer the question of which portion of the semen caused fecundation. Spallanzani placed a small quantity of semen in a watch glass and fixed a number of eggs to another smaller watch glass by means of gluten. The small watch glass was then inverted over the watch glass containing the semen. Although the eggs were kept quite moist and almost one tenth of the semen evaporated during the experiments, the eggs failed to develop when placed in water. To test whether the remaining semen was still active, Spallanzani applied some of it to other eggs; these did undergo development. Further tests showed that other liquids such as blood, urine, wine, vinegar, and so forth could not be substituted for semen. Moreover, the fecundating power of semen could be destroyed by heat, drying, and the admixture of wine. Perhaps the most curious aspect of these experiments, from a modern viewpoint, was Spallanzani's conclusion that it was the liquid portion of the semen and not the *vermicelli spermatici*, or little worms of the sperm, that accounted for fertilization. Various experiments with supposedly sperm-free fluids had convinced him that the spermatic worms were not responsible for fertilization. It has been suggested that when Spallanzani painted allegedly sperm-free fluids onto eggs he had triggered artificial parthenogenesis. Spallanzani concluded that the spermatic worms were parasites passed on from one generation to the next during intercourse or within the uterus during pregnancy; these parasites might even penetrate the egg, find their way into the tiny preformed genital organs, and remain there until puberty. Although he believed that the egg contained the preformed individual, Spallanzani noted that the resemblance of children to both parents raised questions concerning the pre-existence of the germ in the female parent. Unable to resolve these puzzles, or further define the function of the semen, Spallanzani concluded that such questions transcended the sphere of human knowledge.

Despite the authority of Haller, Bonnet, and Spallanzani, preformationist theory and the mechanical philosophy did not go unchallenged even in the eighteenth century. One of first biologists to adopt a new mode of thought known as *nature philosophy* was the embryologist Casper Friedrich Wolff (1733–1794). While studying medicine at Berlin and later at Halle, Wolff absorbed large doses of philosophy from Christian Wolff, professor of mathematics and philosophy, and author of *Thoughts on God, the Soul, and the World.* Caspar Wolff was quite impressed with his mentor's botanical work and his philosophical principle that everything that happens must have an adequate reason for doing so, for it would be absurd for something to come from nothing. In 1759 Wolff completed a dissertation entitled *Theory of Generation,* which became a landmark in the history of embryology, although it brought him little recognition during his lifetime. Unable to obtain a position at the Medical College of Berlin, Wolff emigrated to St. Petersburg in Russia, where he became Academician for Anatomy and Physiology. Although Wolff is generally considered a German embryologist, Russian historians have argued that his work should be considered part of a distinctly Russian tradition that shaped and nourished late eighteenth- and early nineteenth-century embryological theory.

Haller, Bonnet, and Spallanzani saw life and science from a profoundly religious perspective. Consciously or subconsciously, they wanted to preserve and enhance evidence for the biblical account of creation, but scientifically they wanted to place generation within a generally mechanistic physiology. Wolff, in contrast, believed that studies of generation could only be purely descriptive since it was impossible to determine the actual mechanism of development. His theory of generation was founded on the philosophical assumption that development necessarily occurred by epigenesis. When Haller rejected Wolff's theory of generation on largely religious grounds, Wolff retorted that a scientist must search only for truth and could not prejudge biological questions on religious rather than scientific grounds. Yet Wolff's own theoretical preconceptions led him to support and seek out observations favorable to an epigenetic theory of development that was compatible with nature philosophy, that is, a view of a world constructed in conformance with a preestablished harmony. Understanding life according to the precepts of nature philosophy meant attaining full descriptive knowledge of the gradual development of different life forms. It was in the development of the embryo that nature philosophy found its most typical and often most lyrical expression.

In terms that reflect the basic assumptions of nature philosophy, Wolff asserted his conclusion that the organs of the body did not exist at the beginning of gestation, but were formed successively. It was not necessary for scientists to explain exactly how the addition of parts took place, but only to demonstrate that it did indeed occur by means of a process in which each step in development supported the next step. Wolff's essay on generation began with a series of definitions. Generation was defined as the formation of a body by the creation of its parts. Because the actual mechanism of generation is unknowable, a vague descriptive term, the *vis essentialis*, or essential force, was introduced to suggest what brings about change. In addition to observations of the chick egg, Wolff studied the metamorphosis of plants. Basic to all of Wolff's work and conclusions was the assumption that plants and animals are essentially the same in terms of their original undifferentiated materials.

Metamorphosis in plants was an appropriate subject for studies of development in terms of Wolff's program since he was not interested in fertilization or the ultimate origin of the primary undifferentiated materials, but only in the pattern of differentiation once the process had been initiated. Yet while this model system suited his philosophy, it also helped him to circumvent the technological limitations of microscopy. Given the poor quality of microscopes and the lack of staining techniques, plant materials were preferable objects of study because more details could be seen in plants than in unstained animal tissues. Much of what Wolff said about embryogenesis was, therefore, an extrapolation of what he had seen in developing plant parts. According to Wolff, the inner life force of the plant causes liquid to be drawn up from the soil. The liquid collects at the growing point of the stem and becomes a kind of thin jelly. As evaporation of the liquid occurs, small sacs or vesicles are formed; finally, hollow tracks in the jellied mass of vesicles become the ducts of the plant vascular system. Similarly, Wolff asserted, the chick was formed from an undifferentiated mass of little sacs which contained no bodily structures or parts. Eventually the vascular system of the chick formed out of this material just as it did in plants. Indeed, Wolff gave as his first proof of epigenesis in the chick his observation that the blood vessels of the chick blastoderm are not present at the onset of incubation. To lend more credence to his plant-animal analogies, Wolff concentrated on the development of intestines, blood vessels, and kidneys because these parts have a final structure that is tubular. The little vesicles or sacs that Wolff described might have been cells, but many of his contemporaries reported seeing little globules in biological preparations; not infrequently, these globules were simply optical illusions or artifacts. In any case, Wolff insisted that at an early stage of development both plants and animals consist of minute units which are not miniature versions of adult organs.

While Wolff thought that these little sacs could change into other parts, they represented secondary structures in a mass of undifferentiated jelly rather than the primary unit of origin.

Preformation might be thought of as a theory that would ultimately discourage research on the embryo, for if development were simply the expansion of a miniature individual then knowledge of the fully formed organism would be formally equivalent to knowledge of the most minute and fragile precursor at its earliest stage of development. A descriptive epigenetic approach provided a compelling reason to study the course of development, because it asserted that development entailed change as well as growth. To Wolff, the philosophical limitations of preformation rendered it useless as a guide to scientific research. Those who adopt a system of preformation, Wolff warned, do not explain development but simply say that it does not occur.

A change in the climate of opinion late in the eighteenth century made Wolff's philosophy more congenial to naturalists because, with the growing influence of a movement known as nature philosophy, many investigators rejected the application of the mechanistic philosophy to living systems. Nature philosophy was particularly influential among German naturalists and embryology, in the form of developmentalism, was especially attractive to advocates of nature philosophy as a means of demonstrating the unity of nature and the spiritual or immaterial forces that guided development in the material world. Nevertheless, the conflict between preformation and epigenesis was not transcended until the late nineteenth century. Two fundamental changes were involved: first, cell theory provided a framework for a new understanding of egg, sperm, and developing embryo; second, scientists essentially abandoned both mechanistic models and natural philosophy and realized that they need not treat living organisms as machines, nor give up all hope of ever explaining the mechanisms that governed living beings.

CELL THEORY

Cell theory is a fundamental aspect of modern biology, prerequisite to and implicit in our concepts of the structure of the body, the mechanism of inheritance, fertilization, development and differentiation, the unity of life from simple to complex organisms, and evolutionary theory. To say that the cell is the fundamental unit of life is to provide a tremendously powerful generalization integrating the study of structure and function, reproduction and inheritance, growth and differentiation. By presenting the fabric of the body in terms of cells and cell products, cell theory provides one of many possible answers to the question: *What is an organism?* Ancient philosophers and anatomists speculated about the nature of the constituents of the human body that might exist below the level of human vision, but even after the introduction of the microscope, investigators differed as to the level of resolution that might be attainable and appropriate to studies of the human body in health and disease. By the eighteenth century, many anatomists had abandoned humoral pathology and, in analyzing the structure and function of organs and systems of organs, hoped to discover correlations between localized lesions and the process of disease.

Tissue doctrine, elaborated by the great French anatomist Marie François Xavier Bichat (1771–1802), represents an ambitious and influential attempt to systematize the study of the construction of the body. The indefatigable Bichat seemed to live in the anatomical theater and dissection room of the Hôtel Dieu. A prodigious worker, he accomplished at least 600 autopsies in the space of a year. Such a regimen was obviously not conducive to health and longevity; Bichat died when only 31 years old of a fever, which was probably tuberculosis. Working in the autopsy rooms and wards of the hospitals of Paris, Bichat and his colleagues proclaimed their conviction that medicine could only become a true science if physician-researchers adopted the method of philosophical analysis used in the other natural sciences in order to transform observations of general and complex phenomena into precise and distinct categories. This approach and the movement to link postmortem observations with clinical studies of disease were largely inspired by the work of another French physician, Philippe Pinel (1755–1826), author of *Philosophical Nosography* (1798), who argued that diseases must be understood not by reference to humoral pathology but by tracing them back to the organic lesions that were their sources. Because organs were composed of different elements, research must be directed towards revealing these constituents. Organs that manifested analogous traits in health or disease must share some common structural and/or functional components. Failing to find this analogy at the organ level, Bichat assumed that there might be some such analogy at a deeper level, that is, a finer level of resolution. His approach involved studying the body in terms of organs, which could be *decomposed* and *analyzed* into their fundamental structural and vital elements, which he called the tissues. Organs had to be teased apart by dissection, maceration, cooking, drying, and exposure to chemical agents such as acids, alkalis, and alcohol. According to Bichat, the human body could be resolved into 21 different kinds of tissue, such as nervous, vascular, connective, fibrous, and cellular tissues. Organs, which were made up of assemblages of tissues, were in turn components of more complex entities known as the respiratory, nervous, and digestive systems. The actions of tissues were explained in terms of *irritability* (the ability to react to stimuli), *sensibility* (the ability to perceive stimuli), and *sympathy* (the mutual effect parts of the body exert on each other in sickness and health). Obviously, Bichat's simple tissues were themselves complex and compound, except in the sense that they were simpler than organs, organ systems, or the body as a whole. Tissues, as Bichat himself acknowledged, consisted of combinations of interlaced vessels and fibers. Thus, unlike the atom or the cell, Bichat's tissue theory of general anatomy provides no actual unit of basic structure to serve as the locus of the most basic vital phenomena. Embryology was essentially outside the boundaries of Bichat's own research program, and his account of the arrangement of animal tissues generally ignored the problem of tracing the origins of specific organs and tissues back through embryological development. Bichat's goals and guiding principles were thus very different from those that motivated the founders of cell theory. In formulating this doctrine, Bichat's objective was not merely to extend knowledge of descriptive anatomy but to provide a scientific language with which to describe pathological changes. Through an understanding of the specific sites of disease, better therapeutic methods and the means of assessing the efficacy of such treatments should eventually emerge. Although Bichat's work is often regarded as the foundation of the new science of histology, the word *histology* was actually coined about 20 years after his death as part of an effort to distinguish between different approaches to the study of the organization of the body: the static approach found in the work of Bichat and attempts to describe the body in terms of the developmental stages of its tissues as a means of searching for the ultimate origins of life itself.

Having no theoretical objectives for the pursuit of microanatomy beyond the tissues, many of Bichat's followers came to regard the tissue as the body's ultimate level of resolution, even after the cell theory had been well grounded for plants and animals. One reason for this was Bichat's well-known, and often justifiable, skepticism concerning reports of microscopic observations. The microscope was not a trustworthy tool for exploring the structure of the body, Bichat warned, because every person using it gazed into the same darkness and saw a different vision.

Ever since microscopic investigations began in the seventeenth century, observations of various corpuscles, vesicles, sacs, and globules had been reported. In 1665 the word *cell* and illustrations of cellular structures appeared in Robert Hooke's *Micrographia*. Studying sections of freshly cut cork under his microscope, Hooke could see it to be perforated and porous, rather like a honeycomb. Counting the number of little boxes that could be seen at the surface of freshly cut cork, Hooke found more than 1000 per linear inch and estimated that a cubic inch must contain more than one billion. While the cells in cork apparently contained nothing but air, Hooke noted that living plants contained cells full of streaming green juices. Hooke's colleague at the Royal Society, Nehemiah Grew, described the microscopic structures seen in plants as cells or bladders and noted that the cells in the younger parts of plants seemed to be closely packed and very full of juices. Similarly, Malpighi found plants to be made up of little bodies, rather like bags, and Leeuwenhoek reported seeing blood corpuscles, sperm, infusoria, and bacteria. During the early nineteenth century, as instruments and staining techniques improved, microscopists reported an avalanche of discoveries unprecedented since the seventeenth century. Cell theory, however, had been clearly enunciated before the major wave of technical improvements in the microscope and in sample preparation occurred. Since plant and even animal cells are generally large enough to be visible with about 100-fold magnification, it seems likely that the presumed "failure" of the first wave of microscopy to produce cell theory had more to do with intellectual and conceptual limits, attitudes, and expectations than the technical limits of early instruments.

By the beginning of the nineteenth century, questions had been raised about various "globulist theories" concerning the structure of biological materials. While some of these globules may have been cells, in many cases they were probably optical illusions and artifacts. The nature of the cell was also obscured by the connotation of the word *cell* inherited from the seventeenth-century microscopists; the term evoked the idea of walls, as in prisons, surrounding empty spaces. Plants clearly contained cell walls enclosing some fluid, but until observers began to focus on the cell nucleus it was difficult to grasp the fundamental similarity of plant and animals cells. Moreover, some anatomists used the term cellular to refer to tissues in which a network of fibers seemed to enclose cavities or pores. One of the keys to bringing order to these confusing and diverse observations was the study of the cell nucleus conducted by Robert Brown (1773–1858), a naturalist best known for his studies of the random thermal motion of small particles, a phenomenon now called Brownian motion. It was not until 1905 that Albert Einstein proposed an explanation for this phenomenon and thus provided the first proof for the physical existence of atoms. Just as Brown was not the first to see Brownian movement, he was not the first to see the nucleus, but in both cases his investigations were more fruitful than those of his predecessors.

In 1827 while studying pollen grains, Brown saw minute specks that seemed to be in constant motion. He called these particles *active molecules*, but rather than attributing their motion to some innate vital force, Brown decided to look for this phenomenon in nonliving entities of similar size, such as particles of coal dust, powdered glass, and minerals. Four years later, in a report on the reproductive organs of orchids and other plants, Brown described his observations of the germination of pollen grains on the stigma of the flower and the growth of the pollen tube to the ovary. Describing the cell nucleus in his paper on "Fecundation in Orchideae" (1832), Brown noted that each cell in these plants contained a "circular areola," that is, a dark area that appeared to be more opaque than the cell membrane. Similar entities could be found in the cells of other plants.

Thus, studies of plant cells and the juices seen streaming within them began to assume a more coherent pattern in the 1830s; it was more difficult to generalize about the nature of animal cells. But by the end of the 1830s, botanists and zoologists had constructed an explanatory framework for many seemingly diverse phenomena. The naturalists who developed the concept of animal and plant life known as the cell theory were generally influenced by nature philosophy or Romanticism and shared an interest in the large questions concerning the basic elements of life, its organization and historical development. Naturalists inspired by nature philosophy sought to establish a new synthetic science of life, which would uncover the fundamental laws underlying vital phenomena. This enterprise depended upon the assumption that there must be a fundamental unity beneath the superficial but bewildering diversity of the biological realm. For the advocates of Romanticism, the universe was not a great machine, but a living entity in which spirit, will, or God was constantly present. Romanticism overwhelmed art and literature and profoundly influenced the biological sciences, but had relatively little impact on the physical sciences. In the first decade of the nineteenth century, the goals of a new science of life were enunciated by Gottfried Reinhold Treviranus; in contrast to the old *natural history*, *biology* would place all observations of vital phenomena into a new unified and harmonious whole. The reproduction of living beings seemed to offer the greatest challenge to mechanical explanation, for it was clear that if animals were machines like watches, they were quite unlike watches in the sense that male and female animal machines were capable of producing little machines like themselves, whereas mechanical contrivances could not.

Champions of the Romantic approach to science saw nature as a mysterious world of hidden forces, complexity, and continuous but nonrandom change. Thus, because of the assumed underlying unity of the natural world, studies of the life history of the individual, or the species, should ultimately provide an understanding of life itself. The writer Johann Wolfgang von Goethe (1749–1832) helped popularize ideas that intrigued and inspired many

German scientists. In essence Goethe advocated a new pattern of thought, which rejected preformation in favor of a poetic, developmental, evolutionary, dynamic view of Nature that was the opposite of the Newtonian clockwork mechanical universe. Critics who later reevaluated Goethe's claims as a scientist objected that all that he thought he had discovered could be considered "wrong, silly, or already known."

The German naturalist Lorenz Oken (1779–1851) was another highly influential champion of nature philosophy. While visiting a remote island in the North Sea, Oken distracted himself with the study of marine animals, much as Aristotle had done many centuries before. During this period Oken received his inspiration about the structure and evolution of animal life from the archetype, that is, the generalized form which supposedly reflected nature's basic plan and held the key to understanding all living organisms. Combining microscopic observations with philosophical speculations, Oken postulated the existence of a primitive, undifferentiated mucus-like fluid called *Urschleim*. Spherical vesicles were said to arise from this jelly-like material to produce infusoria, the simplest living things. According to Oken, complex organisms were actually aggregates of these simple entities; every animal or plant was, therefore, a colony of living infusoria that had given up their independence to subordinate their lives to that of the organism as a whole. In his play *Faust*, Goethe, Oken's sometime friend and associate, used this concept in describing the primal sea jelly out of which all life arose. Those who rejected mechanical philosophies, which as Goethe complained had tried to extort Nature's secrets by means of "levers and screws," tended to find such ideas very congenial.

Oken's *Urschleim* was eventually replaced by *protoplasm*, an equally vague term for the contents of the cells. The term protoplasm was used by Johannes Evangelista Purkinje (1787–1869) in 1839 to describe the cell substance. Theologians had used the word *protoplast* for Adam, the first formed, but for Purkinje the term covered whatever was first produced in the development of the individual plant or animal cell. According to Purkinje, the body was composed of fluids, such as blood, plasma, and lymph, and fibers, such as those found in the tendons and cells. Devoted to teaching as well as research, Purkinje developed several innovative teaching tools, a knife that was a precursor of the microtome, established the use of balsam-sealed preparations, and adapted Louis J. M. Daguerre's (1789–1851) methods to produce the first photographs of microscopic materials. Because university officials were unwilling to meet his demands for space and equipment, much of his research and teaching was carried out in his own home laboratory, which became known as the cradle of histology. By his example and his teaching methods, Purkinje convinced many scientists that the new achromatic microscopes would prove to be immensely powerful tools for biological research. His observations of nerve cells led to a detailed description of the large flask-shaped cells in the cerebellar cortex that are named for him.

Apparently independently of Purkinje, Hugo von Mohl (1805–1872) used the word protoplasm to describe the part of the plant cell within the cell membrane. In a series of influential articles, von Mohl helped make the term protoplasm a general part of the vocabulary of biology. In 1861 Max Schultze (1825–1874), professor of anatomy at Bonn, inelegantly but succinctly defined the cell as a lump of nucleated protoplasm. According to Schultze, protoplasm was the physical basis of life. In all forms of life—plants and animals, higher and lower forms—protoplasm provided unity of structure and function. The term protoplasm was introduced to the general public at Edinburgh in November 1868 in an address on "The Physical Basis of Life" by Thomas Henry Huxley (1825–1895). As used by Huxley, the term was thoroughly divorced from its previous religious associations. "All vital action," Huxley announced, "may be said to be the result of the molecular forces of the protoplasm which display it." As *Urschleim* had made its way into Goethe's *Faust*, so protoplasm appeared in Gilbert and Sullivan's *Mikado* when Pooh Bah proudly proclaimed that he could trace his ancestry back to a "protoplasmal primordial atomic globule."

Johannes Peter Müller (1801–1858), an eminent physiologist and comparative anatomist, was the mentor of many eminent biologists, including Jacob Henle, Robert Remak, Hermann von Helmholtz, Rudolf Virchow, and Theodor Schwann. Müller served as professor of anatomy and physiology and director of the Museum of Comparative Anatomy at the University of Berlin. When the revolution of 1848 broke out, Müller, who was then serving as the rector of the University of Berlin, was charged with controlling the violent disturbances that broke out among students and staff. Student demands included free education and an end to examinations. After the student rebellion was put down by the military, dealing with the continuing hostility and sabotage that became part of campus life wrecked Müller's physical and mental health. Nevertheless, he retained his position at the University until his death.

A man of broad interests, Müller taught human and comparative anatomy, embryology, physiology, and pathological anatomy and made original contributions to each field. When he died, probably from overwork, three people had to be appointed to replace him. Stimulated by the work of Purkinje, Müller became one of the first to use the new microscopic approach in studies of pathological phenomena. While Müller's work was significant in stimulating interest in a finer level of resolution in pathological anatomy, credit for the formal statement of cell theory is generally attributed to his student Theodor Schwann (1810–1882) and the botanist Matthias Jacob Schleiden (1804–1881). After studying law at the University of Heidelberg, Schleiden attempted to establish a practice as a barrister in Hamburg, but he was such a failure that he attempted suicide. Even with a gun in his hand, aimed at his forehead, Schleiden was unsuccessful. When he recovered from the self-inflicted but superficial wound, he decided to switch from law to natural science and earned doctorates in both medicine and philosophy. For many years he served as professor of botany at the University of Jena. Despite his success in research and teaching, he resigned after 12 years in order to travel and rest his nerves. Perhaps his wanderlust was the result of quarrels with the religious authorities about his rather unorthodox ideas about science and life. One fortunate aspect of Schleiden's travels was his meeting with Schwann in Berlin.

Contemporaries characterized Schleiden as arrogant, temperamental, stubborn, incapable of viewing his own hypotheses objectively, and unsparing in his attacks on the work of his rivals and his predecessors. However, Schleiden did accord considerable respect to the work of Charles Brisseau-Mirbel (1776–1854), an eminent French botanist and microscopist. Brisseau-Mirbel believed that cells were found in all parts of the plant and that they were formed in a primitive fluid in a manner analogous to the formation of a network of membranes and bubbles in the foam of a fermenting liquid, an idea that Schleiden generally accepted. Confronting the practitioners of what he considered archaic systematic botany, Schleiden redefined botany as an inductive science encompassing all aspects of the study of the laws and forms of the vegetable kingdom. Because botany had been under the control of pedantic systematists, he complained, it had established few facts and had discovered no fundamental principles and ideas that could lead to the establishment of its natural laws. Only the chemistry and physiology of plants were truly significant, Schleiden argued. Mere knowledge of the systematic arrangement of plants was dismissed as a waste of time. In order to transform botany into a true science, botanists must study the plant world in all possible ways, including microscopic examination of parts invisible to the naked eye.

In 1838 Schleiden published "Contributions to Phytogenesis" in Müller's *Archives for Anatomy and Physiology*. Taking the important but neglected work of Robert Brown on the nucleus as indicative of a special relationship to plant growth and development, Schleiden focused his attention on this

Matthias Jacob Schleiden

structure. Soon he came to regard the nucleus, which he renamed the *cytoblast*, as a universal elementary organ of the plant world. For Schleiden, all plants of any complexity were aggregates of cells, which he characterized as fully individualized, independent, separate entities. Within the plant, each cell led a double life: one pertaining to its own independent development and another that allowed it to serve as an integral part of a plant. Thus, all aspects of plant physiology were fundamentally manifestations of the vital activity of cells. Finding the origins of life itself in some elementary unit was a primary goal of naturalists influenced by Romanticism. Some researchers believed that the origin of life occurred in a kind of fluid "blastema" out of which organic solids precipitated; the first structures to emerge from this blastema might then be the elementary units of life. Most early nineteenth-century biologists identified these elements as fibers, while others saw them as globules. Early versions of cell theory were very similar in concept to globulist ideas. Schleiden described several possible methods of cell formation in "Contributions to Phytogenesis" and later in *Principles of Botany*, but he favored the hypothesis known as "free-cell formation." According to this doctrine, cell growth was analogous to the process of crystallization: a nucleolus grew by accumulation of minute granules out of the cytoblastema, a fluid composed of sugars and mucus. As mucus particles aggregated, part of the fluid was transformed into a relatively insoluble substance and formed the cytoblast around the nucleolus. When the cytoblast attained its full size, the young cell began to develop as a delicate transparent vesicle which gradually expanded and finally formed a complete cell within a rigid cell wall. Plants could also grow by the formation of cells within cells in the presence of the cytoblastema. In this case the entire contents of a cell were divided into two or more parts and a gelatinous membrane immediately formed around each part. Wood, however, seemed to form by the sudden consolidation of an organizable fluid into a tissue of cells.

Despite his attachment to an oversimplified model of cell formation based on the growth of inorganic crystals, Schleiden rejected the idea that organized life forms could arise through spontaneous generation. Even the lowly algae, lichens, and fungi reproduced their own kind. Confining his work and speculations to the plant world, Schleiden observed that many eminent men had struggled to establish an analogy between the animal and vegetable kingdoms. Failure had been generally acknowledged, although few understood that the precise reason was the impossibility of applying the idea of individuality as used in the animal kingdom to the plant world. Only the very lowest orders of plants, which consisted of single cells, could be called individuals. By clearly enunciating the idea that the plant is a community of individual cells, Schleiden provided the key principle unifying plant and animal life. Moreover, by bringing his ideas to the attention of Theodor Schwann, Schleiden facilitated the extension of cell theory from the plant world to the animal kingdom.

Quite unlike the abrasive, heterodox Schleiden, Theodor Schwann seems to have been a timid, introspective, and extremely pious person. Schwann received his early education at the Jesuits' College in Cologne and then studied medicine at the Universities of Bonn, Würzburg, and Berlin. After graduating in 1834 he became one of Müller's favorite disciples. During the time Schwann had Müller's encouragement, energy, and willpower to keep him at work, he made numerous contributions to histology, physiology, and microbiology in addition to his famous work on cell theory. For example, while making histological preparations for Müller, Schwann discovered the sheath surrounding nerve fibers, which has been named for him, and in studying the process of digestion he discovered the ferment (enzyme) named pepsin. A vigorous program of experiments on fermentation carried out by Schwann challenged the theory of spontaneous generation and suggested that microorganisms are responsible for the chemical changes involved in putrefaction and fermentation.

When Schwann began his microscopic researches, some resemblances between plant cells and certain animal structures had been suggested, but the great variety of forms found in the animal kingdom seemed more significant than any similarities. Even when animal cells, fibers, corpuscles, and so forth were seen, Schwann explained, they remained merely descriptive aspects of natural history. Without a unifying theory, studies of the mode of development of one kind of cell could not be related to that of any other kind. Animals appeared to be more diversified than plants in both their internal and external forms. Moreover, even with the best microscopes, it was difficult to see any detail in animals cells because they were generally transparent and lacked the cell walls found in plant tissues. Schwann had noticed nucleated entities in preparations of notochord, but until he talked with Schleiden, he apparently did not think very much about the implications of these observations. After Schleiden described the role that the nucleus appeared to play in the development of plant cells, Schwann suddenly realized how important it would be to show that the nucleus found in the cells of the notochord performed the same role as that of the nucleus of plant cells.

Like Schleiden, Schwann saw the nucleus as the key to elucidating the relationship between the plant and animal world and the composition of animal tissues. Beneath the myriad forms that animal tissues assumed, forms as diverse as muscles, nerves, and blood corpuscles, there was the unifying

Theodor Schwann

factor of the cell nucleus. Schwann's purpose, as set forth in *Microscopical Researches into the Accordance in the Structure and Growth of Animals and Plants* (1839), was to prove the basic unity of the two kingdoms of organic nature. In this endeavor Schwann was quite successful; moreover, he also provided evidence that even the most physiologically diverse parts of an animal developed according to the same basic principles. In the first section of *Microscopical Researches*, Schwann described the structure and growth of the notochord of the tadpole and cartilage from various sources. Section two provided evidence that cells are the basis of all animal tissues, no matter how specialized. And finally, in the third section, Schwann explicated his theory of cells. Close examination of the notochord, cartilage, and other animal tissues indicated that animal tissues did indeed originate from cells that were analogous to the cells of plants in all important respects. Such findings removed the great barrier thought to have separated the animal and plant kingdoms. Animal tissues, like plant tissues, contained cells, cell membranes, cytoplasm, nuclei, and nucleoli. The generation of cells within cells and the formation of cells around the nucleus seemed to take place in cartilage just as described by Schleiden in plant cells. Having provided proof for the analogy between plant and animal tissue in a particularly favorable system, Schwann undertook the ambitious project of proving that other animal tissues developed from cells. Building on the work of Brown and Schleiden, Schwann introduced a criterion by which true cells could be distinguished: the presence or absence of the nucleus. Microscopic analysis revealed that the whole animal was composed of cells or cell products, but it was frequently necessary to trace a tissue back to an earlier stage of development to observe its cellular origins.

In much the same manner as Schleiden described the double life of plant cells, Schwann proposed that animal cells similarly possessed an independent individual life and a life subordinated to the functioning of the organism as a whole. Two major modifications of cell life were defined. Independent cells were those in which the cell membrane remained clearly distinguishable from those of neighboring structures. Coalesced cells were those in which the walls blended, partially or entirely, with neighboring cells, or intercellular substances, to form a homogeneous substance. Thus tissues could be classified in terms of the degree of development that cells had to undergo in order to form a particular tissue. This analysis was extended to all of the body's most differentiated tissues, including muscles, nerves, bones, teeth, hooves, and feathers. No matter how unique and noncellular any body part might appear, if traced backwards in terms of embryonic development, all the most complex and specialized tissues and parts of the animals were derived from cells.

In some cases, Schwann thought that cells appeared to be formed within previously existing cells, but he also accepted Schleiden's idea that cells could be formed from a structureless fluid, or cytoblastema, by a process analogous to crystal growth. While Schwann took pains to explain that cell growth was only figuratively similar to crystallization, he did seem to find the metaphor of crystallization from the mother-liquor of life very powerful and satisfying, and he urged other scientists to pursue this analogy in their research. Schwann's work has been interpreted as the culmination of the Romantic search for the common origin of all living forms, but Schwann chose to present his conclusions as the result of purely empirical investigations. Whatever his deepest sources of inspiration might have been, Schwann's theory of the cell proclaimed the fundamental unity of anatomical and physiological principles in plants and animals and provided a new framework in which to investigate the origin and development of the embryo.

In the third section of *Microscopical Researches* Schwann summarized all of his research and explicitly elaborated the powerful generalization known as cell theory. His first major proposition states that there is one universal principle of development for all the diverse parts of organisms and that this principle is the formation of cells. His second proposition described the generation of cells from a structureless substance which was present either around or in the interior of existing cells. In a section called "Theory of the Cells" Schwann attempted to deal with the most fundamental aspects of cellular phenomena in organized bodies and the question of whether they should be considered mechanical or vital. Actions carried out altogether blindly in accord with the laws of necessity, Schwann argued, may seem to be adaptive to some higher purpose. Thus, organized bodies which seem to possess powers not found in inorganic nature might actually be acting in terms of physical and chemical laws. Understanding organized bodies in a scientific sense resolved itself into a question of studying the fundamental powers of individual cells. Cellular phenomena were then divided into two natural groups. First, *plastic* phenomena were those related to the combination of molecules to form a cell. Second, chemical changes in the component particles of the cell itself, or in the surrounding cytoblastema, could be called *metabolic* phenomena. The word "metabolic" was coined from the Greek to describe "that which is liable to occasion or to suffer change." Schwann recognized metabolic power as a universal property of cells and, therefore, of life. He suggested that the ancient concept of animal heat could now be explained as the product of cell metabolism.

Fermentation was used as the best known example of the metabolic activities of cells. According to Schwann, the so-called fermentation granules, or yeasts, were actually cells that were particularly amenable to scientific investigations. Schwann's theory of intracellular fermentation and his broader view of the cell as the unit of metabolism was accepted only after a bitter controversy that embroiled the German chemist Justus von Liebig and the French chemist and microbiologist Louis Pasteur. Liebig and his associate Friedrich Wöhler published an anonymous satire of Schwann's theory in which yeast in solution gave rise to eggs that hatched into animals shaped like alembics (special glass vessels used for distillation). These creatures devoured the sugars in solution, digested their meal, and then rudely belched forth carbon dioxide and excreted alcohol. Temperamentally unable to respond to such attacks, Schwann left the field to be defended and conquered by Louis Pasteur, a man who was described as never one to put up his sword until all enemies had been conquered or killed.

While Schwann and Schleiden had provided a powerful new framework for understanding the structure, development, and functions of plants and animals, their theory was different from modern cell theory in several important respects, primarily in terms of their concept of free-cell formation and the notion of the cytoblastema. In subsequent years this concept was attacked by various botanists, zoologists, and microscopists including Karl Nägeli, Hugo von Mohl, and Rudolf Virchow. Karl Nägeli (1817–1891), who further characterized the contents of cells, found that the lower algae were useful models for studies of cell division and for watching the movement and behavior of protoplasm. At first Nägeli, co-editor with Schleiden of a short-lived journal, defended Schleiden's theories, but his comparative studies of the production of cells in various groups of plants eventually convinced him that new cells arose from the division of preexisting parental cells.

The cleavage of eggs had been seen in the seventeenth century by Swammerdam and by Spallanzani at the end of the eighteenth century. Yet when Jean Louis Prévost and Jean Baptiste Dumas saw the segmentation of frog eggs in 1824, they still could not explain the meaning of the phenomenon. In 1854 Martin Barry (1802–1855) published illustrations of cleavage of rabbit eggs, but Barry believed that cells of later stages were derived directly from nuclei. Albert Kolliker (1817–1905), who studied the development of the eggs of the cuttlefish, was probably the first to lay great stress on the division of the nucleus during the process of segmentation of the egg. Kolliker applied Schwann's theory to embryonic development in a great number of animal species and tissues. It was primarily through the work of

Franz Leydig and Robert Remak that the behavior of the nucleus during cell division was finally clarified. From observations of the division of embryonic blood corpuscles in the developing chick, Remak concluded that during normal growth processes cells increased in number by the division of one cell into two new cells. He did not, however, dismiss the possibility that some other process might be involved under pathological conditions. Many scientists continued to accept the possibility of free-cell formation until Rudolf Virchow (1821–1902) convinced them that every cell is the product of a preexisting cell.

Cell theory in its modern form was elaborated and incorporated it into scientific medicine by Rudolf Virchow, a man of many talents and accomplishments. Virchow was a prominent member of the reformist social and intellectual movements of the nineteenth century. Indeed, Virchow and his like-minded colleagues were quite outspoken about the direct relationship between their scientific doctrines and the positions they took in opposing the repressive Prussian state. In 1847 Virchow was assigned to an official investigation of an outbreak of typhus fever in an industrial district of Silesia. Convinced that the fundamental causes of the epidemic could be found in the abysmal social and sanitary conditions prevailing in the district, Virchow prepared a report in which he blamed the government for the misery he had observed. This report led to his immediate dismissal from his official position and helped establish his growing reputation as a radical socialist and medical materialist. Following the revolution of 1848 his reputation for radicalism and sympathy for the opposition made it expedient for him to leave Berlin. Fortunately, he obtained a position at the University of Würzburg, the first chair of pathological anatomy in Germany. Virchow spent 7 very productive years there, until the chair at Berlin became vacant in 1856. After demanding the establishment of an institute of pathology for his research, Virchow worked at the University of Berlin until his death. In 1847 Virchow and Benno Reinhardt (1819–1852) founded the *Archives for Pathological Anatomy*. Editing the journal gave Virchow the opportunity to encourage original scholarship in a variety of fields. Despite the title of the journal, Virchow included articles on comparative anatomy and physiology, anthropology, oriental languages, translations of medieval Greek and Arabic manuscripts, as well as the expected papers on the histology of tumors and infectious diseases. In 1850 Virchow was elected to the Berlin City Council and in 1861 to the Prussian Diet, where he opposed the policies of Otto von Bismarck (1815–1898). He remained on the City Council for the rest of his life and initiated many social, sanitary, and medical reforms. During the wars of 1866 and 1870, Virchow was responsible for military hospitals and the development of the first hospital trains. He served as a member of the Reichstag from 1880 to 1893. In 1869 he founded the Berlin Society of Anthropology, Ethnology and Prehistory, serving as president until his death. He also worked for the founding of the Berlin Ethnological Museum and the Folklore Museum.

In the first volume of the *Archives for Pathological Anatomy*, Virchow reviewed prevailing ideas on the organization and growth of tissues. Reflecting the influence of Schleiden and Schwann, he described the origin of cells from the differentiation of a formless blastema, a fluid exuded from vessels. His studies of inflammation seemed to support this theory; white corpuscles from the blood entered the area of a wound in great numbers and became macrophages. Yet during his early investigations of the healing of the cornea, Virchow had seen phenomena inconsistent with existing views of cell growth, and he began making intensive microscopic studies of pathological processes. He reached the same conclusion that Robert Remak had come to from his embryological studies. Virchow and Remak rejected free-cell formation and spontaneous generation. By 1854 Virchow was convinced that "there is no life but through direct succession," that is, all cells were derived from preexisting cells. In 1855 Virchow published a paper on "cellular pathology," which included the famous motto *omnis cellula e cellula*. Resistance to this fundamental insight, Virchow argued, could be attributed to the fact that on the one hand the microscopic method was still new to medicine and was not deeply ingrained in the thinking of older physicians, while other investigators were overly enthusiastic and uncritical in their use of the instrument. His experiences might have been similar to those of an Oxford University student attending newly established courses in microscopic anatomy in the 1840s. When one of the older faculty members was persuaded to look at the microscopic preparations that illustrated the lectures, the elderly professor proclaimed that he did not believe in such evidence, but even if it were true he did not think that God meant for human beings to know such things.

For Virchow the cell was the fundamental link in the great chain that formed the hierarchy of tissues, organs, systems, and, ultimately, the complete organism. In a series of lectures later published under the title *Cellular Pathology* (1858), Virchow reviewed previous ideas about plant and animal structure, compared plant and animal cells, and analyzed the structure of the tissues that made up human organs in health and disease. Humoral pathology, the ancient concept of general disease, was set aside as Virchow demanded that medical and scientific inquiry focus on a new question: Where in terms of the body's cells is the disease? All disease, according to

Virchow, is simply modified life; there is no essential difference between normal and pathological states. Thus, the study of pathology as a science must be inextricably linked to the study of physiology in order to determine the disturbances that take place in a pathological state. For example, extensive studies of leukemia and various tumors convinced Virchow that cancer cells differ from normal cells primarily in their behavior rather than in their structure. While other scientists had advanced some of these ideas, none had stated them as convincingly as Rudolf Virchow, the man who was often called Germany's Pope of Pathology.

During the last 25 years of the nineteenth century cell theory came to include two more generalizations: cells in animals and plants are formed by the equal division of existing cells, and division of the nucleus precedes division of the cell. Eventually cell theory was able to accommodate the challenge introduced by studies of microbial forms of life and deal with findings that indicated that not all living organisms or tissue components contain cells in the classic sense of a blob of protoplasm containing a nucleus. On the other hand, further studies of the cells of higher organisms have revealed a microcosm full of minute inclusions and organelles, some of which contain their own genetic machinery. In his influential treatise *The Cell in Development and Heredity*, Edmund B. Wilson outlined three stages in the development of cell theory after its enunciation by Schleiden and Schwann. Between 1840 and 1870, scientists labored at the foundations of the theory, marking out the fundamental outlines and principles of genetic continuity. The second period, from 1870 to 1900, witnessed the maturation of cytology and cellular embryology, as well as the development of new concepts of the physical basis of heredity and the mechanism of development. Wilson wrote from the perspective of a scientist in the third stage, after the rediscovery of Mendel's laws when the focus of attention had shifted towards studies of heredity long after improvements in microscopy, especially improved methods of preparing and staining biological samples, made it possible to analyze the fine structure of the cell. Schwann and other biologists usually prepared their materials by teasing out or squashing fresh material into a thin layer and studying this directly under the microscope. The structure of relatively uniform tissues might be adequately studied this way, but the interrelationships among different types of cells in complex organs were essentially obliterated. Fixatives were first used as preservatives for gross specimens. The eminent seventeenth-century chemist Robert Boyle had suggested many preservatives, but it is uncertain whether he actually tested them. Microscopists used various chemicals, such as alcohol, acetic acid, chromic acid and its salts as hardening agents and soaked tissue samples until they became rigid enough to be sliced with razors and knives. Many of the early pioneers of histological technique were English amateurs who were more interested in making beautiful microscopic preparations for display than in learning new facts about nature. English microscopists probably developed the precursor of the microscope slide and the first microtomes. In Germany, in contrast, microscopes were seen as tools for the use of scientists, who had little patience with merely cosmetic, time-wasting techniques. Until German scientists became interested in the subtle details of tissue and cell structure, they continued to use free-hand sections and ridiculed the conceits of British amateurs.

The availability of a whole panoply of synthetic dyes in the late nineteenth century vastly expanded the ability of microscopists to see structural details within biological preparations. Natural dyes had of course been available for hundreds of years. For example, Herodotus had referred to the use of madder for dying goatskin cloaks. Dyes can impart color to fabrics, paper, leather, wood, and to plant and animal tissues prepared for microscopic examination. Staining methods in the first half of the nineteenth century were generally quite crude, and the origins of many common procedures are obscure. Early microscopists used natural dyes such as blueberry juice, red cabbage juice, indigo, carmine, madder, and logwood extract. Carmine is derived from *Coccus cacti*, an insect that had been cultivated and used for dyeing in Mexico long before the Spanish conquest. The female insects, which contain a purple-colored sap, were harvested, killed, and dried just before egg laying, and the pulverized insect preparation was sold as cochineal. Hematoxylin, a very important reagent in histology, is the best known product of logwood.

Crude staining methods, coupled with overactive imaginations and speculative tendencies, led to reports of various phantom structures, as in the case of Christian Gottfried Ehrenberg's (1795–1876) work on the infusoria. Convinced that the infusoria, a category in which he included protozoa and bacteria, had internal organs, Ehrenberg interpreted his stained preparations as evidence for the existence of their digestive organs. He even named these organisms *Polygastrica* to indicate that they had several stomachs in which indigo and other dyes accumulated. Later investigators would reinterpret such observations as evidence of differential staining, that is, the inner portions of the cells Ehrenberg was observing could be stained while the outer parts resisted the stains. Joseph von Gerlach (1820–1896), professor of anatomy and physiology at Erlangen, Germany, has been called the founder of modern staining technique, although he obviously was not the first to use stains in research. His famous *Handbook of General and Special Histology* was published in 1848. For 50 years Gerlach was one of Europe's best known anatomists and most influential microscopists. Unlike some of the other pioneers of staining, Gerlach appears to have carefully controlled his experiments so that they could be described in detail and reproduced by others. In the 1850s he introduced one of the most successful of the early histological stains, a transparent solution of carmine, ammonia, and gelatin generally called Gerlach's stain. Use of this stain produced a noticeable difference in the degree of staining of the nucleus and the intercellular substance. These studies suggested that the uptake of stain indicated that specific chemical reactions had occurred and that different components of cells probably differed in their ability to combine with dyes.

A major addition to the tools available to histologists was the result of the discovery of the aniline dyes by William Henry Perkin (1848–1907) who discovered mauve, the first synthetic dye. Perkin inadvertently prepared this dye while trying, very unsuccessfully, to synthesize quinine. He was awarded a patent for this synthetic dye when he was only 18; the aniline dyes made him rich enough to retire in his thirties. Within a few decades of Perkin's patent, a rainbow of dyes, such as safranin, methyl violet, aniline blue, methyl green, fuchsin, and crystal violet, had been synthesized. By the 1860s many aniline dyes were being used as biological stains, but even in the late nineteenth century commercially available dyes were often impure mixtures and using them systematically and reproducibly was difficult. Further confusion resulted from the fact that the names used for specific dyes varied in different countries.

Generally, the period from 1875 to 1895 was rich in discoveries concerning fundamental cytological phenomena such as mitosis, maturation, and fertilization and important cellular organelles, such as mitochondria, chloroplasts, the Golgi apparatus, and so forth. These methods made possible the discoveries and theories that linked cytology to inheritance and development. By the end of the nineteenth century the minute bodies that appeared during cell division had been investigated. Eduard Strasburger (1844–1912), professor of botany at Bonn, helped to unify the field with his monumental *Cell-Formation and Cell-Division*, first published in 1875. His descriptions of the complex processes taking place in the division of plant cells, however, are not as well known as Walter Flemming's (1843–1905) studies of cell division in animals. In 1876 Flemming became professor of anatomy and director of the Anatomical Institute at the University of Kiel, where he remained until his retirement in 1901. His first papers on the cell and the nucleus were published in 1877. Flemming described the chromosomes in the late 1870s, but the term itself was first used in 1888 by Heinrich W. G.

Waldeyer (1836–1921), who introduced the use of hematoxylin as a histological stain. Flemming had used the term *chromatin* for the nuclear substance, and he gave the name *mitosis* to cell division. Flemming's *Cell Substance, Nucleus, and Cell Division* (1882) established a basic framework for further exploration of the stages of cell division.

EMBRYOLOGY REVISITED: CELLS, ORGANISMS, AND EMBRYOS

Studies of embryology might be summarized in terms of changes in the level of resolution at which scientists were able to work. Embryos could be thought of as minute but complete organisms, assemblages of parts, such as globules, vesicles, and fibers. Once cell theory had been established, embryos could be analyzed in terms of cells and components of cells, such as nucleus, cytoplasm, membranes, chromosomes, intracellular organelles, and macromolecules. Another historical framework would include the general theoretical or metaphysical frameworks that guided research and sparked controversies by confronting scientists with choices between preformationist and epigenetic, mechanistic and vitalistic, or reductionist and organicist paradigms.

Nature philosophy, or Romanticism, which permeated the thinking of European artists, poets, writers, and scientists in the early nineteenth century, emphasized the historical description of growth and change. The Romantic outlook affected the newly defined science calling itself biology by stimulating interest in the forms found in nature and through their claim that variation in nature was a reflection of a limited number of *archetypes*, or ideal forms. The archetype was a kind of generalized form that reflected nature's basic plan. Knowing the archetype was the key to understanding living organization.

Romanticism and nature philosophy generated an intense interest in comparative anatomy and morphology as well as the history of institutions, customs, races, and the individual organism. In terms of its history, the organism could be understood through a detailed description of its development and differentiation; thus many biologists touched by Romanticism pursued embryological studies. In attempting to understand individual growth, some naturalists resorted to a form of explanation that has been called *teleomechanism*. This concept implies a balance between the belief that nature demonstrates purposeful behavior and an attempt to establish mechanical explanations for specific aspects of vital phenomena. Advocates of a new approach to embryology which has been called *developmentalism* saw their work as a branch of morphology concerned primarily with explaining how the fertilized egg gave rise to the structure, form, and apparently goal-directed behavior of the organism.

Many of the scientists involved in the development of cell theory and embryology shared an interest in nature philosophy, sometimes in its most extreme forms, and some seem to have suffered from severe mental and emotional problems. Karl Ernst von Baer (1792–1876), might be seen as the archetype of this species. In his *Autobiography* von Baer specifically recounted many periods of despondency and episodes of depression. One year he had locked himself in his rooms "far beyond winter." When he finally emerged and saw the ripening fields of rye he was so shaken that he threw himself to the ground and vowed to abandon his sedentary and solitary ways. But the next year was the same; obsessed with scientific projects he found himself destroying his nervous system and digestion. Only travel and outdoor activities seemed to help him overcome his bouts of illness and depression. Although von Baer has generally been considered a founder of the new epigenesis, he actually presented a more subtle interpretation of embryological development than a distinction between epigenesis and preformation would suggest. Indeed, it has been said that Aristotle's concept of teleology is deeply embedded in von Baer's biological philosophy. In his *Autobiography* von Baer recalled that as a student at the University of Dorpat, he had discovered that some of the professors were tinged with nature philosophy, while others "suffered from a surfeit of rather useless scholarship." Occasionally students were warned about the bogeyman called nature philosophy, but were not told what harm such ideas might cause. Naturally, this only made nature philosophy more tempting.

When von Baer finally graduated as a doctor of medicine he had little confidence in himself or in medicine as a whole. His brief experience with military medicine and hospital medicine left him with little enthusiasm for a life spent as a practicing physician. The prospect of making his way in the world of science was more exciting, and he turned to the study of anatomy, embryology, physiology, and comparative anatomy. After years of peripatetic poverty he obtained a position at Königsberg, where he carried out most of his embryological work. His major treatise *On the Developmental History of Animals* appeared in 1828. In 1834 he moved to Saint Petersburg. To find relief from depression, he embarked on various expeditions which took him all over Europe, Russia, and even Lapland. When his sight and hearing failed he was forced to retire, but he continued to pursue some aspects of his researches until his death at 84 years of age.

Finding the mammalian ovum was the ultimate aim of von Baer's early embryological research, but in his *Autobiography* von Baer claimed that he would never have had the courage to begin his research as a quest for this goal. The history of this quest was the cause of his caution. After all, even Albrecht von Haller, a man characterized by unsurpassed erudition and prodigious diligence, had been unsuccessful in his research on the development of this enigmatic entity. When Regnier de Graaf (1641–1673) discovered certain swellings on the ovaries of rabbits, ewes, and human females, he assumed they were eggs. Leeuwenhoek had argued that the structures now known as Graafian follicles could not be eggs, but Haller suggested that the egg might be formed by the coagulation of the fluid in the Graafian follicle. Surprisingly, the relatively large mammalian egg had not been discovered earlier, while the sperm had been discovered by the first wave of microscopists. Von Haller had gone to great trouble and expense in his search for the mammalian egg; he dissected about 40 female sheep, but could not find a developing egg in the uterus until about 17–19 days after mating. After examining intact Graafian follicles, burst follicles, the corpus luteum (spent follicles filled with a yellow mass), the oviducts, and the uterus, Haller concluded that a fluid must have been sent to the uterus, where it became more mucus-like, and then produced the egg by some kind of coagulation or crystallization. Haller's theory of the formation of the mammalian and human egg was still being taught at Dorpat University while von Baer was a student. Despite Haller's authority, many scientists continued to believe that the follicles were the actual, original ova and must be received and transported by the oviducts.

Aware of the dispute between Haller and Caspar Wolff, von Baer explained that Wolff had seen his work in opposition to the so-called preformation or emboitement theory, which assumed that a completely formed embryo was present in the ovum. Contrasting his work with that of Wolff, von Baer argued that Wolff's concept of epigenesis, as the truly new formation of all parts and of the entire embryo, had gone too far. While it was certainly true that no limbs or specific parts were present on the ovum, von Baer argued that the parts "do not come into existence by truly new formation, but by a transformation of something already existing." The term *evolution*, in the sense of metamorphosis, von Baer suggested, was more appropriate for embryological development than Wolff's terminology in which epigenesis referred to "new formation." While nothing corporeal was actually preformed, the course of development was in some sense preformed.

Before von Baer discovered the mammalian ovum, several scientists had reported seeing minute entities in the oviducts and had suggested that these bodies might be the mammalian ovum. Von Baer thought that previous investigators had assumed that the mammalian ovum must be transparent and thus did not recognize that the opaque entities seen in the oviduct were ova. In 1827 von Baer told a colleague that he needed a dog that had gone into heat only a few days earlier in order to search for Graafian follicles that were still closed but ready to burst. His friend had such a dog, and she was sacrificed and dissected. At first von Baer was disappointed because he found several follicles already burst and none that seemed close to bursting, but then he noticed that intact follicles all contained a single yellow spot. Curious about what this might be, he opened one of the follicles and with a scalpel carefully lifted the little spot into a water-filled watch glass and put it under the microscope. On seeing a small, well-defined yellow yolk mass, von Baer was so excited that he was almost afraid to look again in case he might have been deluded by a phantom. Establishing a formidable program for his embryological research, von Baer deliberately set about observing, analyzing, and comparing the course of development in a wide range of organisms.

In his *Autobiography* von Baer complained at length about the lack of immediate appreciation for his discovery. In the 1840s, after the establishment of cell theory, he was pleased to see his work honored as the "first building block" of the new embryology. Reflecting on the status of embryology in the 1860s, von Baer noted that it was still impossible to explain how the male contribution rendered the egg capable of developing and how the characteristics of the father passed into the new individual. Von Baer thought that his proof that the mammalian embryo did not coagulate out of some mucoid fluid, but developed by a series of transformations from a previously organized corpuscle, was an important factor in diminishing support for the doctrine of spontaneous generation. While he regarded spontaneous generation as "highly problematic," he did not think that the question had been unequivocally settled.

When later critics accused him of being an opponent of cell theory, von Baer recalled that he had indistinctly observed cell division in fish eggs and mammalian ova. He admitted that he had not understood what was taking place and, in some cases, had assumed that the bulges on the surface of the eggs indicated they were spoiled or defective. His early studies had not referred to the "cell division process," because before the cell theory had been elaborated by Schwann the idea of using the term "cell" for the elements of animal structures was completely foreign to him. He called the globules or elementary particles he had seen in developing embryos "histological elements" and referred to larger organic structures as "morphological elements." In reflecting on the significance of Schwann's theory of the cells, von Baer acknowledged that it certainly had been a fruitful concept since it had evoked great interest and contributed to the development of histology. But von Baer thought that Schwann had given "too great an importance to the so-called life of the cell." How could the cells build the animal organism all on their own, von Baer asked, unless they had "a great deal of morphogenetic intelligence?"

Before embryological development was linked to the theory of the cell, several investigators had called attention to observations that indicated that development does not proceed directly from egg to organ formation, but involves intermediate embryonic structures. The idea that the embryo consists of leaf-like layers out of which new structures develop can be found in the work of Caspar Wolff, but the establishment of the so-called germ layer concept is primarily associated with Heinrich Christian Pander (1794–1865) and von Baer. In 1817 Pander published a well-illustrated treatise in which he called attention to the so-called germ layers of the chick embryo. As a descriptive guide, rather than an attempt to explain the mechanism of development, the germ layer theory simply states that, despite the great differences between adult vertebrates of different species, similar organs are derived from comparable germ layers. In 1855 Robert Remak refined the concept and called the three germ layers *ecotoderm, mesoderm,* and *endoderm.* Ectoderm (outside skin) gives rise to skin and the nervous system; mesoderm (middle skin) produces muscles, the skeleton, and the excretory system; endoderm differentiates to form the notochord, digestive system, and associated glands. The notochord, a transient embryonic structure discovered by von Baer, eventually develops into the backbone. Von Baer thought the term dorsal cord, or vertebral cord, preferable to notochord. While the structure cannot be seen in adult vertebrates, except for certain fish, it serves as an important tool for determining the vertebrate nature of questionable organisms.

Another powerful descriptive generalization formulated by von Baer became widely known as the *biogenetic law,* or the law of corresponding stages. This so-called law essentially states that during development the embryo of a higher animal passes through stages which resemble stages in the development of lower animals or, to put it another way, at early stages of development the embryos of different species resemble each other more than their adult forms do. Thus, development proceeds from a rough sketch of general features to the delineation of the fine details that make each species unique. Johann Friedrich Meckel, an enthusiastic proponent of the recapitulation doctrine, introduced his "law of parallelism" in 1821 to describe the growth of the human embryo. During development, he suggested, the human embryo essentially climbs the hierarchy of animal forms from lowest to highest: fish, reptile, mammal, human. Von Baer challenged Meckel's concept and argued that the human embryo never assumed forms equivalent to the adult forms of lower animals. Embryos became progressively specialized during development even though human and reptile embryos might be difficult to tell apart at early stages of development. Whatever influence Romanticism might have had on von Baer's desire to seek out the origins of developmental processes, his achievements were grounded on systematic microscopic investigations of a great variety of animals and an awareness of differences in adult types as well as similarities in the early embryological stages. Four descriptive propositions associated with the law of corresponding stages were formulated by von Baer. First, during development general characters appear before specialized ones. Second, the most general characteristics gradually develop towards the less general and then to the most specific. For example, limb buds become recognizable as limbs that at later stages differentiate into hands, wings, or flippers. Third, during development the embryo of a given species continuously diverges from those of other species. Fourth, the embryo of a higher species goes through stages that resemble the stages of development of lower animals.

Scientists less cautious than von Baer transformed these general principles into the more dogmatic and deceptive biogenetic law primarily associated with Ernst Heinrich Haeckel (1834–1919). Interpreting embryological history as evidence that all species evolved from common ancestors, Haeckel assumed that during development embryos progressed though stages that were virtually identical to the adult forms of their ancestors. His pithy motto *ontogeny recapitulates phylogeny* confounded descriptions of the development of the individual with evidence of the evolution of all species. Phylogeny, Haeckel insisted, is the mechanical cause of ontogeny. In some sense, despite Lorenz Oken's opposition to the idea of the transmutation of species, Haeckel's biogenetic law shared a deep affinity with the metaphor at the core of nature philosophy that portrayed the universe as a great living being in which the "gestation of nature" established a profound fundamental unity among organic beings. Defining nature philosophy as the study of the "generative history of the world," Oken could see the history of the individual as, in essence, the history of the universe. All other animals, according to Oken, were persistent fetal stages in the production of man, the creature that was the prototype and model of all existence. The individual development of man, therefore, replicated the development of life on earth because the production of man was the goal of the great universal developmental tendency. For Oken and other Romantics, however, the history of development was a reflection of the establishment of archetypes rather than a series of transmutation of types in the evolutionary sense. Many Darwinian evolutionists accepted and exploited the biogenetic law or recapitulation theory as a means of incorporating embryology into the body of evolutionary theory. Charles Darwin, realizing the value of the biogenetic law to his evolutionary theory, called the field of embryology "second in importance to none in natural history." While Thomas Henry Huxley was willing to argue that "in the womb, we climb the ladder of our family tree," despite his admiration for the "excellent Huxley," von Baer never accepted this interpretation of his work.

Embryological researches and the use of comparative methods were stimulated by von Baer's work. Having shown how complex and absorbing embryological development really was, he also provided the guidelines for others to follow. In the 1870s the great Swiss anatomist Wilhelm His urged embryologists to exploit the vast and untapped potential of von Baer's analytical approach to embryology. Von Baer's studies of the mammalian ovum made it possible to see cell multiplication as the basis of embryonic development, and a new focus on the cell made it possible to think about how cell theory might apply to the process of inheritance as well as the study of reproduction and development. Such shifts in conceptual categories and the establishment of new scientific specialties and institutions in the late nineteenth century were reflected in the excitement generated by the research program known as experimental embryology.

The founders of experimental embryology, Wilhelm Roux (1850–1924) and Hans Driesch (1867–1941), were primarily interested in the question of how factors intrinsic or extrinsic to an egg or its parts could govern the development of the embryo. Convinced that descriptive and comparative studies of embryonic development were inadequate, Roux demanded a new approach and saw himself as the founder of a new discipline which he called *developmental mechanics*; the terms *causal analytical embryology* and *developmental physiology* were also used by some of his followers. When celebrating the founding of a new discipline, the appropriate nineteenth-century rite was the establishment of a new journal, in this case the *Archive for Developmental Mechanics of Organisms*; the first volume appeared in 1894. "After sufficient description," Roux exhorted his colleagues, "it is time to take the further step towards knowledge of the processes that produce them." The protocol required for this task involved the resolution of developmental processes into simple though still complex functional processes. These functional processes would then be reduced to their components, so that eventually the truly simple processes could be resolved into their physicochemical aspects. Unencumbered by modesty, Roux predicted that several centuries later students would read his work with the same intensity of interest with which he had studied Descartes. The terminology of the new experimental embryology owed a debt to Charles Darwin and Ernst Haeckel as well as to Descartes, as evident in Roux's 1881 paper "The Struggle of the Parts in the Organism: A Contribution to the Completion of a Mechanical Theory of Teleology." Embryology, Roux argued, must be investigated by means of the tools so successfully developed by the founders of mechanistic physiology. Indeed, Roux's place in the history of embryology must be understood as part of the complex interactions between mechanistic and vitalistic biological theories.

Inspired by Cartesian principles, Roux argued that embryologists must adopt experimental methods as the tools that would make possible the analysis of the immediate causes of development. On a more theoretical plane, Roux provided a way to resolve, transcend, or ignore fruitless debates about preformation and epigenesis. Rather than deal with this historical impasse by traditional means, Roux recast the old embryological questions in terms of new analytical concepts. The primary question Roux posed was whether development proceeded by means of *self-differentiation* or *correlative dependent differentiation*. Self-differentiation was defined as the capacity of the egg or of any part of the embryo to undergo further differentiation independently of extraneous factors or of neighboring parts in the embryo. Correlative dependent differentiation was defined as being dependent on extraneous stimuli or on other parts of the embryo. These were operational definitions, but, Roux insisted, they were powerful because the alternatives could be subjected to experimental tests. Establishing a new conceptual dichotomy had important implications: above all, it removed the adversarial burdens of the old terms, preformation and epigenesis. Now one could ask: To what extent is the differentiation of a given part of the embryo, at a given point in time, self-differentiation, and to what extent is it dependent differentiation?

By defining development as the production of perceptible manifoldness, Roux ignored the ancient conflict between preformation and epigenesis. His concept of development encompassed two principles: first, a true increase in manifoldness, which could be called *neoepigenesis*; second, the transformation of imperceptible manifoldness, which could be called *neopreformation*. As the mechanism of development, *self-differentiation* would involve an independent or mosaic development of each part, whereas *correlative dependent differentiation* would require the interaction of cells or groups of cells. On theoretical grounds, and in keeping with his most famous, but seriously flawed experiment, Roux believed that self-differentiation served as the mechanism of development. In other words, Roux tended to visualize the fertilized egg as similar to a complex machine and development as a process that involved the distribution of parts of the machine to the appropriate daughter cells. One series of experiments refuted earlier suggestions that gravity affected the plane of cleavage of the egg. Some extreme modifications of environmental conditions affected the embryo in a gross way, but more subtle changes in environment had little of no effect on development. Confident that external forces could be neglected, as predicted by the mosaic model, Roux set out to study formative forces within the egg. The first question was whether all the parts of the egg must collaborate to cause normal development or if the separate parts could develop independently. To answer this question, Roux destroyed one of the cells of a frog embryo at the two-cell stage; this is known as the "pricking experiment." When one cell was injured with a hot needle, the undamaged cell developed into a half-embryo. Such experiments encouraged Roux's belief that each cell normally develops independently of its neighbors and that total development is the sum of the separate differentiation of each part. While Roux was convinced of the validity of the mosaic model of development, in other hands the research program he had established rapidly destroyed the experimental base of his theory. Recreating Roux's experiment under different conditions, and using embryos of various species, other scientists obtained quite different results. Nevertheless, long after embryologists had adopted other theories of development, Roux was still cited as the scientist who had formulated the core questions of embryology.

Hans Adolf Eduard Driesch, one of the first to follow Roux's protocol for experimental embryology, provided definitive evidence against Roux's model of mosaic development. According to Driesch, the embryo seemed to develop epigenetically as a harmonious equipotential system. Driesch and Roux differed greatly in philosophical viewpoints and technical approach, but both had studied with Ernst Haeckel. Driesch's interests were very broad, encompassing mathematics, physics, and philosophy. Even as a doctoral candidate, Driesch had questioned the wisdom of his mentors; his work presented a direct challenge to August Weismann as well as to Haeckel and Roux. Eventually relations between Driesch and Haeckel deteriorated to the point where Haeckel advised his former student to take some time off and spend it in a mental hospital. Although Driesch did not take Haeckel's advice, he did eventually abandon biology for philosophy. Restless by nature, Driesch traveled widely, giving lectures in philosophy and science in the Far East and the United States.

Several important differences in experimental conditions led to the rebuttal of Roux's dogmatic extrapolation of results obtained with his injured, but unseparated, frog cells. First, Driesch used sea urchin eggs instead of frog embryos. Second, having discovered that sea urchin embryos at the two-cell stage could be separated into individual cells merely by shaking, he was able to separate the embryonic cells rather than killing one of them. The separated cells formed advanced embryos that were normal in configuration, but half the size of their natural counterparts. This apparently proved that the first two cells of the embryo did not each contain half the determinants of the whole fertilized egg; both cells contained all the information needed for full development. Roux's concept of mosaic development was, therefore, false. With improved techniques, Driesch was able to extend his experiments to the four-, eight-, and even later cell stages.

The implications of Driesch's experiments were momentous, but Driesch was not particularly happy about disproving a nice hypothesis. On seeing his half-embryos developing into typical whole gastrulas, Driesch lamented that the experiment had turned out as it must, but not as he had expected. Still, his experiment seemed a step backward from Roux's initial success in explaining development. Driesch concluded that since a new organism could be generated from parts of the embryo, it could not be regarded as a machine because the parts of a machine, being necessarily simpler than the original machine, cannot reproduce the whole. The process of development is harmonious because the parts normally work together to form one individual, even though each could form an independent individual. In contrast to Roux, Driesch emphasized the epigenetic nature of early development and compared the *presumptive significance* of an embryonic part (what it forms under normal circumstances) and its *prospective potency* (what it might form under altered conditions). The results of experimental manipulations demonstrated that the cell's prospective potency was much greater than its presumptive significance. Therefore, using the terminology of analytical geometry, the fate of a given cell was a function of its relative position in the whole.

Instead of finding that more experimental work rewarded him with a clearer picture of the embryological development, Driesch had apparently reached a more profound level of confusion, which seemed to foreclose all hope of finding a mechanistic explanation for development. There was no way Driesch could picture a machine that could develop into two whole machines, identical to the original, when divided into its component parts. Embryonic development appeared to involve a harmonious-equipotential system. That is, each part had equal capacity or potentiality to substitute for any other part, and the formation of a harmonious whole was the end result of the interplay of parts. This seemed to imply a fundamental structural homogeneity where each part was identical to any other. Such a system, Driesch logically argued, would be devoid of a structural or material cause for change or differentiation; thus, some nonmaterial causal agent had to be involved in setting differentiation into motion. At this point, Driesch turned to Aristotle for inspiration and revived that venerable old mechanic, the *entelechy*, an internal perfecting principle (that which carries the end in itself). Because he ultimately adopted this position, Driesch is generally regarded as a vitalist, but in 1894 when he formulated his analytical theory of organic development, he was still a mechanist. It was not until 1894 that he converted to vitalism; eventually he abandoned experimental embryology and devoted himself to philosophical inquiries, which came to include parapsychology and occultism.

Roux and Driesch were frustrated by the results of their own experiments, but they were sophisticated theorists who established the foundations of a rigorous analytical approach to fundamental problems of embryonic development and an experimental program that allowed researchers to plan and perform increasingly subtle and elegant experiments on living embryos by means of isolation, transplantation, and tissue culture. Driesch's contemporaries Hans Spemann (1869–1941) in Germany and Ross G. Harrison (1870–1959) in the United States were especially important for their roles in refining the techniques of experimental embryology. As the new leader of the science of development, Spemann made his mark through systematic studies of early organ determination, whereas Harrison established the experimental foundations of neurogenesis and tissue culture. In addition to his experimental work, Harrison was involved in the construction of a modern synthetic organicism as a means of understanding the structure and development of organisms. The transformation of embryological paradigms made it possible to separate embryological theories from the old conflicts between vitalism and mechanism and establish links with biochemistry and molecular genetics.

Growing up in a prosperous and cultured family, Spemann's early classical education suited his decision to leave school at the age of 19 and join his father's publishing and bookselling business. Although Spemann had done poorly in his zoology classes, a year spent in military service plus his immersion in the writings of Goethe and Ernst Haeckel led to his decision to resume his education in order to study biology. In 1891 he enrolled at the University of Heidelberg as a medical student, but he was attracted to the study of comparative anatomy and embryology. He was able to continue his studies at the Zoological Institute of the University of Würzburg with Theodor Boveri, a cytologist whose broad interests in heredity and development led to intriguing studies of the role of the nucleus and chromosomes in development. One year after receiving his doctoral degree, while enduring a rest cure for tuberculosis, Spemann read August Weismann's book *The Germ Plasm: A Theory of Heredity* (1892) and became intrigued by the puzzle of the transmission of heredity through the germplasm and the relationship between the reproductive cells and embryological development. On returning to the laboratory in 1897, Spemann began a series of experiments based on those of Roux. Instead of using frogs or sea urchins, Spemann selected the salamander as his experimental system and developed methods of separating and rearranging the cells of the early embryo. These studies led to a series of papers on the "Developmental Physiological Studies on the Triton Egg" (1901–1903) that introduced the technique of manipulating and constricting the egg with a loop of fine baby's hair. Spemann found that if he constricted salamander eggs but did not completely separate the developing cells, he could produce animals with two heads and one trunk and tail. In his autobiography he recalled his fascination with watching the behavior of such "twin embryos" and how the mystery of the split individuality in combination with the pleasure he found in this experimental technique led to his total commitment to embryological research. In 1914, Spemann became co-director and head of the Division of Developmental Mechanics of the Kaiser Wilhelm Institute for Biology in Dahlem, a suburb of Berlin. He found this a congenial environment in terms of research facilities and stimulating colleagues, but World War I delayed construction of his laboratory, and malnutrition had a severe impact on his health. In 1919 Spemann became the director of the Zoological Institute of the University of Freiburg.

The goal of Spemann's program of embryological research was to discover the precise moment when a particular embryonic structure became irrevocably determined in its path towards differentiation. His work was to make the idea of *determination* a dominant theme in development research for many years. For Spemann, embryonic development was the study of the "physiology of development." An acknowledged master of the art of microdissection, Spemann and his associates carried out a series of experiments in which selected parts of one embryo were transferred to a specific region of another. Embryos of different species were used so that color differences would indicate the identity of the transplanted area during development. In 1921 the task of analyzing the effect of transplanting a region known as the dorsal lip was assigned to Hilde Proescholdt, who had joined Spemann's laboratory in 1920. One year later Proescholdt married Otto Mangold, one of Spemann's first students. Hilde Mangold appears as the second author of Spemann's 1924 paper "Induction of Embryonic Primordia by Implantation of Organizers from a Different Species." While Spemann's other students had been allowed to publish their thesis work as sole authors, Spemann insisted on adding his name, as first author, to Hilde Mangold's thesis publication. After the discovery of the organizer region, Hilde Mangold abruptly disappears from accounts of the history of embryology as if she gave up scientific work for marriage and motherhood. Actually, Mangold died in 1924, at the age of 26, from burns caused by the explosion of a gasoline heater in her home, at about the time the organizer paper was published. Mangold's experiment involved grafting the dorsal lip region from an unpigmented species onto the flank of a host of a pigmented species. Three days after the operation was performed an almost complete secondary embryo formed on the host embryo; the secondary embryo was composed of a mosaic of host and donor cells. The mosaic nature of the secondary embryo indicated that both the transplanted tissue and the host embryo participated in its formation. Because the dorsal lip of the donor embryo could cause the formation of a new embryo, it was called the organizer region. Keeping such embryos alive outside of their protective jelly membrane was very difficult; the longest lived embryo in Mangold's experiments survived to the tail bud stage. Mangold described five experimental cases in detail and briefly noted several others. Some years later, using antibiotics and special salt solutions, Johannes Holtfreter and his coworkers were able to keep similar embryos alive into advanced larval stages. The secondary embryos in these recreations of the Mangold organizer experiment were as complete as the primary embryos of the hosts. The products of the experiment looked like twins that had been fused at the flank or belly.

By extending and refining the organizer experiment, Spemann discovered other organizer regions and felt that he was close to achieving an understanding of how the development of the embryo led progressively to the unfolding of vertebrate organization. In many ways Spemann's discovery of the organizer and his concept of induction represents the culmination of the approach epitomized by Roux and Driesch. Spemann's concept of the organizer applied to rather specific stages of development in vertebrate embryos and explained regional differences on a qualitative basis, but Spemann preferred to think of development in terms of supracellular explanation that applied to the embryo as a whole, rather than as a collection of cells. In 1935, the year in which he retired from the University of Freiburg-im-Breisgau, Spemann was awarded the Nobel Prize in Medicine or Physiology. One year later Spemann published his final account of his experiments and ideas in *Embryonic Development and Induction*. Given the state of German politics in the 1930s, and Spemann's own authoritarian proclivities, it is not surprising that historians have pointed out connections between the organizer concept as the mastermind of the cell state and the desire for a dictatorial center of authority to bring order out of the chaos afflicting the social organism.

In the 1930s several embryologists demonstrated that the organizer region was capable of inducing a secondary embryo even after it had been killed by heating, boiling, freezing, or alcohol treatment. Spemann had previously discovered that the organizer was active after its cells had been killed by crushing, but his philosophical approach remained largely that of an organicist seeking supracellular explanations, rather than cellular interpretations of the processes that led to the development of the embryo. The idea that dead embryonic tissues could retain their capacity to induce organized structures had a profound impact on experimental embryology and suggested new biochemical approaches to development and differentiation. Further investigations revealed that many other bits of animal tissues, alive and dead, could induce complex structures, including heads, internal organs, and tails. This new era in embryological research was somewhat disconcerting to the old pioneers, as evident in Spemann's complaint in *Embryonic Development and Induction* that a "dead organizer" was a contradiction in terms. Even as some researchers embraced a new organismic paradigm of embryological development, others adopted more limited, or reductionist approaches in a search for specific chemical agents released by the inducing tissues and attempts to understand how the reacting tissue was able to receive, recognize, and respond to specific chemical stimuli.

Spemann's approach to embryology was built on an organismic, holistic view of embryos and their development. Such attempts to understand developmental processes in terms of the potentialities and interactions of supracellular embryonic parts were largely displaced by the growing force of reductionist trends in post World War II biology. As the focus of biological theory and technique shifted from the organism as a whole to the cellular, subcellular, and molecular level, classical experimental embryology was superseded by developmental biology. Experimental embryology clearly left a legacy of unfinished business, which offered intriguing problems for researchers willing to address old problems with the language and techniques of molecular biology. Following the history of embryology further, therefore, requires a thorough immersion in biochemistry, genetics, and molecular biology. Moreover, for the most part the study of embryological theories and experimental methods becomes overwhelmingly burdened by technical details and prodigious collections of data that seem to resist all efforts to resolve themselves into a satisfactory answer to the central issues posed by embryological development.

During the golden era of induction theory, there was hope that some "magic molecules" would be found with organizing and inducing powers. As no such molecule was discovered, the idea lost much of its appeal. Interest in organizers gradually faded, but organizer theory helped to stimulate a chemical approach to embryology. With further progress in embryology, genetics, and molecular biology, it became apparent that development proceeds in accordance with instructions somehow mapped out in the genetic material, but the molecular biology of the gene proved to be more accessible than embryological development. The exact chain of events by which instruction encoded in the genetic material caused a cell or groups of cells to differentiate in the proper pattern in time and space remained obscure for decades after the broad outlines of the chemistry of the gene had been established. Indeed, many pioneers in genetics and embryology saw the two fields as separate and distinct rather than as different aspects of the same basic problem: *How does the embryo transform cells carrying identical genetic information into an integrated, harmonious population of cells expressing very specific portions of their common heritage?* Not until the 1980s did it once again seem possible to discover the genetic mechanisms that control development, or at least analyze specific steps in the development of flies, worms, and mice. By the end of the 1980s, using the techniques of molecular biology, researchers were able to identify specific genes that become active during embryonic development and thus establish the origin of significant developmental changes. Most surprisingly, it became clear that the same genetic mechanisms were at work in the developmental processes of species as different from each other as fruit flies and human beings. As a problem in the domain of molecular biology, the study of the molecular genetics of developmental programs and pattern formation, or in other words, the genetic switches and proteins that define the destiny of individual cells in embryos, has replaced older themes of generation. Yet the essential core of such studies remains the wonderful coordination of the processes of development and differentiation in time and space. Although macromolecules, genes, homeoboxes, repressors, inducers, organelles, membranes, and receptors have displaced the cell as the central focus of scientific inquiry and theoretical concerns, the cell remains a natural unit of structure and function, the link between generations, and the unit through which the egg becomes embryo and adult.

SUGGESTED READINGS

Ackerknecht, E. (1953). *Rudolf Virchow, Doctor, Statesman, Anthropologist*. Madison, WI: University of Wisconsin Press.

Adelmann, H. B. (1942). *The Embryological Treatises of Hieronymus Fabricius of Aquapendente.* Ithaca, NY: Cornell University Press.

Adelmann, H. B. (1966). *Marcello Malpighi and the Evolution of Embryology.* 5 vols. Ithaca, NY: Cornell University Press.

Amrine, F., Zucker, F. J., and Wheeler, H. (1987). *Goethe and the Sciences: A Reappraisal.* Dordrecht: Reidel.

Anderson, L. (1982). *Charles Bonnet and the Order of the Known.* Boston: Reidel.

Baer, E. von (1986). *Autobiography of Dr. Karl Ernst von Baer.* Ed. and with a preface by Jane M. Oppenheimer. Trans. from the 1886 German edition. Canton, MA: Science History Publications.

Baker, J. (1948–55). *The Cell Theory: A Restatement, History, and Critique.* New York: Garland.

Benson, K. R. (1981). Problems of individual development: Descriptive embryological morphology in America at the turn of the century. *Journal of the History of Biology 14:* 115–128.

Blyakher, L. Ya (Blacher, Leonidas I.) (1982). *History of Embryology in Russia From the Middle of the Eighteenth to the Middle of the Nineteenth Century.* With an Introduction by Jane Maienschein. Trans. from the Russian edition of 1955 by Hosni Ibrahim Youssef and Boulos Abdel Malek. Ed. by G. A. Schmidt. Washington, D.C.: The Smithsonian Institution.

Bonner, J. (1965). *The Molecular Biology of Development.* New York: Oxford University Press.

Bowler, Peter J. (1989). *The Mendelian Revolution: The Emergence of Hereditarian Concepts in Modern Science and Society.* Baltimore: The Johns Hopkins University Press.

Churchill, F. B. (1973). Chabry, Roux, and the experimental method in nineteenth century embryology. In *Foundations of Scientific Method: The Nineteenth Century.* Ed. by R. N. Giere and R. S. Westfall. Bloomington, IN: Indiana University Press, pp. 161–205.

Clark, G., and Kasten, F. H., eds. (1983). *History of Staining.* Baltimore, MD: Williams & Wilkins.

Conn, H. J. (1933). *The History of Staining.* Geneva, NY: Biological Stain Commission.

Cunningham, A., and Jardine, N., eds. (1990). *Romanticism and the Sciences.* New York: Cambridge University Press.

Davidson, E. (1986). *Gene Activity in Early Development.* 2nd ed. New York: Academic Press.

De Duve, C. (1984). *A Guided Tour of the Living Cell.* 2 vols. New York: Scientific American Library.

Driesch, H. (1908). *Science and Philosophy of the Organism.* London: Black.

Farley, J. (1982). *Gametes and Spores. Ideas about Sexual Reproduction 1750–1914.* Baltimore, MD: Johns Hopkins University Press.

Florkin, M., and Stota, E. H. (1967). *Morphogenesis, Differentiation and Development.* New York: Elsevier.

Gaskings, E. (1967). *Investigations into Generation 1651–1828*. Baltimore, MD: Johns Hopkins University Press.

Gould, S. J. (1977). *Ontogeny and Phylogeny*. Cambridge, MA: Harvard University Press.

Hamburger, V. (1988). *Heritage of Experimental Embryology*. New York: Oxford University Press.

Haraway, D. J. (1976). *Crystals, Fabrics and Fields: Metaphors of Organicism in Twentieth-Century Developmental Biology*. New Haven, CT: Yale University Press.

Harrison, R. G. (1969). *Organization and Development of the Embryo*. New Haven, CT: Yale University Press.

Harvey, W. (1965). *The Works of William Harvey, M.D.* Trans. with *A Life of the Author* by Robert Willis. (London: The Sydenham Society, 1847.) New York: Johnson Reprint.

Haymaker, W., ed. (1970). *Founders of Neurology*. 2nd ed. Springfield, IL: Charles C. Thomas.

Horder, T. J., Witkowski, J. A., and Wylie, C. C., eds. (1986). *A History of Embryology*. New York: Cambridge University Press.

Hughes, A. (1959). *A History of Cytology*. New York: Abelard-Schuman.

Lee, A. B. (1885). *The Microtomist's Vademecum*. Philadelphia, PA: Blakistons.

Lenoir, T. (1982). *The Strategy of Life: Teleology and Mechanics in Nineteenth Century German Biology*. Boston, MA: Reidel.

Maienschein, J. (1983). Experimental biology in transition: Harrison's Embryology, 1895–1910. *Studies in the History of Biology 6:* 107–127.

Manning, K. R. (1983). *Black Apollo of Science: The Life of Ernest Everett Just*. New York: Oxford University Press.

Meyer, A. W. (1936). *An Analysis of William Harvey's Generation of Animals*. Stanford, CA: Stanford University Press.

Meyer, A. W. (1939). *The Rise of Embryology*. Stanford, CA: Stanford University Press.

Nakamura, O., and Toivonen, S., eds. (1978). *Organizer: A Milestone of a Half Century From Spemann*. New York: Elsevier.

Needham, J. (1959). *The Rise of Embryology*. Cambridge: Cambridge University Press.

Needham, J. (1968). Organizer phenomena after four decades: a retrospect and prospect. In *J. B. S. Haldane and Modern Biology*. Ed. by K. R. Dronamraju. Baltimore, MD: Johns Hopkins Press, pp. 227–298.

Oppenheimer, J. (1967). *Essays in the History of Embryology and Biology*. Cambridge, MA: MIT Press.

Oppenheimer, J. (1970). Cell and organizers. *American Zoologist 10:* 75–88.

Poynter, F. N. L. (1968). Hunter, Spallanzani, and the history of artificial insemination. In *Medicine, Science and Culture*. Ed. by L. G. Stevenson and R. P. Multhauf. Baltimore, MD: Johns Hopkins University Press.

Roe, S. A. (1981). *Matter, Life, and Generation. Eighteenth Century Embryology and the Haller-Wolff Debate*. New York: Cambridge University Press.

Rudnick, D., ed. (1958). *Cell, Organism and Milieu*. New York: Ronald.

Sandler, I. (1973). The re-examination of Spallanzani's interpretation of the role of the spermatic animalcules in fertilization. *Journal of the History of Biology 6:* 193–223.

Schleiden, M. J. (1968). *Principles of Scientific Botany; or, Botany as an Inductive Science*. Trans. by E. Lankester with a new introduction by J. Lorch. New York: Johnson Reprint.

Schwann, T. (1969). *Microscopical Researches into the Accordance in the Structure and Growth of Animals and Plants*. Trans. by Henry Smith. New York: Kraus Reprint.

Slack, J. M. W. (1983). *From Egg to Embryo. Determinative Events in Early Development*. New York: Cambridge University Press.

Spemann, H. (1938). *Embryonic Development and Induction*. New Haven, CT: Yale University Press.

Twitty, Victor C. (1966). *Of Scientists and Salamanders*. San Francisco, CA: W. H. Freeman.

Virchow, R. (1971). *Cellular Pathology*. New York: Dover.

Virchow, Rudolf (1958). *Disease, Life and Man: Selected Essays*. Trans. and introduction by L. J. Rather. Stanford, CA: Stanford University Press.

Willier, B. H., and Oppenheimer, J. M., eds. (1974). *Foundations of Experimental Embryology*. 2nd ed. New York: Hafner.

Wilson, E. B. (1925). *The Cell in Development and Heredity*. New York: MacMillan.

Winsor, Mary P. (1976). *Starfish, Jellyfish, and the Order of Life. Issues in Nineteenth-Century Science*. New Haven, CT: Yale University Press.

6

PHYSIOLOGY

Physiological studies from the most ancient times to the present have been guided implicitly or explicitly by a philosophical framework that has been either *mechanistic* or *vitalistic*. The mechanistic philosophy asserts that all life phenomena can be completely explained in terms of the physical-chemical laws that govern the inanimate world. Vitalists claim that the real entity of life is the soul or vital force and that the body exists for and through the soul, which is incomprehensible in strictly scientific terms.

Although the activities that distinguish living from nonliving things might be considered more exciting than their morphology, anatomical inquiries could be pursued with only the naked eye and a few simple tools. An understanding of physiological phenomena, in contrast, depends on subtle inferences drawn from chemistry and physics. Thus, although Aristotle, Galen, and Vesalius attempted to analyze the relationship between structure and function, their explanations rarely transcended obscure references to cooking, fermentation, animal heat, spirits, and faculties. It was not until the seventeenth century that new ways of dealing with the dynamic functions of the body were realized. By the end of that remarkable century, a new concept of the cosmos was well established. Through the work of Galileo, Kepler, and Newton, the universe emerged as a great law-bound machine with the earth merely one of the planets circling the sun. Gravity, not crys-

The private laboratory of a professor of physiology at Altanta Medical College in 1913

talline spheres and angels, kept the heavenly bodies moving in their orbits. Although the Scientific Revolution may not have transformed the life sciences as profoundly as it did physics, it was not without effect. Two different ways of exploring vital phenomena emerged with new clarity in this period. These might be called the *metric experimental*, best exemplified by the work of William Harvey, and the *rational-philosophical* approach used by the philosopher and mathematician René Descartes (1596–1650).

The method exploited so well by Harvey could be seen as the culmination of the work of Aristotle, Galen, and Vesalius. Harvey said that through vivisection he was led to truth, that is, his studies of the structures and activities of the living body gave him clues to explain vital phenomena in a new way without any special knowledge drawn from chemistry or physics. Indeed, Harvey had essentially no tools or instruments that had not been available to his famous predecessors. On the other hand, the career of Santorio Santorio, who sat so patiently in his balance and measured what could be measured, proves that measurement and patient experimentation are not enough to answer fundamental physiological questions. Quantitation alone gives little insight into vital phenomena without an appropriate theoretical framework for guidance.

In the seventeenth century physiologists dealt most successfully with problems of a mechanical nature. Recognizing the power of mechanical explanation in Harvey's work, other physiologists tried to force vital phenomena to fit mechanical analogies. The jaws were explained as pincers, the stomach as a mill, the veins and arteries as hydraulic tubes, the heart as a spring, the muscles and bones as a system of cords and pulleys, the lungs as bellows, the viscera and kidneys as sieves and filters. Digestion and metabolism, muscle and nerve action are problems requiring an understanding of chemical and electrical phenomena. In contrast to the physiologists known as *iatromechanists*, who believed that all functions of the living body could be explained on physical and mathematical principles, the *iatrochemists* and Paracelsians attempted to explain vital phenomena as chemical events.

The writings of René Descartes provided the most influential philosophical framework for a mechanistic approach to physiology. While Descartes

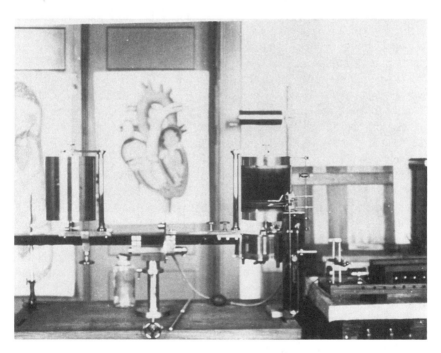

Demonstrating the action of the heart in a physiology laboratory

made no original physiological discoveries and dissected only to support his preconceived ideas, he played a critical role in assimilating and integrating the work of Galileo, Kepler, and Harvey into a new system of philosophy and provided his followers with an apparently complete and satisfying mechanistic system. His writings on physiology, which include *On Man* and *On the Formation of the Fetus*, were not meant as particular contributions to that science, but were part of a general system of philosophy. The fundamental platform of the mechanical philosophy set forth in a bold and comprehensive manner in the *Principles of Philosophy* (1644) was that all natural phenomena could be explained in terms of matter and motion.

Descartes was born into a wealthy family, but his early life was marked by tragedy and illness. His mother died when he was born, and his own pulmonary weakness seemed to doom him to an early grave. During the years he spent at a Jesuit school, he was drawn to mathematics as the only area of truth and reality to be found in a world where Aristotelian philosophy was under attack and the Copernican theory was being widely adopted by enlightened thinkers. Doubt may have been the major product of his Jesuit education, but the profound impact of this religious indoctrination was at the core of a system that has been described as the metaphysics of a Roman Catholic mathematician. Although convinced that he could erect a philosophical framework that would unite all the physical and natural sciences, when Descartes heard of the persecution of Galileo, he suppressed his own *Treatise on the World*, which owed too much to the work of Galileo, in order to avoid the appearance of heresy. Descartes destroyed some of his manuscripts and turned his attention to safer subjects; parts of his great treatise *On the World* were finally published after his death.

Even a superficial introduction to philosophy must include Descartes's system and his famous phrase, *cogito, ergo sum* (I think, therefore, I am). This was the one idea that Descartes was sure of even as he went about systematically doubting everything else; that is, mind must exist. Because it was inconceivable that God would deceive us, clear and distinct ideas about matter must also be true. The world, therefore, must be composed of two distinct entities: *mind* and *matter*. Ideas that were true and self-evident became the premises by which other truths were established. Although Descartes acknowledged the importance of observations, his approach was a form of rationalism which subordinated the crude facts of observation and experiment to the test of reason and *a priori* principles. His universe was a great mechanical system in which God was the first cause of all motion. Matter had the essential qualities of extension in space, divisibility, and motion. Rejecting the atomic theory of Democritus and Lucretius, Descartes ruled out the existence of their great void. "Give me motion and extension," Descartes said, "and I will construct the world."

Cartesian doctrine treated animals as automata whose activities were explained in purely mechanical terms as the motions of material corpuscles and the heat generated by the heart. The Galenic doctrine of the three spirits was replaced by one where only the animal spirits remained, but were transformed into a subtle fluid utilized in the brain and nerves. Even human beings emerged from Descartes's exposition as another kind of earthly machine which differed from animal automata by virtue of the rational soul that governed their actions. Serving as the agent of thought and governing volition, conscious perception, memory, imagination, and reason, the rational soul was the only entity exempted from a purely mechanical explanation. The purpose of Descartes's *Treatise of Man* was to extend his concept of the universe as a machine to the explanation of human beings as machines working in accordance with physical laws. To do this, he sifted through the anatomical knowledge of his time, carried out his own dissections, added certain self-evident ideas, and concocted a theory of the action of the body as a machine. He was quite enthusiastic about Harvey's work on the motion of the heart and the blood, the perfect example of a mechanical system, but he imposed his own interpretation on fundamental aspects of the workings and purposes of the circulatory system.

According to Descartes, the heart was a heat machine rather than a pump. Thus, in contrast to Harvey, Descartes endorsed the ancient idea that the action stroke of the heart was the expansion phase, rather than the contraction. Because the heart was so fiery and hot, as soon as the blood entered its chambers it was immediately dilated, warmed, and subtilized. The purpose of the fire in the heart was, therefore, to heat and expand the blood so that it continually fell drop by drop through a passage from the vena cava into the right chamber, from which it was driven into the lungs. It might have been possible to obtain estimates of the temperature of the heart and other organs with the thermometers available in the seventeenth century, but Cartesian physiology did not call for testing self-evident ideas against such crude experimental verification. In contrast to the fiery heart, the lung was described as a delicate, soft organ in which the hot vapors of the blood were cooled by fresh air so that they condensed and fell drop by drop into the left cavity of the heart. Similar mechanical explanations were proposed for other physiological phenomena, but the nervous system presented a formidable challenge. According to Descartes, all other systems were subservient to the nervous system, which carried out the commands of the rational soul. The strongest and subtlest parts of the blood were, therefore, carried to the cavities of the brain by arteries taking the most direct route from the heart. These special parts of the blood nourished and sustained the substance of the brain and produced the animal spirits, which Descartes compared to a subtle wind or very pure flame. Through conduits in the brain the animal spirits were able to enter the nerves and affect the muscles and cause movement. The animal spirits of the Galenic system were transformed into a special subtle fluid which could be described in purely mechanical terms. Thus, the nerves were tubes in which purely hypothetical Cartesian valves governed the direction of flow of the nervous fluid, just as the venous valves governed the flow of the blood.

As a means of explaining how the rational soul interacted with the earthly machine, Descartes selected the pineal gland as the seat of the soul and the only site in which a direct interaction between the immaterial soul and the corporeal machine took place. Descartes believed that the pineal gland was soft and mobile and that it served as a reservoir for the animal spirits. The pineal gland was chosen because it is located at the base of the brain, is a single rather than a paired organ, and because it was (erroneously) thought to be present in humans but absent in lower animals. In Descartes's system nerves not only possessed valves, but also delicate threads along the length of their cavities which connected the brain and the sense organs. The tiniest motion along the thread exerted a pull upon the site of the brain where the thread originated, opening up the orifices of certain pores on the internal surface of the brain, allowing the animal spirits to flow into the muscles, causing the machinery to move. The concept is reminiscent of Plato's description of men as the puppets of the gods. The nerves, muscles, and tendons of the body were compared to the engines and springs that moved the devices in the elaborate fountains and grottoes of the French royal gardens. The gardens were rather like an early version of Disneyland where mechanical gods, goddesses, and monsters moved when hidden triggers were activated by the visitor. Indeed, in the Cartesian system all physiological functions might be as mechanistic as the workings of clocks, mills, and other machines. Although Descartes was not always consistent as to whether ideas were differentiations impressed upon matter or aspects of the soul alone, he attempted to explain thoughts and emotions in mechanical terms. Ideas were impressions received by animals spirits as they left the pineal gland. Memories were apparently physically recorded, rather like permanent press, into the fabric of the nerves. Since human beings were a combination of soul and body, emotion was experienced both in the soul and in the machinery of the body. For example, the joy experienced in the soul had its counterpart in the body because in this state the blood would become finer, expanding more readily in the heart, and thus facilitating the excitement of the nerves.

Descartes's work was widely read, imitated, and honored. The Cartesian system challenged scientists to treat the physical and mental aspects of human beings in the same manner as all other scientific problems. Disciples of Descartes saw him as the first philosopher to dare to explain all the functions of human beings, even the brain, in a purely mechanical manner. Like many other systems of thought, Descartes's physiology began as heresy and ended as dogma. The mechanistic approach was adopted by philosophers and experimentalists, especially those like Giovanni Alfonso Borelli (1608–1697), who had a special interest in the problem of muscle action.

Like Descartes, Borelli favored the mechanical mode of explanation for physiological processes, but if Descartes may be regarded as the founder of iatromechanism as a philosophy, Borelli was the founder of iatrophysics as an experimental science. Little is known about Borelli's early life other than his precocious mastery of mathematics. Inspired by the work of Galileo, Borelli obtained the consent and financial support of the University at Messina to take a prolonged leave from his position as professor of mathematics to study with Galileo and Evangelista Torricelli in Florence. Unfortunately, Galileo died in 1642 and Borelli returned to Messina. In 1656 Ferdinand, Duke of Tuscany, invited Borelli to serve as professor of mathematics at the University of Pisa. Although Borelli was involved in establishing the Accademia del Cimento and making Pisa a respected center of research in mathematics, medical science, and the new experimental approach to natural science, after 12 years at Pisa and many quarrels with his colleagues, he left the university to seek quieter and healthier surroundings. Returning to the University of Messina, Borelli became involved in literary and antiquarian studies, investigated an eruption of Mount Etna, and continued to work on the problem of animal motion. In 1674 he was accused of involvement in a political conspiracy to free Sicily from Spain. Forced into exile, he fled to Rome. Overwhelmed by financial problems and illness, Borelli died before his great work *De motu animalium* (*On Motion in Animals*) was published. An introduction was added by an ecclesiastic dignitary who commended Borelli for upholding the authority of the Church in his lectures on astronomy. In physiological matters, too, Borelli remained a faithful son of the Church by acknowledging that all the mechanical phenomena described in his book were ultimately governed by the soul.

De motu animalium was an attempt to apply mathematical and mechanical principles to the study of muscle action. The text was a sustained analysis of the mechanics of muscle contraction which dealt with the movements of individual muscles and groups of muscles treated geometrically in terms of mechanical principles. Animal movements were divided into *external movements*, such as those carried out by the skeletal muscles, and *internal movements*, such as those of the heart and viscera. Borelli's method of study involved a progression from the simplest element of the motor system, the independent muscles, up to the more complicated organs and organ systems, and finally the power of movement of the organism considered as a whole. According to Borelli and his contemporary Nicolaus Steno (1638–1686), a pioneer in the microscopic study of muscle structure, the fleshy muscular fibers played a fundamental role in muscle contraction; the fibers of the tendons, in contrast, were merely passive agents which did not take part in contraction.

The action of the heart particularly intrigued Borelli. Unlike Descartes, he recognized that the heart was a muscular pump rather than a heat engine and confirmed this by simple experiments. Measuring the temperature of the heart and other internal organs in a vivisection experiment on a deer, he proved that the temperature of the heart was not significantly different from that of other parts of the body. During the eighteenth century several scientists took up the question of the regulation of body temperature. Of course the ancients had noted that body temperature was apparently constant in healthy people, but varied in pathological states. It was also apparent that no matter what the ambient temperature, members of certain species are warm to the touch when alive and become cold when dead. Other species were relatively cold, or varied in temperature and activity, depending on ambient temperature. With the clinical thermometer, experimenters could demonstrate that human body temperature remained very constant, except during attacks of fever. In the eighteenth century, Charles Blagden (1748–1820) and John Hunter (1728–1793) performed an ingenious series of experiments, which proved that body temperature was constant at a broad range of ambient temperatures. Small rooms were maintained at temperatures of 110–120°F, 180–190°F, and even 260°F, while the outdoor temperature was about 32°F. Blagden and Hunter worked in the hot rooms, noted their own reactions, and measured body temperature. As controls, nonliving materials and foods, such as eggs, meats, wines, and water, were tested. From these experiments Blagden concluded that temperature regulation was a fundamental characteristic of life.

While measuring the weight that different muscles could support, Borelli found that the muscles on the two sides of the jaw could support a weight of more than 300 pounds when acting together. He assumed that the force of contraction of all healthy muscles would be the same for a given unit of bulk. Thus, he extrapolated from his measurement of some exterior muscle to the forces that presumably could be generated by the heart and other internal muscles. He was particularly impressed by the strong muscles of the stomach and their ability to crush foods and other objects. To test the grinding capacity of the stomach, he introduced hollow glass spheres, hollow lead cubes, wooden pyramids, and so forth into the stomachs of turkeys. The next day the objects were crushed, eroded, and pulverized. Borelli admitted that chemical reactions also played a role in digestion, but some iatromechanists refused to admit that digestive juices played any significant physiological role. In explaining the action of the heart, Borelli compared the ventricles to a wine press or piston and noted that during systole (contraction) the walls of the ventricles obliterated the cavities from which the blood has been driven. He concluded that when the heart muscles contracted they increased in bulk and that this was a general phenomenon in muscle action. The hardening and tension apparent during muscular contraction must be caused by inflation of the muscular substance by something flowing into the muscle, or a sudden fermentation in the muscle itself, triggered by animal spirits traveling from the brain through the nerve and into the muscle. Trying to formulate a strictly mechanical explanation, Borelli suggested that muscular fibers are chains of rhombuses and that contraction was due to inflation caused by the sudden insertion of a number of wedges.

In his attempts to analyze the interaction between nerves and muscles, Borelli tested theories involving the movement of an incorporeal influence, vapor, or air. If air rushed into the cavities of muscles, experiments could be designed to reveal this movement. A simple experiment in which the muscles of a living animal were divided length-wise while the animal was held under water appeared to refute this hypothesis. If some spirituous gas had entered the muscles, it should burst out of the wound and bubble up through the water. Borelli concluded that inflation of the muscles must result from something that occurred in the muscles themselves rather than the movement of an "air" transmitted by the nerves. A fermentation or ebullition might occur within the muscle itself to cause its sudden inflation in response to some influence from the brain. This seemed an admirable explanation, but Nicolaus Steno, Jan Swammerdam, Francis Glisson (1597–1677), and other seventeenth-century physiologists were able to demonstrate that muscles do not increase in volume when they contract. Steno believed that it should be possible to apply the methods and philosophy of Galileo to the biological problem of muscular contraction. While his approach was primarily geometrical, his analysis of the nature of muscle contraction confirmed the fact that the apparent swelling of a working muscle was due to shortening of the fibers and did not involve an increase in volume. Some of Steno's ideas, and his interest in the microscopic structure of various tissues, were influenced by his collaboration with Jan Swammerdam. The two met while at Leiden and they traveled together to Paris in 1665. Steno and Jan Swammerdam conducted experiments to show that muscles do not swell during contraction and that a frog can continue to move for some time after its heart has been cut out. Similarly, if the head of a tortoise was cut off, the feet and tail continued to move, and the muscles of a dog convulsed when the nerves were stimulated. Furthermore, an isolated heart could continue to beat for some time although no new blood was flowing in and no new spirits could travel from the brain into the nerves. Thus, Steno argued that all previous writings on the relationship between spirits or subtle fluids and muscular action were merely attempts to disguise a profound ignorance. Under the microscope muscles appeared to be collections of motor fibers. Each motor fiber was in turn a complex of very minute fibrils arranged lengthways, with a middle part that differed from the ends in its consistency, thickness, and color. The only part involved in contraction was the fleshy part of the motor fibers, which became shorter, firmer, and more corrugated on the surface. The tendons did not change during muscle contraction. Such

Apparatus for measuring the contraction of a frog's muscle

observations clarified the differences between tendons and nerves and refuted the ancient assumption that tendons caused motion and that muscles were merely passive, fleshy material.

Francis Glisson was interested in various medical and physiological problem, such as rickets, the fine structure of the liver, and the physiological property referred to as *irritability*. After earning the doctoral degree in medicine at Cambridge in 1634, Glisson was elected a Fellow of the Royal College of Physicians. In 1636 he was appointed Regius Professor of Physic at Cambridge, but because he was a staunch Presbyterian he was forced to leave this center of Royalist influence at the outbreak of the Civil War. He moved to Colchester, where he played a notable role during the siege of the city by Royalist forces in 1648. After the war he settled in London where he practiced medicine and became active in the College of Physicians; from 1667 to 1669 he served as president. Glisson was a member of the remarkable group of intellectuals and natural philosophers who met informally at Gresham College and later founded the Royal Society. Although he rarely returned to Cambridge, Glisson later petitioned the university for 5 years' back salary for the time he spent at Colchester. During his studies of the liver Glisson tried to explain why the bile was not discharged into the intestines continuously, but only when needed. He discovered that the gallbladder and biliary duct discharge more bile when they were irritated. This occurs, he argued, because they have the capacity to be irritated. Glisson's doctrine of *irritability* became very influential in physiology, largely through the work of Albrecht von Haller. While Glisson had used the term irritability to describe a broad range of phenomena, Haller tried to confine the concept to the property that caused muscles to contract in response to an external stimulus. Other workers extended the concept to mean any kind of change in a living organism.

In the eighteenth century, as in the seventeenth, physiology remained a mixture of speculation and experimentation, but the assimilation of the ideas that emerged from the revolution in chemistry provided new ways of understanding digestion, respiration, and the role of the circulation of the blood. Much of this work was initiated and promoted by teachers at Europe's leading medical schools, such as the indefatigable Hermann Boerhaave (1668–1738), who taught chemistry, physics, botany, ophthalmology, and clinical medicine at the University of Leiden. Although Boerhaave is not associated with any particularly striking discovery or theory, he was so famous in his day that a letter from Asia addressed to "The Greatest Physician in the World" was supposedly delivered directly to Boerhaave. His *Elements of Chemistry* was still lauded some 80 years after its publication as the "most learned and luminous treatise on chemistry" ever written. Boerhaave's approach to anatomy, physics, and chemistry was widely disseminated by devoted students and disciples, such as Albrecht von Haller. Physiology, like embryology, owes a great debt to Haller, who carried out an enormous number of experiments despite an almost paralyzing melancholy, bad health, and a profound revulsion for the pain caused by the vivisection of animals. Haller's *Elements of Physiology* summarized the state of physiology in a concise and accessible form, but the range of his work was so extensive that the great French physiologist François Magendie (1783–1855) complained that whenever he thought he had performed a new experiment he always found it had already been attempted or described by Haller.

In his studies of the form and function of various organs and organ systems, Haller attempted to link anatomical knowledge to physiology by means of experiment, that is, to see physiological research as a new discipline of *vitalized anatomy*. Reviving Glisson's concept of irritability, Haller refined the concept and contrasted the *irritability* of muscle to the *sensibility* of nerves. The task of the physiologist was to determine which parts of the body were irritable and which were not. The irritable parts were defined as those that contract when touched, while the sensible parts were those that conveyed a message to the mind when they were stimulated. The property of irritability in muscles was shown to be due to the nerves. For example, the diaphragm could be made to contract by irritating the appropriate nerves. According to Haller, a special contractile force existed in muscles. In a living animal, or one that had just died, he often found spontaneous contractile movement in muscle tissue. This contractile property could also be induced by applying some stimulus such as pinching or pricking. Haller argued that there was a fundamental difference between the living contractile force and spontaneous contractions. Irritable tissues were those in which the special force, the *vis insita* (inherent force), resided. Irritability, he concluded, was the definitive characteristic of muscle fibers. While others had called this special force the *vital force,* Haller rejected this doctrine because he found that the force inherent in muscles survived for some time after death. Another force that was carried to the muscles from the brain by means of the nerves was called the *vis nervosa*. This force initiated muscular contractions, but it could not be called the vital force either because it also remained for a time in dead animals.

Experiments indicated that the nerves alone served as the instruments of sensation so that only those parts of the body served by nerves experienced sensations. The nerves also elicited the power of contraction in the muscles that served as the instruments of movement. While using evidence from pathological lesions and vivisection experiments to determine whether specific parts of the brain had particular properties and functions, Haller concluded that the question was too complex to be answered satisfactorily at the time. He rejected the idea that the nerves act as solid bodies like elastic strings conveying vibrations and concluded that the material substrate of the nerves must be a subtle and unique fluid known to us only by its effects. The nerves and fibers of the brain, therefore, must be hollow in order to transmit the nervous fluid which was responsible for sensation, movement, and the preservation of life. While denying the possibility that the soul was diffused over the whole body, Haller argued that the soul had nothing in common with the body, except for sensation and movement. Because both sensation and movement seemed to have their source in the medulla, it must be the seat of the soul. Despite his success in research and teaching, Haller was the victim of poor health and chronic depression, which he tempered with opium. He died with his finger on his pulse after telling his friends: "The artery no longer beats."

The life and work of Julien de La Mettrie (1709–1751) offers a sharp contrast to the pious Haller. La Mettrie was quite willing to let his materialistic scientific theories conflict with or contradict Christian dogma, despite the fact that he was himself a priest. His father was a rich merchant who demanded that his son study for the priesthood instead of allowing him to be a poet. Advised by a friend that it would be more rewarding to be a mediocre physician than a good priest, La Mettrie decided to study medicine at Leiden with Boerhaave. In addition to translating the works of Boerhaave into French, La Mettrie deliberately antagonized the Parisian medical community by publishing a series of pamphlets satirizing his conservative opponents. His own unorthodox ideas about the nature of vital phenomena appeared in a controversial book called *A Natural History of the Soul*. According to La Mettrie, we cannot know what the soul truly is, but neither do we know what matter is. Since we never find a soul without a body, to study the properties of the soul we must study the body, and to do this we must investigate the laws of matter. His own experience while suffering from a fever made La Mettrie realize that the clearness of his thinking varied with the severity of his illness. This proved to him that thought is a function of the brain, dependent on physical conditions. His natural history of the soul traced the evolution of an active principle in matter through plants and animals to humans. These materialistic views were clearly in disagreement with Church dogma. La Mettrie launched another attack on his fellow physicians in a book called *The Politics of Physicians*. With his practice and reputation in ruins, La Mettrie sought refuge in Holland, where he amused himself by writing another attack on physicians.

Feeling relatively safe in Leiden, La Mettrie composed another unorthodox work entitled *Man, the Machine* (1748), but he discreetly published it anonymously. Without Descartes's veneer of piety, La Mettrie presented man as a machine whose actions were entirely due to physical-chemical factors. In contrast to Descartes, La Mettrie rejected the idea that humans were essentially different from animals. For La Mettrie, the human being was a special variety of monkey, superior mainly by virtue of the power of language. Indeed, he argued, animals were not merely machines, although their ability to reason was less developed than that of humans. Experiments on animals indicated that peristalsis continued after death and that isolated muscles could be stimulated to contract. Presumably, if this occurred in animals, it must also be true for humans because both were essentially the same in terms of bodily composition. Dismissing the mind-body dualism of Descartes, La Mettrie argued that even the mind must depend directly on physiochemical processes. Substances such as opium, coffee, and alcohol affected both the body and the mind, affecting thoughts, mood, imagination, and volition. Moreover, diet could affect character and diseases attacked mind as well as body. As a rather sardonic joke, La Mettrie added to the attacks on his book by publishing his own satirical critique called *Man, More than a Machine.* Eventually, La Mettrie found refuge at the court of Frederick II of Prussia, where he was able to write, lecture, and practice medicine. The physician who had called death the conclusion of a farce provided a demonstration that this was true at a feast given in his honor by a grateful patient. Immediately after eating an enormous quantity of a truffle pastry, La Mettrie fell ill and died. The cause of death was unclear, but Voltaire pronounced it a great occasion since, for once, the patient had killed the doctor.

THE CHEMICAL APPROACH TO LIFE

Despite the general preference for iatromechanism, some natural philosophers were attracted to chemical explanations of life. Like the Renaissance alchemist Paracelsus, Johannes Baptista van Helmont (1579–1644) believed that the workings of the universe could be explained in chemical terms. Unlike Paracelsus, who has been labeled a quack and a mystic, van Helmont has a secure place in the history of chemistry for his pioneering studies of gases and fermentation. It was van Helmont who introduced the term *gas* to replace the Paracelsian word *chaos* (which indicated what typically happened in the alchemist's laboratory when substances being heated released great quantities of gas).

As a student of philosophy and theology at the University of Louvain, van Helmont found so-called higher education entirely empty and unsatisfying. He referred to the Master of Arts degree as a meaningless symbol of scholasticism rather than a sign of learning. Rejecting philosophy, he turned to botany, law, and finally medicine. At the age of 22 he earned the degree of Doctor of Medicine and then spent 10 years traveling through Europe. Eventually, he settled near Brussels in order to spend his time performing chemical experiments while practicing medicine as a form of charity. Although van Helmont was a devout Catholic, he came into conflict with Church authorities who thought that his naturalistic exposition of magnetic cures in *De magnetica vulnerum curatione* (1621) conflicted with orthodox interpretations of cures as miracles. Van Helmont published very little after this experience, but just before his death he entrusted his manuscripts to his son for posthumous publication.

While striving to reconcile the chemical view of life with a vitalist philosophical outlook, van Helmont conducted chemical experiments of an exact and quantitative nature in which the idea of the indestructibility of matter was implicit. For example, he proved that a metal could be dissolved in acid and then recovered without loss of weight. Intrigued by the transformation of liquid water into air and the way in which this vapor could again be transformed into liquid water, van Helmont planned and executed an experiment to prove that water is the source of all things. In a large earthenware vessel he placed 200 pounds of dried earth and a willow tree that weighed 5 pounds. After watering the tree with nothing but pure rain water for 5 years, he found that the tree now weighed almost 170 pounds, but the weight of the soil was virtually unchanged. Van Helmont was satisfied that his experiment proved that water was the primary element. This experiment certainly proves something fundamental. It proved once again that meticulous measurements will not illuminate the secrets of nature where theory throws a false light.

To confirm his belief that earth could be formed from burning vegetation, which in turn had come from water, van Helmont carried out an experiment in which he burned 62 pounds of charcoal. Having determined that only one pound of ash remained, he assumed that the rest of the material had been driven off as a form of air, which he named *gas sylvestre*, the same gas released when organic materials were burned, during fermentation, and from shells and limestone treated with acids. While van Helmont apparently realized that different gases existed, he lacked the special apparatus required to collect them for further study and characterization. However, his research on gases led him to reject the Paracelsian concept of the three elements—sulfur, mercury, and salt. Instead, van Helmont proposed that there are only two elements: air (the natural atmosphere) and water (everything that is not air).

According to van Helmont, all physiological phenomena could be explained in terms of chemical processes that were governed by a hierarchical series of *archaei* in various organs. Subordinate to the *archaei* was an entity that he named the *blas* which performed specific functions. A more prosaic way of explaining these concepts is to say that all changes in the body were due to the action of *ferments*; the *blas,* or *archaeus* acted on matter through ferments. For example, according to van Helmont digestion involved a series of conversions that transformed food into living flesh. The first stage was said to occur in the stomach by means of an acid ferment produced by the spleen. The acidic chyle prepared in the stomach then passed into the duodenum where very complicated ferments from the bile were at work. Another stage of digestion was caused by the liver as a prelude to the conversion of chyle into crude blood. After the nutritious chyle was absorbed, the useless refuse passed into the large intestine where another ferment converted it into feces. Another digestion supposedly took place in the heart and arteries to convert the darker and thicker blood of the vena cava into blood that was lighter in color and more volatile. In explaining the work of the heart, von Helmont suggested that there were minute pores in the septum that allowed the passage of *vital spirit* from the left side to the right side. The pores allowed blood to pass from the right side of the heart to the left side but prevented movement in the opposite direction. Finally, each part of the body took the nutrients it needed from the blood and transformed them into its own special components. Although much of this theory was obscure and confusing, von Helmont's emphasis on the relationship between fermentation and physiological processes was of considerable value to other researchers, such as the iatrochemist Franciscus Sylvius (1614–1672), who established a vigorous program of chemical and experimental approaches to physiology at the University of Leiden.

While bringing the work of Harvey and van Helmont to the attention of his students, Sylvius encouraged them by emphasizing how much remained to be discovered. Fascinated by the properties of acids, bases, and salts, Sylvius transformed his observations into a theory that explained disease as the result of an excess of *acridity,* which could be either acid or basic. Therapy could then be rationally designed. A very important and bold insight expounded in Sylvius's work is the idea that the chemistry of living things is the same as the chemistry of nonliving things. Thus, at least in theory, it should ultimately be possible to reproduce in the laboratory chemical events supposedly peculiar to the living body. Digestion, for example, could be explained as a fermentation involving the saliva, the bile, and the pancreatic juice that had been discovered by Regnier de Graaf (1641–1673) in 1664. All mystical *archaei* could, therefore, be banished from the digestive process. Like Borelli, Sylvius was the founder of a distinct school of thought. While Borelli insisted that physiological phenomena could be explained in purely mechanical terms, Sylvius rested his case on the science of chemistry. Both men had great confidence in their methods and theories, and both thought that they had the exclusive key to the explanation of physiological phenomena. The question of digestion was one where the advocates of iatrochemistry had a significant advantage.

Based on studies of the pancreatic juice carried out by Regnier de Graaf and René Antoine Ferchault Réaumur's (1683–1757) studies of the gastric juice, the intrepid Lazzaro Spallanzani (1729–1799) used himself and his students as guinea pigs in attempts to understand the mechanism of digestion. Spallanzani began his investigation of digestion by testing the action of saliva on various foods; he went on to verify Réaumur's work on the power of the gastric juices of birds of prey to digest food and prevent putrefaction. To determine whether observations made with other animals were applicable to humans, Spallanzani valiantly swallowed tubes and bags containing selected foods despite the possibility that such materials could cause a fatal obstruction of the alimentary canal. Independently of Spallanzani, Edward Stevens published a thesis in 1777 which contained one of the first descriptions of the isolation of human gastric juice and *in vitro* studies of its properties. Little is known about Stevens other than the date on which he defended and published his thesis. An English translation of part of his thesis was appended to the English edition of Spallanzani's *Dissertation Relative to the Natural History of Animals*. Stevens conducted his experiments with the cooperation of a man of rather weak intellect who made his living by swallowing stones for the amusement of spectators. After 20 years of swallowing stones, this human stone-swallowing regurgitator might have found Stevens's hollow silver spheres a pleasant delicacy. Various foods were placed inside the perforated spheres so that the action of the digestive juices could be compared. Knowledge of human digestion was not substantially improved until 1833 when the surgeon William Beaumont (1785–1853) published the results of studies conducted through the gastric fistula of Alexis St. Martin.

A very different approach to chemical phenomena led to an understanding of the mechanism of combustion and the physiology of respiration. This was the doctrine of phlogiston, derived from the alchemical theories of Johann Joachim Becher (1635–1681). Despite his lack of a formal education,

Becher was able to attain great academic and financial success, alternating with periods of exile and infamy. In 1666 he was appointed professor of medicine at the University of Mainz, but his interest in economic issues led to a position at the Commercial College of Vienna. By 1678 he was forced to seek refuge in Holland; 2 years later, after the failure of his alchemical enterprises, he fled to England. According to Becher, bodies consist of air, water, and three kinds of earth: *terra pinguis* (fatty earth), *terra mercurialis* (metallic earth), and *terra lapidia* (stony earth). During combustion, *terra pinguis* was released as fire.

Through the work of George Ernst Stahl (1660–1734), *terra pinguis* was transformed into *phlogiston*, the key to a new chemical system that seemed to explain many bewildering phenomena while guiding chemical research for at least a century. After studying medicine at the University of Jena, Stahl held a teaching position there until 1693, when he was invited to teach at the new University of Halle. His colleague Friedrich Hoffmann (1660–1742), a distinguished physician and founder of one of the famous medical systems of the eighteenth century, taught chemistry, physics, anatomy, surgery, and medical practice. Stahl's responsibilities included botany, physiology, dietetics, pathology, and materia medica. In 1703 Stahl published a new edition of Becher's *Physica subterranea* and appended his own exposition of the mechanism of combustion. Phlogiston was defined as a material principle, a component of combustible materials, and the material and principle of fire; phlogiston, however, was not fire itself. Phlogiston was found in the vegetable, animal, and mineral kingdoms, but was more abundant in the first two, which accordingly left little residue when they burned and released their phlogiston. One problem for phlogiston theory was the apparent gain in weight that occurred when metals were burned. But because phlogiston could not be trapped and analyzed, there was no compelling reason to assume that it was an ordinary substance with weight. Like heat, light, magnetism, or electricity, it might be a weightless subtle fluid. Indeed, instead of being attracted to the earth like other elements, phlogiston might have a tendency to rise. This would explain the increase in the weight of metal oxides and the weight loss that occurred when the oxides were reduced, but it had little relevance to Stahl's physiological theories. For Stahl, all chemical changes in the living body were fundamentally different in nature from ordinary chemical events. In the living body, he asserted, all chemical changes were controlled by the *anima sensitiva* (sensitive soul). Rejecting Cartesian dualism and claims that the body was a machine, Stahl contended that physiological phenomena obeyed laws entirely different from those governing the inanimate world. Although phlogiston theory implicitly accepted the alchemical concept that substances were composed of matter plus intangible spirits, essences, or principles, it retained the loyalty of some of the most distinguished eighteenth-century chemists, such as Joseph Black, Henry Cavendish, and Joseph Priestley.

While the connection between respiration and combustion had been long suspected, the chemical basis of this relationship was not demonstrated until the end of the eighteenth century. Solving the riddle of respiration required the confluence of studies of the circulation of the blood, the exchange of gases in respiration, the microanatomy of the lungs, and the chemistry of the gases. It was primarily the invention of the pneumatic trough by Stephen Hales (1677–1761) that made the isolation and characterization of the gases possible and led to a revolution in chemistry. Reverend Hales, the Perpetual Curate of Teddington, investigated the hydraulics of the vascular system, the chemistry of gases, the respiration of plants and animals, and the measurement of blood pressure. Much of this work was reported in *Vegetable Staticks* (1727) and *Statistical Essays, Containing Haemastaticks* (1733). Hales was interested in the study of natural history as well as the physics of fluids, gases, and liquids. Hales was a Fellow of the Royal Society and served as a member of various commissions investigating matters of public health and alleged wonder cures. As a result, Hales became interested in practical problems such as ventilation and ways of introducing fresh air into confined quarters, such as ships, prisons, and hospitals. He invented several practical ventilation devices. It has been said that Harvey demonstrated the logical necessity of his theory of the circulation of the blood, but that it was Hales who provided rigorous experimental demonstrations through his analysis of the hydrodynamics of the vascular system. Hale's scientific works were outlined in a poem by Thomas Twining, who noted that in his serene retreat in Teddington, the good pastor searched for Nature's secrets by methods that ranged from weighing "moisture in a pair of scales" to stripping the skins from living frogs. The "lingering death" suffered by mares and dogs used in the vicar's studies of blood pressure was also alluded to in the poet's assessment of these philosophic studies. By 1724 Hales had begun the series of experiments that established the general outlines of the physiology of plants: the circulation of the sap and the interactions between the plant and its environment; the uptake of water by the roots and its transport to the leaves and its transpiration; the growth of the parts of plants and the proportional aspects of growth. He also extended previous work on the relationship between respiration, or combustion, and air.

Microscopic examinations of plants in the seventeenth century, especially those of Nehemiah Grew, revealed that plants contained ramifying systems of tubes which appeared to run from the roots, through the stem and branches, and into the leaves. Since some of the tubes appeared to be filled with liquid and others with air, Grew and others thought that there might be a circulation in plants like that discovered by Harvey in animals. This raised many questions: Since plants had no muscular heart or pump, what force could cause the flow of sap? Was there a closed system of tubes in plants? What were the functions of the different parts of the plant? Was plant life analogous to animal life? Nehemiah Grew had suggested many important aspects of plant physiology by about 1670. It was apparent that sap moved upwards from the roots to the leaves, distributing fluid to the whole plant. Also, the formation of some elements of plant substance must take place in the leaves and such substances must move from the leaves to other parts of the plant. Some substances seemed to be produced and stored for later use, perhaps in the tubers associated with the roots. Hales demonstrated that it was the evaporation of water from the leaves rather than pressure of water in the roots that caused the upward movement of the sap. Many naturalists thought that the movement of sap in plants might be analogous to the circulation of the blood in animals. Advocates of this theory generally assumed that sap moved up via the inner part of the stem and down through the outer parts. Hales proved that such a hydraulic cycle or circulation does not occur in plants. Having clarified the "water economy of plants," Hales realized that plant nutrition and its relationship to the exchange of gases required further study. By separating his reaction vessels from the collection vessel and collecting gaseous products over water, Hales was able to separate and store several different gases, including carbon dioxide and oxygen. Applying more rigorous qualitative methods to the study of plants and animals, Hales returned to van Helmont's plant growth experiment. Rather than accepting the assumption that water alone entered into new plant material, Hales measured the amount of water taken up by roots and given off by leaves. This led to the conclusion that in addition to water, plants must take some "secret food of life" from the air. Plants, Hales suggested, might take part of their nourishment from the air.

The pneumatic trough invented by Hales was well exploited by a group of scientists known as the pneumatic chemists, which included Joseph Black (1728–1799), Henry Cavendish (1731–1810), Carl Wilhelm Scheele (1742–1786), Joseph Priestley (1733–1804), and Antoine-Laurent Lavoisier (1743–1794). In 1754 Joseph Black received the doctorate of medicine for a dissertation entitled "On the acid humour arising from food and on magnesia carbonate alba." Stones formed in the bladder and gravel passed in the urine were of considerable interest to medical researchers at the time. A heated controversy was raging over the use of caustic agents as solvents for such stones. To express his satisfaction with Mrs. Joanna Stephens's secret remedy for the stone, Prime Minister Walpole paid £5000 for her recipe and had it published in the *London Gazette* (1739). When Black discovered carbon dioxide, the gas he called "fixed air," he was testing the major component of Mrs. Stephens's powders and pills—calcined snails. According to phlogiston theory, limestone became caustic quick lime by taking up phlogiston; when quick lime was slaked, it supposedly gave off phlogiston. Black discovered that when chalk was calcined or burned it gave off a gas called *fixed air*. Driving this gas through a clear solution of limewater caused the formation of "mild lime," which precipitated out of solution. Fixed air was found in expired air, was produced by burning charcoal, and was released during fermentation; this gas was deadly to animals and would extinguish a flame. At first Black thought that the entire portion of the atmospheric air unsuited for respiration must be fixed air, but other chemists discovered another noxious gas, later identified as nitrogen.

A new species of gas was discovered by Henry Cavendish, a man said to be as wealthy as he was eccentric. Cavendish had left Cambridge without a degree because he objected to the strict religious tests applied to candidates. After studying physics and mathematics in Paris, he settled in London, where he used his large inheritance to purchase the finest instruments and reagents for his scientific experiments. In 1766 Cavendish published an account of his experiments on "factitious airs" in the *Philosophical Transactions* of the Royal Society. By dissolving zinc, iron, or tin in vitriolic acid Cavendish produced hydrogen, which he called *inflammable air*. At first he thought that inflammable air might be phlogiston itself. In one of his most significant experiments, Cavendish exploded a mixture of hydrogen and oxygen and thus proved that water was not an element, but a compound of two gases.

Oxygen was discovered almost simultaneously by Carl Scheele and Joseph Priestley, but both were staunch advocates of phlogiston theory. Conducting a series of experiments to prove that air is made up of two components, "foul air" and "fire air," Scheele determined the ratio to be one part of fire air to three parts of foul air on a volume/volume basis. Scheele believed that the function of fire air was to absorb the phlogiston given off by burning substances. When air became saturated with phlogiston, it could no longer support combustion. Scheele and Priestley knew that fire air was used up in combustion and that the remaining air had less weight and volume than the original mixture. This was ingeniously explained by postulating that the combination of air and phlogiston produced a compound so subtle that it passed through the pores of the glass and disappeared into the atmosphere.

Although mainly remembered as an experimental chemist, Joseph Priestley was also a minister, theologian, author, and educator. During his lifetime, he was more famous as a radical theologian and political thinker than as a scientist. His experiments on gases began in the public brewery next to his house in Leeds. The gas that bubbled out of the beer making vats was, of course, carbon dioxide, Joseph Black's fixed air. Although the gas was largely insoluble in water, Priestley discovered that it produced a very pleasant effervescent beverage, which he likened to the best mineral waters. The College of Physicians suggested that Priestley's soda water might be used as a cure for scurvy. In July 1791 a mob burned down Priestley's house during a series of attacks on dissenters and radicals. Priestley barely escaped the riots with his life, but his home, library, laboratory, and many manuscripts were destroyed. Reluctantly, he decided to emigrate to America where he settled in Northumberland, Pennsylvania.

An enthusiastic and independent experimentalist, Priestley was curious about everything, but unsystematic in his researches. Indeed, he often said that if he had known any chemistry he would never have made any discoveries. Chance and careful observation, he argued, were more significant in making discoveries than preconceived theories. Among the gases Priestley discovered were ammonia, sulfur dioxide, carbon monoxide, nitric oxide, and hydrogen sulfide. As Sir Humphry Davy (1778–1829) said: "No single person ever discovered so many new and curious substances." Priestley did not think that it was possible to prepare air that was purer than the best common air, but this was what appeared to have happened in 1774 when he extracted a new air from mercuric oxide. The new air caused a candle to burn with a very vigorous flame and allowed mice to survive for a longer period than a similar limited quantity of ordinary air. Priestley called this gas *dephlogisticated air*. After breathing this air himself, Priestley predicted that it would be useful in medicine, although it might be dangerous as well. While investigating the relationships among the various gases and the nature of respiration, Priestley discovered that air that had been "injured" by animal respiration or by the burning of candles could be restored by plants. This observation puzzled him at first because he had assumed that both plants and animals should affect air in the same manner. While the French chemist Antoine-Laurent Lavoisier explained respiration and combustion in terms of oxidation theory, Priestley believed that air became saturated with phlogiston from combustion, animal respiration, and other chemical processes. Plants, in contrast, took up phlogiston and thus purified the air. Dark venous blood, laden with phlogiston, released phlogiston in the expired air. Bright red dephlogisticated arterial blood was, therefore, capable of absorbing phlogiston from body tissues. Venous blood exposed *in vitro* to dephlogisticated air became as bright as arterial blood. Although Priestley acknowledged that no one had been able to ascertain the weight of phlogiston, he did not see this as a reason for rejecting the theory. No one had weighed light or heat, he argued, but scientists did not doubt their existence.

The intriguing but apparently paradoxical results of Priestley's studies of the effects of plants and animals on the various gases led Jan Ingen-Housz (1730–1799) to undertake a series of experiments on plant nutrition that led to his discovery of photosynthesis. Well known as a physician and an advocate of inoculation against smallpox, Ingen-Housz conducted the experiments that clarified the nature of respiration and photosynthesis while living in a country house near London in 1779. In 1771 Priestley had reported that plants restore air that had been "spoiled" by burning candles or the death of an animal through suffocation. Although Priestley noted that in some experiments plants were able to restore such vitiated air, some chemists were unable to duplicate these results. Carefully repeating these experiments, Ingen-Housz demonstrated that only the green parts of plants could improve "spoiled" air and, most importantly, they could do this only when they received sunlight. As he reported in his book *Experiments Upon Vegetables, Discovering Their Great Power of Purifying the Common Air in the Sunshine and of Injuring It in the Shade and at Night* (1779), plants, like animals, performed respiration, the process that produced fixed air (carbon dioxide). But when green plants received the visible part of the sunlight, they released dephlogisticated air (oxygen). Plants and animals, therefore, mutually supported each other; animals consumed oxygen and produced carbon dioxide, whereas plants consumed less oxygen by respiration than they produced by photosynthesis. As a physician, Ingen-Housz was concerned with the possible usefulness of his observations. Although he designed an apparatus that could be used to administer pure dephlogisticated air to patients with respiratory diseases, he probably never used it.

Unlike Priestley, Antoine Lavoisier approached chemistry as a scientist who had mapped out a protocol for the complete reformation of chemistry at the outset of his career. While studying law, Lavoisier became interested in the natural sciences. At the age of 21 he submitted the first of many articles to the Royal Academy of Sciences. Four years later, when he became a member of the Academy, he decided to devote himself to science. To support himself in a manner that would not interfere with his research, Lavoisier became a member of the network of private, and much hated, tax collectors. He married 14 year old Marie Anne Pierrette Paulze in 1771. Marie illustrated his scientific works, assisted him in the laboratory, kept his notes, translated the works of English chemists into French for him, and entertained his famous visitors. By the time Lavoisier published his *Treatise on the Elements of Chemistry* (1789), he was quite satisfied that he had established a revolution in chemistry. Astute as he was in science, Lavoisier did not seem to recognize the danger of the social and political changes fermenting all around him. Along with many other tax-farmers, Lavoisier was arrested, tried, and condemned to the guillotine. Madame Lavoisier married another chemist, Benjamin Thompson (1753–1814), later known as Count Rumford.

In contrast to Priestley, Lavoisier's work was remarkable not for its experimental originality or the discovery of new entities, but for its precision, planning, and explanatory power. On November 1, 1772, Lavoisier deposited a sealed letter with the Secretary of the Academy of Sciences. This was to serve as his claim to priority against future publications, providing all went as he expected and his investigations confirmed his hypotheses. His plan was to bring about a revolution in physics and chemistry by repeating previous work with new precautions. The revolution began to take shape in 1772 when Lavoisier met Priestley and learned about his dephlogisticated air. After repeating Priestley's experiments, Lavoisier renamed the new gas *oxygen*. Further experiments demonstrated that "eminently respirable air" was converted into "fixed air" by both combustion and respiration. To exorcise the spirit of phlogiston from the body of chemical theory, in 1777 the Lavoisiers ceremoniously burned the writings of Georg Ernst Stahl while chanting a requiem for phlogiston. In "Experiments on the Respiration of Animals and on the Changes Which the Air Undergoes in Passing Through the Lungs" (1777), Lavoisier explained the process of respiration as a slow combustion or oxidation that used oxygen and released carbon dioxide. To provide quantitative measurements of the production of animal heat by the slow combustion that occurred within the animal body, Lavoisier and Pierre Simon Laplace (1749–1827) designed a calorimeter. Tragically, the violence of the French Revolution truncated Lavoisier's career just at the point where he was learning to apply the new chemistry to physiological phenomena. The idea that physiological phenomena could be analyzed and explained in chemical terms had, however, been well established.

FROM ANIMAL CHEMISTRY TO BIOCHEMISTRY

Studies of the *animal economy*, or animal chemistry, became a prominent aspect of the research program of early nineteenth-century scientists such as Jöns Jacob Berzelius (1779–1848), Friedrich Wöhler (1800–1882), and Justus von Liebig (1803–1873). After completing his medical studies, Berzelius discovered that he preferred the study of chemistry to the practice of medicine. Building on the work of Lavoisier and John Dalton (1766–1844), the founder of modern atomic theory, Berzelius carried out the tedious analytical work involved in preparing a table of the 50 elements then known and determining their atomic weights. He also discovered several elements, established a journal, developed new analytical methods, discovered pyruvic acid, developed the concepts of isomerism and catalysis, and trained so many chemists that many of the major nineteenth century chemists were either his students or students of his students. A prolific writer, Berzelius unleased a torrent of articles and books that helped unify and direct chemical studies for almost a century. Since Berzelius was trained in medicine, it was not surprising that he would eventually turn to the analysis of complex organic substances such as bile, blood, and feces.

Beyond his specific discoveries, Berzelius was important in bringing the chemicals of life within the purview of the atomic theory. Not surprisingly, Friedrich Wöhler, the first man to synthesize an organic chemical in the laboratory, was one of Berzelius's disciples. As a student Wöhler was, according to his own admission, distinguished neither by special zeal nor broad learning, but at Heidelberg University he was allowed to abandon routine course work in order to devote himself to research. Wöhler's most famous experiment involved the production of urea from ammonium cyanate; this was the first time that a chemical produced by living beings had been synthesized from materials that were available, at least in principle, from nonliving matter. In a letter to a colleague, Wöhler wrote with great excitement that he could "make urea without kidney of man or dog." Although the preparation of urea is often seen as proof of the theoretical equivalence of inorganic and organic chemicals, many other demonstrations were needed to abolish the conceptual abyss separating "animal chemistry" from inorganic chemistry. Hermann Kolbe (1818–1884) is generally considered the first chemist to perform the complete *in vitro* synthesis of an organic compound, acetic acid. While Kolbe may have excelled as a chemist, he appears to have had little tolerance for the play of the imagination as a stimulus to scientific thought. In 1877, Kolbe attacked contemporary chemists for causing a decline in chemistry by their advocacy of a stupefying nature philosophy just barely separated from spiritualism. Attempts to think about the geometrical arrangement of atoms within molecules struck Kolbe as the most disgusting pseudoscientific fantasies.

Prior to 1750 nutrients were regarded as necessary for the animal as lubricants for the muscles and joints and as replacements for parts worn out by the wear and tear of daily life. Chemists assumed that these components were present in foodstuffs and were directly assimilated into the tissues. Organic chemicals seemed to be overwhelmingly complicated, but by the nineteenth century chemists were confident that knowledge of chemical composition would provide a new scientific basis for agriculture and nutrition. Believing that empirical formulae adequately characterized organic chemicals, chemists still found themselves unable to deal with all the chemical species found in foodstuffs. They turned instead to an analysis of bulk foods, body fluids, solids, and excrements and created simplicity from complexity by seeing all foods as mixtures of saccharine, oleaginous, and albuminous materials (that is, carbohydrates, fats, and proteins).

Chemists and physiologists differed as to the proper approach to elucidating the vital phenomena of the body. Many physiologists saw vivisection, or animated anatomy, as the only way to unravel the complex phenomena of life. Test tube chemistry, in their view, could contribute nothing to understanding life. The work of the German chemist Justus von Liebig (1803–1873) and the French physiologist Claude Bernard (1813–1878) illustrates significant aspects of this dispute. Struggling to create a community of chemists in Germany, Liebig saw it as his mission to prove that the science of chemistry would revolutionize agriculture, industry, and nutrition. During the first phase of his career, Liebig investigated classical organic chemistry, developing and improving methods of analysis and identification. He later turned to more complex problems of agricultural chemistry and the chemistry of living things. This new interest brought him his greatest fame and triumphs, but also revealed his tendency to formulate grandiose theories from a shaky and meager base of data. For example, Liebig rejected evidence that yeast is a living organism and ridiculed the fermentation studies of Schwann and Pasteur. According to Liebig, yeast was the product of fermentation rather than the cause. Liebig believed that only green plants could build up complicated organic substances from the simple inorganic elements they took from the air and the soil. While plants were synthetic chemical factories, animals were degradative chemical processors which ultimately took the components they needed for their tissues from plants and used the rest as fuel. Assuming that the law of conservation of matter was also applicable to living things, Liebig attempted to determine what chemical events occurred in the living body by measuring and balancing all the components ingested and excreted by the body. This approach to metabolic phenomena was not unlike that of Santorio Santorio. Moreover, Liebig thought that muscle activity was directly dependent on protein degradation and that urea output could be used as an index of physiological activity. Claude Bernard said that this attempt to deduce the invisible metabolic phenomena from such analyses of input and output was like trying to deduce what happened inside a house by measuring what goes in the door and out the chimney. Liebig's ideas about nutrition stimulated much food faddism, including that of Sylvester Graham, the American health reformer whose name is immortalized in the graham cracker.

Experimental physiology owed much to Claude Bernard's mentor François Magendie (1783–1855). Although his work did not culminate in any broad or all inclusive theory, Magendie virtually founded the field of experimental pharmacology. Magendie served as professor of medicine at the College of France, president of the French Academy of Sciences, and president of the Advisory Committee on Public Hygiene. In these offices and through his researches and private lectures, Magendie left his imprint on every branch of physiology.

The foundation of Magendie's special method of investigation was vivisection; reports of his ruthless experimentation stimulated much antivivisectionist activity. His goal was to establish physiochemical explanations of vital phenomena, and he approached this through what critics called an orgy of experimentation. Although he attempted a purely mechanical explanation of most vital phenomena, he seems to have allowed for a vital force, at least in the nervous system. Studies of the nervous system, resulting in an acrimonious priority battle with Sir Charles Bell, included the experiments that led to the discovery of the separate motor and sensory functions of spinal nerves. Perhaps Magendie's most important contribution to science was liberating Claude Bernard from his chronic apathy and inertia. Once aroused and engaged in experimental physiology, Bernard surpassed his mentor in many ways. Even Magendie had to admit that Bernard's skill in vivisection was superior to his own; time has affirmed Bernard's ability to formulate significant generalization.

Born into a family of poor peasants, Bernard was fortunate to have received instruction in classical subjects from the parish priest. After more advanced studies, Bernard taught language and mathematics at a Jesuit school while tutoring private pupils. Financial difficulties forced him to abandon teaching to take a position as assistant to an apothecary at Lyons. Here he carried out menial tasks such as sweeping the floors and washing glassware.

Delivering prescriptions was the only enjoyable part of his work, because it allowed him to stop and watch the operations carried out at a nearby veterinary school. Finding many aspects of medicine quite ridiculous, Bernard wrote a short play revolving around theriac, a popular medicine so haphazardly compounded from over 60 ingredients that no two batches were ever the same. The success of this play made him aspire to greater things, and he began work on an ambitious five-act historical drama. Soon after arriving in Paris with his play, Bernard was advised by a literary critic to learn another profession if he wanted to eat. Bernard chose medicine and did well enough at the Parisian School of Medicine to be awarded an internship in 1839. Bernard was not by any measure regarded as a brilliant student. Indeed, he ranked twenty-sixth out of the 29 taking the examinations. Unable to obtain a professorship and unwilling to practice medicine, Bernard served as Magendie's assistant for many years. Trying to rescue himself from financial difficulties, Bernard contracted an arranged marriage, which became a source of intolerable tension because his wife was adamantly opposed to vivisection. Despite persistent ill health and recurrent attacks of gastritis, Bernard's research program was increasingly successful. He became a member of the Academy of Sciences in 1854 and then assumed Magendie's professorship. At 47 years of age, exhaustion and illness forced him to retreat to his birthplace. During this period of enforced rest and reflection, he wrote about the broader implications of his work in *Introduction to the Study of Experimental Medicine*. This remarkably lucid and perhaps deceptively simple text has been more widely read and discussed than his numerous research papers and 14-volume *Lessons in Experimental Physiology Applied to Medicine*. When Bernard died, the French Chamber of Deputies voted to give him the first state funeral to honor a scientist. The novelist Gustave Flaubert (1821–1880) described the event as more impressive than the ceremonies held at the funeral of the Pope.

In the course of his research, Bernard illuminated physiological phenomena in new ways, demonstrating that many vital functions might better be seen in terms of chemistry than as aspects of animated anatomy. His most significant discoveries included the glycogenic function of the liver, the role of the pancreatic juices in digestion, the functions of the vasomotor nerves, and the nature of the action of curare, carbon monoxide, and other poisons. More importantly, Bernard placed his observations in a theoretical framework based on his concept of *determinism*, faith in the experimental method and its applicability to physiology, the science of life. For Bernard, vitalism and mechanism were both essentially worthless distractions leading to endless disputes that obstructed scientific progress. Bernard insisted that there

A portrait of Claude Bernard in 1866

were always real physical-chemical bases for all vital phenomena, no matter how diverse and mystifying they might appear. Thus, serious physiological research could only proceed when scientists banished that capricious agent, the "vital force," that resisted the laws governing the inanimate world and put all acts performed by living organisms beyond the scope of science. While praising doubt and an open mind, Bernard warned against excessive skepticism. Scientists must believe in *determinism*, that is, in science itself, based on complete and necessary relationships among phenomena in living beings and the inanimate world. The study of the science of life, he wrote, was like coming into a superb and dazzlingly lighted hall which could only be reached "through a long and ghastly kitchen." While it was impossible to define precisely what life is, the scientist must not allow speculations to interfere with true scientific work, which was to analyze and compare the manifestations of life. If forced to define life in a single phrase, Bernard would say, "life is creation." An organism might then be described as a machine that works by means of the physico-chemical properties of its constituent parts. The way to understand vital phenomena was through vivisections. Indeed, Bernard argued, the science of life could only be established by experimentation and, since medicine was ultimately an aspect of experimental physiology, human beings could only be saved from death through the sacrifice of other beings. The experimentalist envisioned by Bernard was a man so totally absorbed by the pursuit of scientific ideas that he was oblivious to the cries and blood of animals and could see only "organisms concealing problems which he intends to solve."

Perhaps then it is surprising that Bernard believed that his demonstration of the glycogenic function of the liver was his most important piece of work. To do this he had to divorce himself from prevailing theories of plant and animal metabolism. At the time, scientists believed that sugar in the blood of carnivores must be supplied entirely from ingested foods. Animals, whether carnivores or herbivores, supposedly used materials originally synthesized by plants in order to support a combustion that took place either in the blood or the lungs. Revolutionizing ideas about metabolism, Claude Bernard proved that animal blood contains sugar even when it was not supplied by foodstuffs. In tests of the theory that sugar absorbed from food is destroyed when it passed through the liver, lungs, or some other tissue, Bernard put dogs on a carbohydrate diet for several days and then killed the animals immediately after feeding. Large amounts of sugar appeared in the hepatic veins. To his surprise, animals in the control group, which had been fed only meat, had large amounts of sugar in their hepatic veins, but not in the intestines. Before concluding that current theories were wrong, Bernard did many additional experiments to ascertain the location of the tissue that supposedly served as the site of the destruction of carbohydrates. Through these experiments, Bernard discovered gluconeogenesis, the conversion of other substances into glucose in the liver. Further work led to the discovery of glycogen (the carbohydrate storage polymer of animals), as well as the synthesis and breakdown of glycogen. All animal tissues appeared to have ferments (enzymes) that acted on glucose. Thus, the investigation of glucose metabolism led to the concept of the "internal secretions," which were products transmitted directly into the blood instead of being poured out to the exterior of the gland or organ secreting them.

French scientists generally ignored Theodor Schwann's metabolic theory of the cell, with its emphasis on the importance of the nutritive medium bathing the cells, but Bernard saw that this concept was applicable to the fundamental problem of physiology, the relationship between the cells and their immediate environment. Bernard believed that he was the first scientist to insist that complex animals had two environments: an *external environment* in which the organism lived and an *internal environment* in which the cells functioned. Ultimately, vital phenomena occurred within the fluid internal environment bathing all the anatomical elements of the tissues. This was the basis of Bernard's well known dictum: "The constancy of the internal milieu is the condition for free and independent life." Higher animals were not totally independent of their external environment, but in close and intimate relation to it, so that their equilibrium was the result of continuous and exact compensatory adjustments. *Introduction to the Study of Experimental Medicine* portrays Claude Bernard as the ideal scientist and sage, always lucid and rational, never at a loss for a reasonable hypothesis. A close examination of his research notebooks indicates that the path to each of his discoveries was much more arduous, confused, and tortuous than the published accounts admit. Although Bernard's skill in experimental surgery was superlative, deficiencies in his mastery of chemical techniques seem to have hindered him throughout his career. Nevertheless, Bernard truly believed in himself as a reformer of physiology and a highly individualistic researcher despite long years of struggle, obscurity, and failure.

Elucidating the complex pathways of metabolism and the enzymes, hormones, and neurotransmitters that regulate them is still an enormous and incomplete task. Bernard's work marks an epoch in which biochemistry nearly superseded animated anatomy as the key methodology of physiological research. In the United States, Lawrence J. Henderson (1878–1942) extended Bernard's work and called attention to the concept of the constancy of the internal environment. Through his voluminous and popular writings,

A portrait of Walter Bradford Cannon in 1918

Walter Bradford Cannon (1871–1945) helped to disseminate ideas about physiological regulatory mechanisms beyond the academic community. In 1926 Cannon coined the word *homeostasis* to describe the conditions that maintained the constancy of the interior environment. According to Cannon, the term did not mean something fixed and unchanging, but a relatively constant, complex, well-coordinated, and generally stable condition. Since Bernard and Cannon made homeostasis the guiding principle of physiological research, the concept has broadened considerably. New terms such as feedback loops, servomechanisms, transfer functions, and cybernetics indicate that the concept is also applicable to problems studied by engineers, sociologists, economists, ecologists, and mathematicians.

SUGGESTED READINGS

Astrup, P., and Severinghaus, J. W. (1986). *The History of Blood Gases, Acids, and Bases*. Copenhagen: Munksgaard International Publications.

Benison, S., Barger, A. C., and Wolfe, E. L. (1987). *Walter B. Cannon*. Cambridge, MA: Harvard University Press.

Benzinger, T. H., ed. (1977). *Temperature: Part I, Arts and Concepts; Part II, Thermal Homeostasis*. Stroudsburg, PA: Dowden, Hutchinson & Ross.

Blasius, W., Boylan, J. W., and Kramer, K., eds. (1971). *Founders of Experimental Physiology*. Munich: Lehmanns.

Brooks, C. McC., and Cranefield, P. F., eds. (1959). *The Historical Development of Physiological Thought*. New York: Hafner.

Brooks, C. McC., Koizumi, K., and Pinkston, J. O., eds. (1975). *The Life and Contributions of Walter Bradford Cannon, 1871–1945*. Albany, NY: State University of New York Press.

Cannon, W. B. (1932). *The Wisdom of the Body*. New York: Norton.

Cannon, W. B. (1945). *The Way of an Investigator: A Scientist's Experiences in Medical Research*. New York: Norton.

Cranefield, P. F. (1974). *The Way In and the Way Out: François Magendie, Charles Bell, and the Roots of the Spinal Nerves*. Mount Kisco, NY: Futura.

Descartes, R. (1972). *Treatise of Man*. Trans. by T. S. Hall. Cambridge, MA: Harvard University Press.

Foster, M. (1899). *Claude Bernard*. New York: Longmans, Green.

Foster, M. (1970). *Lectures on the History of Physiology*. Reprint. New York: Dover.

French, R. D. (1975). *Antivivisection and Medical Science in Victorian Society*. Princeton, NJ: Princeton University Press.

Friedmann, H. C., ed. (1981). *Enzymes*. Vol. 1. Benchmark Papers in Biochemistry. New York: Academic Press.

Fulton, J. F., and Wilson, L. G., eds. (1966). *Selected Readings in the History of Physiology*. Chicago, IL: Charles C. Thomas.

Fye, W. B. (1987). *The Development of American Physiology*. Baltimore, MD: Johns Hopkins University Press.

Geison, G. L. (1978). *Michael Foster and the Cambridge School of Physiology: The Scientific Enterprise in Late Victorian Society*. Princeton, NJ: Princeton University Press.

Goodfield, G. J. (1960). *The Growth of Scientific Physiology*. London: Hutchinson.

Grande, F., and Visscher, M. B., eds. (1967). *Claude Bernard and Experimental Medicine*. Cambridge, MA: Schenkman.

Graubard, M. A. (1953). *Circulation and Respiration: The Evolution of an Idea*. New York: Philosophical Library.

Hales, S. (1961). *Vegetable Staticks*. London: Oldbourne Science Library.

Hall, T. S. (1975). *History of General Physiology 600 B.C. to A.D. 1900*. 2 vols. Chicago, IL: University of Chicago Press.

Haller, A. von (1966). *First Lines of Physiology*. Trans. by W. Cullen. New York: Johnson Reprint.

Holmes, F. L. (1974). *Claude Bernard and Animal Chemistry. The Emergence of a Scientist*. Cambridge, MA: Harvard University Press.

Holmes, F. L. (1985). *Lavoisier and the Chemistry of Life: An Exploration of Scientific Creativity*. Madison, WI: University of Wisconsin Press.

Ihde, A. J. (1964). *The Development of Modern Chemistry*. New York: Harper & Row.

Jonas, H. (1966). *The Phenomenon of Life. Toward a Philosophy of Biology*. Chicago, IL: University of Chicago Press.

Langley, L. L., ed. (1973). *Homeostasis, Origins of the Concept*. Stroudsburg, PA: Dowden, Hutchinson & Ross.

Leake, C. D., ed. (1956). *Some Founders of Physiology: Contributions to the Growth of Functional Biology*. Washington, DC: American Physiological Society.

Lenoir, T. (1982). *The Strategy of Life: Teleology and Mechanics in Nineteenth Century German Biology*. Boston: D. Reidel.

Lomax, E. (1979). Historical development of concepts of thermoregulation. In *Body Temperature: Regulation, Drug Effects and Therapeutic Implications*. Ed. by P. Lomax and E. Schonbaum. New York: Marcel Dekker, pp. 1–24.

McKie, D. (1952). *Antoine Lavoisier: Scientist, Economist, Social Reformer*. New York: Schuman.

Melhado, E. M. (1981). *Jacob Berzelius. The Emergence of His Chemical System*. Madison, WI: University of Wisconsin Press.

Mendelsohn, E. (1964). *Heat and Life: The Development of the Theory of Animal Heat*. Cambridge, MA: Harvard University Press.

Needham, D. (1971). *Machina Carnis. The Biochemistry of Muscular Contraction in Its Historical Development*. Cambridge: Cambridge University Press.

Olmsted, J. M. D. (1938). *Claude Bernard, Physiologist*. New York: Harper.

Olmsted, J. M. D. (1944). *François Magendie, Pioneer in Experimental Physiology*. New York: Schuman.

Pagel, W. (1982). *John Baptista Van Helmont: Reformer of Science and Medicine*. Cambridge: Cambridge University Press.

Parascandola, J., and Whorton, J., eds. (1983). *Chemistry and Modern Society*. Washington, DC: American Chemical Society.

Partington, J. R. (1961). *A History of Chemistry*. 2 vols. New York: St. Martin.

Pauly, P. J. (1987). *Jacques Loeb and the Engineering Ideal in Biology*. New York: Oxford University Press.

Ritterbush, P. C. (1964). *Overtures to Biology. The Speculations of Eighteenth-Century Naturalists*. New Haven, CT: Yale University Press.

Rossiter, M. W. (1975). *The Emergence of Agricultural Science: Justus Liebig and the Americans, 1840–1880*. New Haven, CT: Yale University Press.

Rothschuh, K. E. (1973). *History of Physiology*. Trans. and ed. by G. B. Risse. New York: Krieger.

Schultheisz, E., ed. (1981). *History of Physiology*. Elmsford, NY: Pergamon Press.

Shea, W. R. (1991). *The Magic of Numbers and Motion. The Scientific Career of René Descartes*. Canton, MA: Science History Publications.

Smith, B., and Hill, M. (1983). *Joseph Priestley, 1833–1804, Scientist, Teacher and Theologian: A 250th Anniversary Exhibition*. Oxford: Manchester College.

Woolf, H., ed. (1981). *The Analytic Spirit. Essays in the History of Science*. Ithaca, NY: Cornell University Press.

7

MICROBIOLOGY,
VIROLOGY, AND IMMUNOLOGY

It is virtually impossible to imagine the development of the science of microbiology without the microscope, but the relationship between the instrument and the conceptual framework of modern microbiology is rather ambiguous. Seventeenth-century microscopists were certainly able to see a new
world teeming with previously invisible entities, including protozoa, molds,
yeasts, and bacteria. While Antoni van Leeuwenhoek was quite sure that he
had discovered "little animals," with parents like themselves, others took
exception to this conclusion. Indeed, questions concerning the nature, origin,
and activities of the citizens of the microbial world were not clarified until
the late nineteenth century. Several accounts of infusoria were, however,
published in the eighteenth century. For example, Louis Joblot (1645-1723),
remembered primarily for his opposition to the doctrine of spontaneous generation, confirmed the existence of some of Leeuwenhoek's animalcules. In
1718, Joblot published an illustrated treatise on the construction of microscopes that described his observations of the animalcules that could be found
in various infusions. Carl von Linnaeus (1707-1778), the great arbiter of the
names, classification, and (almost) the existence of all species of living
things, was quite skeptical of microscopic studies. Believing that an orderly
arrangement of all species was the supreme achievement of a naturalist,
Linnaeus found the creatures discovered by Leeuwenhoek, Joblot, and oth-

ers rather a nuisance. All such creatures were tossed into the all-purpose miscellaneous category known as *Vermes*, in a class called *Chaos*. Sorting out these minute but apparently infinitely diversifed creatures provided a special challenge for nineteenth-century microscopists.

During the nineteenth century, microbes were studied by scientists investigating problems as fundamental and profound as the origin and evolution of life and as pragmatic as fermentation and putrefaction. Theodor Schwann had predicted that fermentation could be successfully exploited as a model system for investigating the vital processes occurring in each cell of the higher animals and plants. Studies of microbes were also seized upon in the battle over evolutionary theory for insights into the great question called *biogenesis*, the origin of life. If microorganisms were the product of spontaneous generation, they might even fill the gap between living and nonliving things. Microorganisms could also be studied from the medical point of view. Infection and disease could be analyzed as phenomena analogous to fermentation and putrefaction, processes already suspected of being caused by microorganisms.

The idea that corruption, impurity, or disease could be transmitted by means of contact is an ancient folk belief, but it was generally ignored in Hippocratic doctrine. The establishment of the germ theory of disease, however, is often cast in terms of a conflict between *contagion theory* and *miasma theory*. Closer inspection of the evolution and usage of these terms suggests that in earlier periods they were not necessarily seen as mutually exclusive. Thus, sharp distinctions between contagion and miasma theory might be considered rather misleading and anachronistic when applied to the period between the publication of *On Contagion* (1546) by the Renaissance physician and poet Girolamo Fracastoro (1478–1533) and the golden age of microbiology in the late nineteenth century. During this period medical writers often switched back and forth between the two terms, or used them interchangeably; when contagion was defined loosely enough to include harmful material that was indirectly as well as directly transmitted, it was not incompatible with equally vague definitions of miasma as disease-inducing, noxious, contaminated air. Fracastoro is generally regarded as the earliest champion of the germ theory of disease, but it was Giovanni Cosimo Bonomo (d. 1697) who provided the first convincing demonstration that a contagious human disease was caused by a minute parasite close to the threshold of invisibility. Bonomo proved that scabies, commonly known as "the itch," was caused by a tortoise-like mite just barely visible to the naked eye. The mites could be transferred directly from person to person or by means of bedding or clothing used by infested persons. The itch mite, however, was regarded as merely an interesting curiosity rather than a paradigm that might apply to other diseases. Even after the microscope had revealed a world of living creatures invisible to the naked eye, most physicians and scientists regarded the notion of "disease-causing animalcules" as little better than ancient superstitions about disease-causing devils and demons. There was, moreover, the obscure issue of whether the minute entities observed under the microscope were the *product* of putrefaction, fermentation, and disease or the *cause* of such phenomena. Further evidence for contagion theory appeared in studies of the silkworm disease known as muscardine. Agostino Bassi (1773–1857) argued that the disease was caused by a minute parasitic fungus and that other diseases might have similar causative agents. Intrigued by Bassi's work, Johann Lucas Schönlein (1793–1864) demonstrated in 1839 that a fungus could be found in the pustules of ringworm.

Living ferments have been used for thousands of years to produce beer, wine, and bread, but the nature of fermentation was obscure. Originally, the term *ferment* was used to refer to an active substance that could transform a passive (fermentable) substance into its own nature; for example, when yogurt (*ferment*) is added to milk (*fermentable substance*), the milk is transformed into yogurt. More than 150 years after Leeuwenhoek had described yeast, several investigators suggested almost simultaneously that yeast was a living organism and that its activities caused alcoholic fermentation. In the 1830s Theodor Schwann and Charles Cagniard-Latour (1777–1859) called attention to the association between the growth of yeast and the process of alcoholic fermentation and suggested that the development of yeast might be the cause of fermentation. Despite the lucidity of Schwann's reasoning and the ingenious nature of his experiments and observations, the great chemist Justus von Liebig (1803–1873) ridiculed the concept of cellular fermentation. To Liebig, Schwann and Latour were guilty of injecting occult vital phenomena into a purely chemical process. According to Liebig, the transformations known as decay, putrefaction, fermentation, and rotting were caused by the ability of certain chemical substances, when in contact with other substances, to cause them to undergo decompositions. In his earliest papers on fermentation, the French chemist Louis Pasteur (1822–1895) defended the position taken by Latour and Schwann that organized beings, that is, microorganisms, appeared to be associated with fermentation and attacked the position taken by Liebig.

Experiments on the nature of fermentation and the question of spontaneous generation eventually led Louis Pasteur to problems in medical microbiology, a very large and complicated field, perhaps best illustrated by examining the work of Pasteur and Robert Koch (1843–1910), the scientists who have come to exemplify the establishment of the theoretical, methodological, and ideological principles of the new science of microbiology. Pasteur's research program can serve as a case study for the interplay between researches devoted to practical problems and so-called pure or basic scientific knowledge and established the fundamentals of disinfection, sterilization, and immunization. Convinced that "the sciences gain by mutual support," Pasteur was generally involved in several research problems at the same time. The interactions among his various interests make it impossible to discuss his work as a neat chronological progression.

As a youth, Pasteur was a diligent student and a talented artist. His father, a tanner and ex-soldier of Napoleon's army, hoped that his bright and talented son would distinguish himself as a scholar or artist. Many of the portraits the young Pasteur made of family and friends are considered quite good, but at 19 years of age he gave up painting and devoted himself strictly to science. Not all of Pasteur's teachers were impressed with his intellect; in chemistry he was rated as mediocre. (Stories about such obvious errors in judgment are commonly encountered in the hagiographies of great scientists, perhaps to make teachers more humble and to offer hope to the truly mediocre.) In the competitive examination for the École Normale Supérieure of Paris, Pasteur ranked sixteenth. Admission to the school was considered an honor in itself, but Pasteur refused to enroll until he could prepare himself better. In 1843 he competed again and advanced to fifth place. His first attempt at student life in Paris, however, ended in homesickness so acute that he had to return to his family. The most important lesson Pasteur learned as a student of chemistry and physics was the applicability of the experimental approach to a broad range of questions, even to problems in biology and medicine, areas in which he had no specific training.

While still a student, Pasteur became intrigued by new studies of crystal structure, stereoisomerism, and molecular dissymmetry. These issues may seem quite remote from the biological and medical researches that made Pasteur's name a household word, but they seem to have given him the impetus, inspiration, and insight that guided him through the labyrinth of his later research problems. Jean Baptiste Biot (1774–1862) had demonstrated that even in solution some organic chemicals such as sugars and tartaric acid rotate the plane of polarized light. Another chemist, Eilhard Mitscherlich (1794–1863), found that in addition to the common large crystals of tartaric acid, there were smaller crystals which he called paratartaric acid or racemic acid. Although both types of crystals had the same chemical properties, racemic acid was inactive to polarized light. Pasteur seized upon this apparent inconsistency and devised experimental means of answering the question:

How could chemicals be the same and yet different? Having prepared and crystallized 19 different salts of the tartrates and paratartrates, Pasteur could separate out two kinds of crystals, left-handed and right-handed mirror images, which had equal and opposite polarizing properties. As he pursued this remarkable trait from the behavior of crystals to that of microorganisms, he came to see molecular dissymmetry as a fundamental criterion that distinguished the chemical processes carried out by living organisms from those of the inanimate world.

Among the aphorisms attributed to Pasteur, the most quoted have to do with the importance of theory and the role of chance in discovery. Pasteur liked to say that chance favors only the mind that is prepared. Moreover, he insisted that the theoretical was as important as the practical, although he accepted the idea that, for the good of France, scientific education should be made relevant to industrial and commercial needs. "Without theory," Pasteur argued, "practice is but routine born of habit." When asked about the value of a purely scientific discovery, Pasteur liked to answer: "What is the use of a new-born child?" The path that led Pasteur from the theoretical question of molecular dissymmetry to the behavior of microorganisms involved a chance discovery that took place in his laboratory in 1857. It was common enough in warm weather for molds to grow on tartrate solutions; most investigators would note this and throw away the spoiled preparations in disgust. But Pasteur wondered whether the effect of molds would be the same upon the two forms of tartaric acid. Discovering that the molds used only the D (dextrorotatory) form, Pasteur developed a simple and ingenious method of separating stereoisomers by using living agents. More important, this chance observation provided a key to broader issues. The conclusion Pasteur drew was that molecular dissymmetry was a test that discriminated between the chemistry of life processes and ordinary chemical reactions. This association between dissymmetry and life permeated Pasteur's concepts of fermentation and the role of microorganisms in the balance of nature. Every aspect of the universe, living and nonliving, seemed to reflect some fundamental aspect of molecular dissymmetry. Pasteur speculated that the universe was a dissymmetrical whole and that life itself was a function of the dissymmetry of the universe and the consequences this entailed.

In 1849 Pasteur began teaching at the University of Strasbourg. While continuing his research on crystals and preparing his lectures in chemistry, Pasteur met and married the daughter of the Rector of the Academy of Strasbourg. Five years later, having achieved considerable recognition for his chemical researches, Pasteur was appointed professor of chemistry and dean of sciences at the University of Lille in northern France. The mission with which he was charged included assistance to local industries. Pasteur tried to make education relevant to the needs of the district by encouraging more laboratory experience and creating a new diploma for those who wished to enter an industrial career at the level of foreman or overseer.

Pasteur became directly involved in problems of alcoholic fermentation, a major industry in the region around Lille, and applied the principles of experimental chemistry to the problems of fermentation. These investigations essentially began when a local industrialist came to Pasteur for advice on his problems with the manufacture of alcohol from beet roots. Visiting the factory every day and studying samples of the fermentation juices under the microscope, Pasteur discovered that changes in the population of microorganisms were associated with healthy and spoiled fermentations. In addition to these changes in the population of microorganisms, Pasteur also noticed the appearance of optically active compounds. From the clues obtained through his previous studies of organic crystals, Pasteur formed the hypothesis that fermentation was a process carried out by living ferments. This hypothesis was opposed by the most renowned chemists of the period, Justus von Liebig, Jöns Jacob Berzelius, and Friedrich Wöhler, who argued that fermentation was a purely chemical process and that microorganisms were the product rather than the cause of fermentation. After analyzing many kinds of fermentations, Pasteur suggested that all fermentations are caused by microorganisms and that each living ferment is specific for a particular kind of fermentation. Changes in environment, temperature, acidity, composition of the medium, and poisons affect different ferments in particular ways. In 1857 Pasteur left Lille and joined the École Normale in Paris as assistant director in charge of administration and direction of scientific studies. When the Academy of Sciences awarded the Prize for Experimental Physiology to Pasteur in 1860, the great French physiologist Claude Bernard wrote the report and emphasized the physiological significance of Pasteur's studies on alcoholic, lactic, and tartaric acid fermentation. By this time, Pasteur had begun his studies of spontaneous generation and was reporting his preliminary experiments to the Academy. His old friend Jean Biot warned Pasteur to put aside this topic, because, as both Biot and Pasteur knew, attempting to prove a universal negative is logically absurd and thus neither a scientifically nor a philosophically rewarding proposition.

The question of spontaneous generation was an ancient and controversial one, but Pasteur was convinced that microbiology and medicine could only progress when the idea of spontaneous generation was totally vanquished. Belief in the spontaneous generation of life had been almost universal from the earliest times up to the seventeenth century. As suggested by the term

"Mother Earth," many peoples conceptualized the world as a living organism, supporting the possibility of living creatures arising out of seemingly inanimate substrates. The lowest creatures, parasites, and vermin of all sorts, which often appeared suddenly from no known parents, seemed to be the result of some kind of transmutation of lifeless materials into organized beings. Aristotle had supported the doctrine of spontaneous generation, even cataloging the particular species that would be generated from various substrates. According to Aristotle, heat was necessary for generation by any mechanism: sexual, asexual, or spontaneous. While higher creatures reproduced by virtue of their animal heat, the lower forms arose from slime and mud in conjunction with rain, air, and the heat of the sun. A combination of morning dew with slime or manure produced fireflies, worms, bees, or wasp larvae, while moist soil gave rise to mice. Seventeenth-century physician and alchemist Johannes Baptista van Helmont (1579–1644) believed that mice could be produced by incubating a flask stuffed with wheat and old rags in a dark closet. But the Italian physician Francesco Redi (1626–1698) initiated an experimental attack on the question of spontaneous generation. Redi was a member of the Academy of Experiments of Florence. In keeping with the principles espoused by the Academy, Redi tested various kinds of substrates, raw or cooked, including flesh from animals as varied as lions and lambs, fishes and snakes. Noting the way that different flies behaved when attracted to these substances, Redi suggested that maggots might develop from the objects deposited on the meat by adult flies. He followed the development of larvae from eggs and observed that different kinds of pupae gave rise to different species of flies. These studies were published in 1668 as *Experiments on the Generation of Insects*. While these experiments did not altogether discredit the concept of spontaneous generation, they did shrink the field of battle from the generation of macroscopic creatures to the small new world of infusoria and animalcules discovered by van Leeuwenhoek.

While most seventeenth- and eighteenth-century preformationists rejected spontaneous generation as a contradiction of encapsulation theory, and a dangerous materialistic theory, mathematician and philosopher Gottfried Wilhelm Leibniz (1646–1716) suggested the existence of living molecules, or monads. The French naturalist Georges Buffon (1707–1788) and the English microscopist John Turbevill Needham (1713–1781) combined their considerable talents to disprove the work of Louis Joblot (1645–1723). To prove that infusoria are not spontaneously generated, Joblot boiled his medium and divided it into two parts. One half was sealed off and the other left uncovered. The open flask was soon teeming with life, but the sealed vessel was free of infusoria. To prove that the medium was still susceptible to putrefaction,

Francesco Redi

Joblot exposed it to the air and showed that infusoria were soon actively growing. Joblot concluded that something from the air had to enter the medium to produce microorganisms. When Needham repeated Joblot's experiments, whether the flasks were open or closed, the water boiled or not boiled, the infusions placed in hot ashes or not, all vessels soon swarmed with microscopic life. Needham, therefore, concluded that there was a vegetative force in every microscopic bit of matter and every filament that had been part of living beings. When animals or plants died they slowly decomposed to one common principle, which Needham thought of as a kind of universal semen from which new life arose.

Published in the *Philosophical Transactions of the Royal Society* (1748), Needham's views were quite well known. The claims of Needham and Buffon did not, however, stand unchallenged for very long. The indefatigable Lazzaro Spallanzani exposed the fallacious assumptions and poorly contrived experiments that seemed to support spontaneous generation. By heating a series of flasks for different lengths of time, Spallanzani determined that various sorts of microbes differed in their susceptibility to heat. Whereas some of the larger animalcules were destroyed by slight heating, other very minute entities seemed to survive in liquids that had been boiled for almost an hour. Further experiments convinced Spallanzani that all these little animals entered the medium from the air. Convinced that a great variety of animalcular "eggs" must be disseminated through the air, Spallanzani concluded that the air could either convey the germs to the infusions or assist in the multiplication of those already present.

Although Spallanzani's experiments answered many of the questions raised by advocates of spontaneous generation and proved the importance of sterilization, his critics claimed that he had tortured the all-important "vital force" out of the organic matter by the cruel treatment of his media. The vital force was, by definition, capricious and unstable, rendering it impossible to expect reproducibility in experiments involving organic matter. A more serious objection was posed in 1810 by the French chemist Joseph-Louis Gay-Lussac (1778–1850), who showed that the sterile vessels lacked oxygen and concluded that oxygen was necessary for fermentation and putrefaction. Fortunately for the food industry, Nicolas Appert (1750–1841), a French chef, ignored the theoretical aspects of the dispute and applied Spallanzani's techniques to the preservation of food. He placed foods in clean bottles, corked them tightly and raised them to the boiling point of water. His techniques were described in a book published in 1810, which heralds the beginnings of the canning industry.

During the nineteenth century, the design of experiments for and against spontaneous generation became increasingly sophisticated, as proponents of the doctrine challenged the universality of negative experiments. Theodor Schwann repeated many of Spallanzani's experiments, but added the refinement of heating the air as well as the medium. To prove that the vital principle had not been tortured out of the heated air, he proved that a frog could live quite happily when supplied with such air. Still, Schwann's results were not wholly consistent or reproducible. In another attempt to purify the air used in such experiments, Franz Schulze (1815–1873) passed air through concentrated potassium hydroxide or sulfuric acid. Efforts to repeat his experiments gave equivocal results and critics still claimed that his methods tortured the vital principle. Heinrich Schröder (1810–1885) and Theodor von Dusch (1824–1890) adopted a new approach; they filtered the air entering a flask of putrefiable medium through a long tube of cotton wool. While this treatment should have been gentle enough to satisfy the opposition, it was not uniformly successful. Indeed some test substrates invariably spoiled. Because any single case of apparent spontaneous generation could allow the proponents of the theory to maintain that it only occurred under special, favorable, perhaps even sympathetic conditions, the opponents were always on the defensive.

One of the most vigorous defenders of spontaneous generation was Félix Archiméde Pouchet (1800–1872), director of the Natural History Museum in Rouen, a member of many learned societies, and a respected botanist and zoologist. In 1858 he began to present a series of papers to the Academy of Sciences of Paris purporting to show proof for spontaneous generation, which he renamed *heterogenesis*. In 1859 he published *Heterogenesis*, a massive treatise devoted to his philosophy and experiments. Since Pouchet believed it was self-evident that nature employed spontaneous generation as one of the means for the reproduction of living things, his experiments were designed not to determine *whether* heterogenesis took place, but only to discover the *circumstances* under which it occurred. Pouchet's assumptions were similar to those of Buffon and Needham; he rejected the idea that life arose de novo from accidental aggregations of molecules or in solutions of mineral substances. Heterogenesis required the existence of a vital force coming from preexisting living matter and could only occur in solutions of organic matter that retained some vital properties. According to Pouchet, the factors that promoted heterogenesis were organic matter, water, air, and the proper temperature. Any alterations in these prerequisites could affect the kind of organisms produced; thus, each factor had to be examined systematically in a series of experiments. In Pouchet's hands, all the standard experiments on spontaneous generation gave positive results. Just as some gardeners seem to be blessed with a green thumb, Pouchet apparently had a heterogenetic thumb. Moreover, like his main opponent, Louis Pasteur, Pouchet was a vigorous and enthusiastic experimenter.

While many scientists criticized Pouchet's work, Pasteur forced his adversary to admit that the existence of germs in the air was the critical issue in establishing the experimental basis of the debate. Work on fermentation had shown Pasteur that the so-called ferments were microorganisms and that the air was the source of these entities. Rigorous experimentation, Pasteur argued, forced reasonable people to conclude that spontaneous generation is a chimera. All purported evidence in support of the doctrine resulted from experimental artifacts and careless technique. Demonstrating the germ-carrying capacity of air was, therefore, Pasteur's first priority. If the microbes associated with fermentation were brought to their substrate by the air, it should be possible to intercept them. This was demonstrated by sucking air through filters and trapping the dust particles. After washing the filter in a mixture of alcohol and ether, the dust was collected and examined under the microscope. The number and kind of microbes that were found varied with temperature, moisture, air currents, and other environmental factors. For example, the germ content of hospital air was quite high, while that of mountain air was low. Both Pasteur and Pouchet explored mountains and glaciers to test different kinds of air. Pasteur even considered a balloon ascent to literally rise above Pouchet, who had taken dust samples from the the roof of the Cathedral of Rouen and the tomb of Rameses II. Despite the theatricality of many of these maneuvers, one of Pasteur's most convincing experiments was a simple demonstration using flasks with peculiar long necks drawn out and bent into a curve resembing the neck of a swan. If liquids were sterilized in these flasks, the medium would remain sterile even though ordinary air could enter via the swan neck. Germ-laden dust particles were trapped in the curve of the neck. If the medium was tipped to slosh through the bend in the neck, or if the neck was broken off and dust was allowed to enter, the medium soon seethed with microbial life. Experimenting with various media, Pasteur proved that even the most easily decomposable fluids, such as milk, blood, or urine, could remain sterile if proper precautions were taken. He also showed that microbes could be grown on a simple defined medium, that certain microbes grew only in the absence of oxygen, and that microbes remained true to their original type and were not transmuted into different species. These years of experimentation on the germ-carrying capacity of the air apparently caused some of Pasteur's peculiar mannerisms. Whether at home or dining out, he never used a plate or a glass without examining it for some speck of dust and wiping it carefully.

Clearly, Pasteur's experiments did not deal with the question of the ultimate origin of life, a process Thomas Henry Huxley named *biogenesis*; they demonstrated that microbes do not arise de novo in properly sterilized media under conditions prevailing today. In 1862 Pasteur was awarded a prize for this work by the Academy of Sciences, but Pouchet issued another challenge. A commission was appointed to settle the debate. Curiously, Pouchet withdrew from the showdown scheduled for 1864. When echoes of the old debate appeared again in the 1870s, Pasteur undertook a new series of experiments to prove that living ferments were necessary for fermentation. In August, before yeasts become associated with the grapes, he had some hothouses built to seal off parts of his vinyard. When these grapes were harvested in October, they did not ferment unless yeasts were added. The last major champion of spontaneous generation was Henry Charlton Bastian (1837–1915), professor of pathological anatomy at University College Hospital, London, and author of a 1000-page tome entitled *The Beginnings of Life*. Going beyond other heterogenesists, Bastian claimed he had evidence for *archebiosis*, the production of life de novo from inanimate matter even in the simplest solutions. In a letter to Bastian written in 1877, Pasteur explained that he felt it absolutely necessary to fight and conquer the advocates of spontaneous generation because their doctrine was an obstacle to progress in the art of healing. As long as doctors and surgeons continued to believe in the spontaneity of all diseases, their efforts at treatment and prevention would remain largely futile.

Further experimental refinements were introduced by John Tyndall (1820–1893), an Irish physicist who had studied the optical properties of particulate matter, a phenomenon now known as the light-scattering or Tyndall effect. Inspired by Pasteur's work, he carried out experiments on the germ-carrying capacity of the air and the efficacy of various sterilization methods. These studies were described in his *Essays on the Floating Matter of the Air in Relation to Putrefaction and Infection* (1881). Using a chamber in which light scattering by dust particles served as a measure of the optical activity, or "biological purity" of the air, Tyndall trapped dust particles with a layer of glycerine until the air was optically empty and, therefore, biologically pure. Test tubes filled with various infusions remained sterile when kept in the Tyndall Box. In tests designed to probe the sensitivity of microbes to heat, Tyndall discovered that some germs formed remarkably heat-resistant spores. These tests led to a method of sterilization involving cycles of heating and cooling known as fractional sterilization, or tyndallization. After destroying the active bacteria in the first step of sterilization, Tyndall allowed the medium to cool and then heated it again. Discontinuous boiling for one minute in five separate steps could sterilize medium containing microbial species that had resisted one hour of continuous boiling. Tyndall, who had been attacked by Bastian for intruding into the domain of biologists and physicians, was awarded an honorary Doctor of Medicine degree by the University of Tübingen. Further studies of the nature of microorganisms made it increasingly obvious that they were not simple blobs of organic material that might be spontaneously generated during the putrefaction of organic matter. Even the smallest bacteria are too complex to simply precipitate out of solution like crystals. The discovery of viruses stimulated a brief revival of the idea, but even viruses, while very small, are quite complex and endowed with definite mechanisms for replication.

Advocates of the doctrine of spontaneous generation have argued that some form of the doctrine is necessarily true in the sense that if life did not always exist on earth, it must have been spontaneously generated at some point. Alternatively, while avoiding the question of spontaneous generation on earth, some scientists and philosophers have argued that the seeds of life and epidemic disease have come to earth from outer space in the form of spores, bacteria, or viruses. In contrast, scientists who have devoted their attention to the question of the origin of life have suggested that when the earth was young, a process like spontaneous generation must have occurred in the primordial chicken soup. The idea that living beings were formed by means of the mixing of the primary elements in the nourishing environment of the ancient oceans of the earth also appears in the ancient Indian writings known as the *Rig Veda* and the *Atharva Veda*. Modern attempts to understand the origin of life, however, can be traced to the work of A. I. Oparin, a Russian scientist, who in 1924 proposed an apparently plausible explanation for the chemical evolution of life in the enriched organic soup that could have accumulated in the ancient seas when the earth's atmosphere was very different from that of today. According to Oparin, the origin of life was not the result of some "happy chance," but was "a necessary stage in the evolution of matter." Intrigued by such ideas, in 1957 scientists organized the first of many international symposia on the origin of life. To some degree Oparin's speculations were vindicated by experiments performed by Stanley Miller in 1953. Since then Miller, Sydney Fox, and others have been able to demonstrate the formation of important organic chemicals, including amino acids and components of the nucleic acids, under experimental conditions simulating the earth's probable prebiological environment.

THE GERM THEORY OF DISEASE

Primitive ideas about contangion are not directly related to the germ theory of disease, but to the general notion of transfer through contact. Just as heat and cold could apparently be transferred to neighboring bodies, so were putrefaction, uncleanliness, corruption, and disease. Epidemics were probably rare in small bands of primitive peoples, but would have been terrifying events once population density increased sufficiently to produce and sustain them. Many peoples believed diseases were sent by the gods as punishment for their sins, but the Hippocratic physicians rejected such supernatural agents of disease and had little interest in the idea that disease was spread by contagion. Nevertheless, some Greek and Roman philosophers, poets, and architects thought that disease might be caused by tiny animals dwelling in swampy places or by contact with the sick and with contaminated articles. For example, in his discussion of hygienic regulations for selecting building sites, the Roman architect Marcus Terentius Varro (177–27 B.C.) warned against swampy locations. Disease-causing animals living in swampy places, so small as to be invisible, might enter the body through the mouth and nose and cause grave illnesses. Alternatively, epidemic diseases might be caused by comets, eclipses, floods, earthquakes, or major astrological disturbances that charged the air with poisonous vapors known as miasmata.

The kinds of evidence used to support the concept of pathogenic miasmata and the alternative concept that disease was transmitted by contagion were analyzed by the Renaissance physician and poet Girolamo Fracastoro (1478–1553). In 1530 Fracastoro published a classic study of the venereal disease known to Italians as the French disease. His medical poem *Syphilis, or the French Disease* gave the disease its modern name. Having observed the epidemics of syphilis, plague, and typhus that had ravaged Italy in the sixteenth century, Fracastoro compared the implications of the miasmatic theory of disease with those of contagion theory. As part of his analysis of the ancient term contagion and the way in which various epidemic diseases appeared to be disseminated, he distinguished three different modes of contagious diseases. The first kind infected by direct contact only. The second kind infected by contact and also by means of *fomites*, inanimate articles such as clothing, linens, and utensils that had been in contact with the sick and were capable of carrying the germs of contagion to another victim. In the third category he placed contagion transmitted not only by contact and fomites, but also capable of being transmitted at a distance; tuberculosis, certain eye diseases, and smallpox seemed to fall into this category. The first category of diseases could be compared to the putrefaction or infection that occurred in fruits, spreading from grape to grape or apple to apple. That is, infection appeared to be a form of putrefaction or fermentation.

Girolamo Fracastoro

Although Fracastoro has been called the founder of the germ theory of disease, his writings were more ambiguous than this title would imply. In general, the idea of contagion existing in tiny germs or seeds of disease did not prove very useful in guiding medical practice. The contagion theory led to attempts to stop epidemics by quarantines, isolation and disinfection, but these methods did not seem to be successful enough to justify their inconvenience. While such measures might have mitigated the spread of bubonic plague to a limited extent, they were ineffective against typhus fever, typhoid fever, and cholera. When such diseases were epidemic, isolation of the sick appeared to have no effect. Public health reformers, therefore, seemed to have good reason to believe that such diseases were spread by filth and poisoned air.

In 1840 Jacob Henle (1809–1885), a prominent German pathologist, physiologist, and anatomist, revived the contagion theory and published his examination of the relationships among contagious, miasmatic, and miasmatic-contagious diseases. His *On Miasmata and Contagia* was generally ignored by his contemporaries, but after the establishment of the germ theory of disease, it was retrospectively recognized as a landmark. Analyzing the patterns of transmission of various diseases seem to prove that malaria was a purely miasmatic disease, while smallpox, measles, scarlet fever, typhus, influenza, dysentery, cholera, plague, and puerperal fever were miasmatic-contagious. Diseases such as syphilis, foot-and-mouth disease, and rabies were acquired only from contagion. Moreover, Bonomo, Bassi, and Schönlein had provided evidence that mites and mircoparasites transmitted scabies, muscardine, and ringworm. Clearly separating the concept of *disease* from the concept of *parasite*, Henle defined the material basis of contagion as an organic entity capable of living a separate existence, or a parasitic existence within the diseased body. That is, the contagion was not the disease but the inducer or cause of disease. Critically evaluating the experimental evidence, Henle discussed the nature of the proofs that would be required to establish a causal relationship between microbes and disease.

According to Henle, physicians blamed disease on miasma, which they defined as something that mixed with and poisoned the air, but, he argued, no one had ever demonstrated the existence of miasma. It was simply assumed to exist, by exclusion, because no other cause could be demonstrated. A more likely hypothesis, according to Henle, was that *contagia animata* (living organisms) caused contagious diseases because whatever the morbid matter of disease might be, it obviously had the power to increase in the afflicted individual. The natural history of epidemics could be explained by assuming that disease was caused by an agent excreted by sick individuals.

If this agent was excreted by the lungs, it might pass to others through the air; if excreted by the intestines, it would enter sewers and wells. Given the fact that the pus from pox pustules could be used to infect a multitude of people, the contagion must be an animate entity that multiplies within the body of the sick person. Chemicals, organic or not, remain fixed in amount; only living things have the power to multiply themselves. While Henle's discussion of the question of contagion and miasma is superficially similar to that of Fracastoro, the context in which they worked and the centuries that separated them infused very different meanings into their use of the terms miasma and contagion.

Using Schwann's metabolic theory of the cell as a point of departure, Henle compared the action of contagion to fermentation. Well aware of the difficulties confronting the contagion theory of disease, Henle warned that finding some microorganism in the sick did not prove that it had a causal role. The agent must be isolated and cultured, free from the diseased individual. Indeed, Henle came close to outlining what are usually called Koch's postulates. Acknowledging the lack of rigorous evidence for the germ theory of disease, Henle argued that science could not wait for unequivocal proofs because scientists could only conduct research "in the light of a reasonable theory." A major problem in establishing the truth of germ theory was methodological. Obtaining pure cultures was a difficult and tedious procedure, almost impossible in the hands of any but the most meticulous experimentalists. Working out proper sterilization and culture procedures required a prior understanding of and commitment to the germ theory of disease. Given the state of the methods available until quite late in the nineteenth century, it was no wonder new animalcules seemed to appear spontaneously in the tempting broths set out for them, as well as in the raw wounds of surgical patients. Confused by the claims of the heterogenesists, physicians tended to reject the germ theory or dismiss it as a laboratory curiosity unrelated to clinical medicine. But as Pasteur's associate Émile Duclaux contended, "the great merit of a new theory is not to be true, because there is no such thing as a true theory, but to be fruitful."

The study of crystals had led Pasteur to the nature of fermentation and the diseases of wine, beer, and vinegar. His next practical challenge was a mysterious disease threatening the silk industry of France, causing great despair and hardship in many households and villages. Although Pasteur had never before seen a silkworm and knew nothing about these invertebrates, he expected to resolve the problem by finding a germ. The problem was more complex than expected, because it involved two different disease organisms, as well as nutritional and environmental effects, and thus the epidemic was the result of complex interactions among host, germ, and environment. From silkworm diseases, Pasteur progressed to the riddle of disease in higher animals, and finally, to human rabies. During this phase of his career, Pasteur was devastated by the deaths of two of his daughters, his son's war experiences, and a series of strokes that left him partially paralyzed. As part of the treatment for cerebral hemorrhage, Pasteur's physicians prescribed the application of 16 leeches behind the ear.

Louis Pasteur dictating an article on the diseases of silkworms to Madame Pasteur

Medical microbiology owes much to Pasteur's studies of chicken cholera, a disease unrelated to human cholera except in the virulence of the infection and the rapidity of death. Chickens that pecked at foods soiled with the excreta of the sick quickly acquired the disease. Affected birds become drowsy and showed signs of anoxia, that is, oxygen starvation. Based on reports by veterinary surgeons that certain bacteria were present in the tissues of sick chickens, Pasteur isolated a microbe that could be cultured in a medium made of chicken gristle. A small drop of the fresh culture would quickly kill a chickens, but the microbe only caused a small abscess in guinea pigs. If new cultures were made every day, small samples could be used to demonstrate the virulence of laboratory strains of the microbe. When Pasteur went away on his summer holiday in 1879, chicken broth cultures were left in the laboratory. In October, Pasteur tried to resume his experiments by injecting some of the old cultures into chickens, but they were essentially unaffected. When he obtained fresh samples of virulent microbes and inoculated the hens, once again they remained healthy. Realizing that his old cultures had protected the experimental animals from the new virulent microbes, Pasteur announced that it would now be possible to create artificial vaccines that acted against virulent diseases much as Edward Jenner's cowpox vaccine protected people from smallpox. That is, the virulence of pathogenic bacteria could be modified by manipulating laboratory cultures. Changing the conditions of culture would make it possible to decrease virulence progressively and obtain a vaccine that caused only mild disease but provided protection from the natural disease. The microbe that caused chicken cholera could be attenuated by storing the cultures for variable intervals, but it was more difficult to attack the microbe that caused anthrax, because the bacillus was able to form protective spores. Anthrax, a disease that primarily attacked sheep and cattle, also infected farmers and butchers. In many French provinces, as many as 20% of the sheep succumbed to this disease. Some farms seemed especially cursed by the scourge and experienced even greater losses. Afflicted sheep often died within hours after the apparent onset of symptoms. Autopsies revealed thick black blood and a black and liquid spleen. Several microscopists noted the presence of rod-shaped bacteria in the blood of anthrax victims, but Robert Koch was the first to provide unequivocal proof that the disease was caused by a specific pathogenic microbe.

Human rabies is a very rare condition, but the image and the threat of the rabid animal once exerted a powerful grip on the popular imagination. Pasteur himself said that he remembered the howls of wolves in the countryside surrounding his village and the screams of a neighbor cauterized af-

Louis Pasteur studying rabies

ter the bite of a rabid animal. Generally, no matter what the treatment, the disease was fatal and gruesome in its progress. In his previous studies of animal diseases Pasteur had begun by identifying and isolating the microbe that appeared to be the etiological agent. But no visible agent could be found in prepartions of materials capable of rabies. The invisible agent of disease was referred to as a *virus* according to the traditional use of the term as an unknown agent of disease. Pasteur and his associates found that the long incubation period between the bite of a rabid animal, or subcutaneous inoculations, and the onset of disease could be shortened by inoculating material from rabid animals directly into the brain of dogs or rabbits. Although the "microbe of rabies" was invisible, a vaccine could be prepared by suspending the spinal cords of rabid rabbits in a drying chamber. After about 2 weeks the material became virtually harmless. By proceeding from harmless to more virulent materials in a series of inoculations, Pasteur could protect dogs that had been bitten by rabid animals or inoculated with virulent preparations.

On July 6, 1885, 9-year-old Joseph Meister was brought to Pasteur after an attack by a rabid dog left him with deep wounds on his hands, legs, and thighs. Advised by physicians that the case was hopeless, Pasteur began a series of inoculations. Meister recovered completely and within a year more than 2000 people had received the Pasteur rabies vaccine. The rabies vaccine brought Pasteur international acclaim. The Pasteur Institute was dedicated in 1888, but by then Pasteur was too weak to even speak at the ceremonies. The Pasteur Institute began with microbiology laboratories and gradually added more buildings and laboratories, a hospital, and outpatient clinics. The first Institute served as the model for a network of Pasteur Institutes that were established in the former French colonies. For twentieth century researchers like molecular geneticist François Jacob, who first began work there in 1949, the Institute was still the mecca of French biology, a place where innumerable discoveries had been made by almost legendary scientists. It remained a unique and highly flexible institution, unlike the universities and other organizations that invariably staggered under the burdens of bureaucracy. Workers at the Institute participated in an annual tribute commemorating the death of Pasteur by assembling at the Institute's oldest building, which housed Pasteur's crypt, and marching past Pasteur's tomb. The ostentatious mausoleum contains mosaics depicting aspects of Pasteur's work on alcoholic fermentation, silkworm diseases, chicken cholera, anthrax, and rabies, and four angels depicting Faith, Hope, Charity, and Science. Marble panels on the walls of the tomb contain inscriptions recording

Pasteur's victories: molecular dyssymmetry, fermentation; so-called spotaneous generation; studies on wine; silkworm disease; studies on beer; virulent diseases; vaccines; prophylaxis against rabies. After the ceremonial visit to the tomb of Pasteur, the Pastorians marched towards the simple tomb of Émile Roux (1853–1933), who had succeeded Émile Duclaux (1840–1904) and served as director of the Institute for about 30 years.

Very different from Pasteur in training, temperament, and philosophical approach, Robert Koch was most successful at formulating the principles and techniques of modern bacteriology. While he lacked Pasteur's gift for dramatic presentations, his contemporaries considered the rather phlegmatic Koch "a man of genius both as technician and as bacteriologist." Unlike Pasteur, Koch approached bacteriology as a physician, and his research was primarily motivated by medical questions. Koch was born in a small village in the Harz Mountains to hard-working, diligent, parents; he was the third of 13 children. Koch acquired his interest in research as a medical student at the University of Göttingen, but none of the faculty members, including Jocob Henle, seemed to have any interest in the possible relationship between bacteria and disease. When he graduated in 1866, Koch spent several months observing clinical medicine at the Charité hospital in Berlin and attending a course of lectures by Rudolf Virchow. Dreaming of travel and adventure, Koch hoped to find work as a ship's doctor or military surgeon, but after becoming engaged to Emmy Fraatz, he seemed to be doomed to spend his life as a country doctor and district medical officer. Despite the demands of his private practice and official duties, he also explored new ideas in natural history, archaeology, photography, public health, hygiene, and bacteriology. A small laboratory was set up next to his consultation room; his equipment included a Zeiss oil immersion microscope and the illuminating apparatus invented by Abbe. Experimenting with the new aniline dyes, he was able to get clear images of bacteria and distinguish between what he called the "structure picture" and the "color picture."

When an outbreak of anthrax occurred in his district, he was able to make his own tests of the relationship between the anthrax bacillus and the disease. Although anthrax is primarily a disease of sheep and cattle, it can cause severe, localized skin ulcers known as malignant pustules in humans, a dangerous condition known as gastric anthrax, or a virulent pneumonia known as wool-sorters' disease. Several investigators had found microbes in the blood of animals with anthrax, but their evidence for a causal association between bacteria and disease remained largely circumstantial. Rather large for a bacillus, *Bacillus anthracis*, the causal agent of anthrax, appears in the bloodstream in such large numbers that it can be found by direct examination of blood. By 1860, Franz Pollender (1800 1879), Pierre Rayer (1793–1867), and Casimir Joseph Davaine (1812–1882) had observed bacteria in the blood of anthrax victims. Moreover, several attempts had been made to transmit the disease to experimental animals by inoculating them with blood from diseased animals.

Like his predecessors, Koch began his work by attempting to transfer anthrax from infected farm animals to rabbits and mice. When injected with fresh anthracic material, infected mice died within 24 to 30 hours. Moreover, the disease, with all its characteristic symptoms, could be transmitted through a series of experimental animals. To provide further proof that the disease was caused by *Bacillus anthracis* and not some poison in the blood, Koch cultivated tiny bits of spleen from infected animals in the aqueous humor from the eye of a cow. This material served as the first link in a chain of cultures transferred through fresh aqueous humor. Even after 10 to 20 transfers, a bacterial culture grown in aqueous humor could kill a mouse just as quickly as blood taken directly from a diseased sheep. These experiments ruled out the possibility that the disease was caused by some poison from the original animal; only an entity capable of multiplying in experimental animals and in laboratory cultures could create such a chain of infection. While studying anthrax bacilli trapped on microscope slides, Koch noticed that as the medium evaporated the thread-like chains of growing bacteria were transformed into bead-like spores. By adding fresh medium, he could reverse this transformation.

In 1876, when Koch was convinced that he had isolated the microbe that caused anthrax, established its life cycle, and explained the natural history of the disease, he contacted Ferdinand Cohn (1828–1898) for advice and criticism. Primarily a botanist, Cohn was the first prominent German scientist to take a special interest in bacteriological research. Certainly Koch was not the first amateur microbiologist to approach Cohn with "unequivocal evidence" that would solve the riddle of infectious disease. Despite his initial skepticism, Cohn invited Koch to come to the Institute of Plant Physiology at the University of Breslau and demonstrate his findings. Impressed by the rigorous nature of Koch's experiments, Cohn arranged for the publication of Koch's paper "The Etiology of Anthrax Based on the Developmental Cycle of *Bacillus anthracis*" in the journal *Contributions to Plant Biology*.

An understanding of the complex life cycle of *Bacillus anthracis* immediately explained the mystery of the persistence of anthrax in pastures that farmers came to think of as cursed by the disease. Because spores could survive under harsh conditions, one contaminated carcass dumped in a shallow pit could serve as a reservoir of spores for many years. Spores were the most general means of disseminating the disease among animals, but people who slaughtered, butchered, or skinned infected animals could contract the disease by means of still active bacilli in fresh tissues. Having satisfactorily elucidated the role of *Bacillus anthracis*, Koch proposed various prophylatic measures to prevent the dissemination of the disease. The first step was the proper disposal of carcasses. Sheep buried in shallow wet graves served as excellent foci for the development and dissemination of spores, because spore formation required air, moisture, and a temperature above 15°C. Since the average soil temperature at a depth of 1 to 10 meters in Germany was below the critical temperature, carcasses buried in deep dry trenches far from farms would become harmless. Surveys of the disease pattern in sheep, cattle, and horses suggested that sheep were the normal reservoir of infection. Simply separating other animals from sheep interrupted the chain of infection. Koch predicted that similar modes of dissemination would be found for typhoid fever and cholera, but it was difficult to prove this because no animal models seemed to be available for research on these diseases. Despite such obstacles, Koch confidently predicted that progress in bacteriology would lead to the control of the epidemic diseases that threatened human beings. To overcome the skepticism of the conservative medical profession, Koch urged advocates of the germ theory of disease to abandon careless and speculative work, develop methods of cultivating pure strains of bacteria, and demonstrate the value of microbiology in the prevention and treatment of human disease.

Even after Koch's classic work on the anthrax bacillus, the controversy concerning the etiology of the disease and the role of *Bacillus anthracis* continued. In 1877 the French physiologist Paul Bert (1833–1886) announced that oxygen could destroy the anthrax bacillus, but the blood could still cause the disease. Therefore, Bert claimed, the bacillus was not the cause of anthrax. This argument drew Pasteur into the battle. Carrying out a series of 100 transfers from culture to culture, Pasteur purified the bacteria in a medium of broth or urine. Only a living organism capable of multiplying in the course of these transfers could be the virulent agent; the dilution had been so great that no inanimate poison from the original sample could remain. In explaining Bert's observations, Pasteur noted that much confusion had arisen because some experimenters used fresh materials while others took samples from long-dead carcasses. In such cases, their experiments involved the germs of putrefaction rather than anthrax. Another peculiarity of the anthrax bacillus was discovered in connection with a dispute as to whether chickens could be affected. Under normal conditions, chickens were not susceptible to the anthrax germ. To test the hypothesis that chickens did not acquire anthrax because their body temperature was several degrees above that of susceptible species, Pasteur immersed an inoculated hen in a cold water bath. The chilled chicken took the disease and died. One control chicken had been inoculated, but not placed in a similar bath; another had been placed in a bath, but not inoculated. A hen that had been inoculated and kept chilled until it showed signs of the disease recovered completely after it was removed from the bath. Later research suggested that Pasteur's hypothesis about resistance in chickens was incorrect, but the dispute drew attention to the complex phenomenon of differential susceptibility among various species.

Although it was Koch who demonstrated the bacterial etiology of anthrax and the critical problem of spore formation, it was Pasteur who developed a preventive vaccine based on his studies of chicken cholera. By growing bacteria under controlled laboratory conditions, Pasteur argued, a series of attentuated cultures could be established. In the case of the anthrax bacillus, cultivation in broth at 42–44°C was critical; spores could not develop at a temperature above 42°C, but the vegetative form died at 45°C. After several weeks of cultivation under these conditions, the anthrax bacillus gradually lost its ability to cause disease and could be used as a preventive vaccine. With the cooperation of the Agricultural Society of Melun, a public demonstration of the anthrax vaccine was held in May 1881 at Pouilly-le-Fort. The editor of the *Veterinary Press*, who called Pasteur the "prophet of microbiolatry," had initiated a campaign for subscriptions to test Pasteur's anthrax vaccine. On May 5, 1881, before a crowd of supporters and skeptics, Pasteur and his associates inoculated 24 sheep, 6 cows, and 1 goat with his anthrax vaccine; 24 sheep, 4 cows, and 1 goat were set aside as controls. On May 17 the experimental animals were vaccinated again with a more virulent laboratory-attenuated culture. Late in May all the animals were inoculated with a highly virulent strain. In June the contest was over and Pasteur was judged to have achieved a stunning success; the vaccinated animals remained free of anthrax, but the unvaccinated sheep and goat died and the unvaccinated cows became ill. Even among historians and philosophers of science 100 years later, there is little doubt that the demonstration at Pouilly-le-Fort was an event of high drama, deliberately planned and orchestrated by Pasteur as part of his strategy for effecting a revolution in biology, medicine, and hygiene. Understanding how Pasteur was able to mobilize support from the French public and much of the scientific establishment during what has been called the "high point of the scientific religion" is more problematic.

By 1894, millions of sheep and cattle had been vaccinated against anthrax. Despite the apparent success of Pasteur's approach to the development of preventive vaccines, Koch used the pages of his journal *Record of the Works of the German Sanitary Office* to argue that Pasteur was incapable of cultivating pure strains of microbes. Koch ridiculed Pasteur's experiments on anthrax in chickens, his account of the role played by earthworms and spores in propagating anthrax, and even the preventive value of vaccination. In response to the honors awarded to Pasteur as a "second Jenner," Koch called attention to the obvious fact that his French rival was not a physician and noted that Jenner's vaccine saved human lives rather than sheep. Unfortunately, bitter and disruptive disputes between the French and German pioneers of microbiology were not uncommon and, at least in part, reflected the hostilities between their countries. Deeply troubled by the French defeat in the Franco-Prussian War, Pasteur blamed the neglect of science for the loss and saw his work as part of a much needed effort to raise the level of French science and industry. In retrospect, it is clear that Koch was correct in the claims he made concerning the superiority of the methods he had developed to identify and purify the bacterial agents of disease. Nevertheless, it was Pasteur who established practical methods of preventing disease, and, while quantity need not correlate with quality, it is interesting to note that Pasteur published 31 papers on anthrax, whereas Koch published only two.

After his successful studies of anthrax, Koch turned to the general problem of wound infection and traumatic infective diseases. By this time, Joseph Lister (1827–1912) had introduced his antiseptic system as a means of applying Pasteur's germ theory to the prevention of postsurgical infection, and scientists had generally accepted the idea that traumatic fever, purulent infection, putrid infection, septicemia, and pyemia were conditions that shared the same essential nature. Nevertheless, disputes continued as to whether the microorganisms found in these infections were the cause or the result of disease processes or simply nonspecific entities. Part of the problem was the difficulty of determining whether microbes were present in the blood of healthy animals. With improved illumination and suitable stains, Koch was able to demonstrate that bacteria were not found in the blood or the tissues of healthy animals, but when blood from infected animals was injected into healthy mice, various septic diseases were produced as bacteria proliferated. Having observed the subtleties of gangrene in mice, spreading abscess, pyemia, septicemia, and erysipelas in rabbits, Koch concluded that the criteria for associating the diseases with the microbes had been adequately met and that his research had obvious clinical importance in the control of human traumatic infectious diseases. Many critics of the doctrine of specific

Robert Koch

etiology dismissed this work as a series of laboratory curiosities, but Joseph Lister believed that Koch had provided a general proof of the applicability of germ theory to wound infection and promoted the publication of an English translation of Koch's work *The Aetiology of Traumatic Infective Disease*. Finally, in recognition of the importance of his work to German science, Koch was appointed professor of hygiene at the University of Berlin and director of the University's Institute of Hygiene. The Institute for Infectious Diseases was created for him in 1891.

Each disease Koch studied was invariably associated with a definite form of microorganism. While many distinguished scientists believed that bacterial forms were easily interconvertible, Koch maintained that pathogenic bacteria existed as distinct species. His studies of the traumatic infective diseases were meant, at least in part, as a defense of the concept of bacteria as distinct, fixed species against an attack by Karl von Nägeli (1817–1891), who is primarily thought of as the distinguished botanist who failed to appreciate Gregor Mendel's (1822–1884) theory of inheritance. The apparent generation of new forms in the laboratory, Koch contended, was not proof of transformation of type, but evidence of technical problems that allowed the introduction of contaminants. Clearly, it would be impossible to say that a specific bacillus caused a specific disease if bacteria did not exist as distinct species. Using pure cultures was essential, but unfortunately the animal body seemed to be the only cultivation apparatus available for the growth of many pathogenic microbes.

Growing bacteria outside the animal body and obtaining pure strains became Koch's obsession. After many unsuccessful experiments, including attempts to grow bacteria on slices of potato, Koch concluded that it would be impossible to construct a universal medium on which all microorganisms could grow. Therefore, he attempted to find a means of converting the usual nutrient broths into a form that was firm and rigid so that individual bacteria would be fixed in place and would grow into colonies like little islands on a sea of jelly. Medium solidified by the addition of gelatin was useful for cultivating many different organisms, but gelatin has several disadvantages; it melts at body temperature and can be digested by some bacteria. An ancient Far Eastern culinary technique provided a practical substitute. The wife of one of Koch's associates suggested agar-agar; her mother had learned to use agar-agar in making jellies from a Dutch friend who had been introduced to it while in Java. Nutrient agar remains solid up to 45°C and resists bacterial digestion. A simple but very effective device for culturing microbes on solid media was introduced by Richard Julius Petri (1852–1921), the curator of Koch's Hygiene Institute. Although the use of solid media was initially called "Koch's plate technique," thanks to the universal adoption of the petri dish, Petri's name is probably more familiar to students of microbiology than Koch's.

While attending the Seventh International Medical Congress in London in 1881, Koch enjoyed the opportunity to demonstrate the superiority of his plate technique to an audience that included Louis Pasteur and Joseph Lister. Shortly after returning from this meeting, Koch committed all of his considerable energy and skill to the formidable task of proving that tuberculosis was an infectious disease, identifying the causal agent, and finding a means of prevention or cure. In contrast to Pasteur, who had taken rabies, one of the most dramatic but rare threats to human life, as his greatest challenge, Koch chose tuberculosis, a commonplace, indeed almost ubiquitous scourge, for his greatest work. Working in strict secrecy, Koch was ready to present his preliminary results concerning the discovery of the tubercle bacillus, *Mycobacterium tuberculosis*, in little more than a year. Accounts of Koch's discovery, accompanied by speculation about a possible cure, appeared in newspapers throughout the world shortly after his presentation in March 1882 at a meeting of the Berlin Physiological Society. To understand the excitement generated by Koch's discovery, it is essential to appeciate the way in which this dreaded disease permeated the whole fabric of nineteenth-century life. When Koch began his work, epidemiologists estimated that tuberculosis was the cause of death in one of every seven human beings. Even in the 1940s, pathologists commonly found evidence of tuberculosis when performing postmortem examinations of the lungs and lymphatic glands of persons who had presented no evidence of the disease during life. The impact of the disease on society was amplified by the fact that it was particularly likely to claim victims in what should have been their most productive adult years. Not unlike AIDS in the 1980s, tuberculosis seemed to claim a disproportionate number of young artists, writers, composers, and musicians. The romantic myth of a link between the fire of genius and the fever of tuberculosis was dispelled by the discovery of what Franz Kafka saw as the "germ of death itself" in the sputum of its many impoverished victims and the dust of their filthy, dark, crowded tenements. As in the case of AIDS, the connection between creativity and the disease was fortuitous, while the association between the dissemination of the pathogen and poverty was indicative of fundamental inequities in society.

Other pathogens, Koch noted, had fallen into the hands of investigators "like ripe apples from a tree," while the tubercle bacillus remained out of reach. Special fixation and staining methods were needed before the fine rod-like forms could be distinguished from the nonspecific debris found in tuberculous tissues. The tubercle bacillus is only a tenth the size of the anthrax bacillus and is covered by a waxy layer, which made staining very difficult. Great patience and a firm conviction that tuberculosis was caused by a specific bacterial species were needed in these studies because the microorganism grew exceptionally slowly. Although tuberculosis can be found in many animals, including cattle, horses, monkeys, rabbits, and guinea pigs, not all strains of the bacillus are pathogenic for every species. Fortunately, the microbe that caused tuberculosis in humans could be transferred to the guinea pig. *Mycobacterium tuberculosis* was identified in the tissues of patients suffering from a bewildering array of disorders known as phthisis, consumption, scrofula, miliary tuberculosis, and so forth. Indeed, the tuberculosis bacillus can attack virtually every part of the body. Thus, after decades of controversy as to the nature of tuberculosis and whether it was a contagious disease, Koch's discovery vindicated the unitary theory of tuberculosis. Identification of the tubercle bacillus helped dispel decades of perverted romantic sentimentalism. The pale, thin, "angel of phthisis" with eyes bright from fever discreetly coughing up blood stained sputum into her white lace handkerchief before her inevitable, but redemptive, death was actually filling the air around her with clouds of bacteria. Worse yet, the tubercle bacillus was, as Koch noted, very similar in size, shape, and staining properties to the microbe associated with leprosy.

Working feverishly, Koch was ready to present his preliminary report to the Physiological Society in Berlin on March 24, 1882. Paul Ehrlich later described the presentation as a masterpiece of scientific research and a deeply moving experience. In addition to identifying the microbe that caused tuberculosis, Koch had considered the problem of the dissemination of the disease. Although the bacillus could form spores, it did not do so as easily as the anthrax bacillus. The tubercle bacillus could only grow in specially constituted medium at temperatures greater than 30°C, but even under ideal conditions several weeks were needed for good growth on solid media. Tubercle bacilli were, therefore, true parasites, dependent on the animal body for their survival, rather than facultative parasites like the anthrax bacillus. Pulmonary tuberculosis, the most common form, provided the most efficient means of transmission, because its victims coughed and spit up large quantities of germ laden sputum. Infectious germs disseminated through the air contaminated floors, furniture, clothing, and so forth. In poorly ventilated, dirty, dusty tenement rooms rarely illuminated by the cleansing rays of sunlight, the germs could remain viable and virulent for days or even months. Although the milk of infected animals could be a source of infection, Koch underestimated this danger, largely because of his interest in the role of sputum in spreading the pulmonary form of the disease.

During his studies of tuberculosis, Koch formalized the criteria needed to prove unequivocally that a particular microbial agent causes a specific disease. These criteria are now known as *Koch's postulates*. First, it is necessary to find the microbe invariably associated with the pathological condition. A thorough study of the relationship of the microbe to the natural history of the disease provided good, albeit circumstantial evidence. Skeptics, however, were entitled to argue that even if the microorganism appeared in association with the disease, it might not be the cause of the disease. Rigorous proof of a causal relationship required complete separation of the microbe from the diseased animal, tissue fragments, and all possible contaminants. After the putative pathogen had been grown as a pure laboratory culture, it should be introduced into healthy animals. If the disease was then induced with all its typical symptoms and properties, the investigator could then assert that the microbe was truly the cause of the disease. Ideally, the researcher would carry out all these steps before making claims to have solved the riddle of the etiology of infectious diseases, but this might be impossible for human diseases lacking animal models. Although Koch's postulates were formulated to prove a causal relationship between bacteria and disease, his general approach has been extended to guide studies of other disease-causing agents, such as viruses, asbestos, lead, mercury, and other chemicals that are thought to induce disease, trigger allergic responses, or depress the immune system.

Despite the rigorousness of Koch's investigations of the tubercle bacillus, not all scientists were willing to accept the idea that a single pathogen could be responsible for an enigma as complex as tuberculosis. Although systematic postmortem examinations by René Laennec (1781–1826) and other French physicians had found evidence that tuberculosis caused morbid effects throughout the body, Rudolf Virchow, the German "pope of pathology," insisted that pulmonary tuberculosis and miliary tuberculosis were different diseases. Like Koch, Virchow was given to nationalistic attacks on his French counterparts, but Virchow also tended to denigrate Koch's work on the "so-called tubercle bacillus." In any case, Virchow argued, if the microbe were as widespread as Koch asserted, but only certain individuals became consumptives, the true cause of tuberculosis could not be a contagion. Advocates of the germ theory of disease objected that one might as well say that bullets did not kill because not every soldier on the battlefield was killed by a barrage of bullets. Some critics argued that Koch had not discovered anything new because others, such as the English epidemiologist William Budd (1811–1880) and the French physician Jean Antoine Villemin (1827–1892), had previously claimed that tuberculosis was contagious. Villemin had demonstrated that tuberculosis could be transmitted from humans to rabbits by means of sputum, blood, and bronchial secretions. Given the fact that the disease seemed to "run in families," many physicians were convinced that the disease was due to a noncontagious, hereditary "tubercular diathesis." In other words, people who are susceptible to tuberculosis, are susceptible to tuberculosis. Whether the disease was caused by a microbe or a hereditary weakness, in the absence of a therapeutic agent or preventive vaccine, its victims faced a long, lingering, inexorable decline that inevitably terminated in death.

For those who accepted the relationship between the tubercle bacillus and the disease, Koch's discovery stimulated hope that appropriate medical guidance might break the chain of transmission and help victims in the early stages of disease. On the other hand, the ability of physicians to detect infections that would have gone undiagnosed previously might exacerbate public perceptions about the threat of the disease. Of course a remedy for such a highly prevalent illness would be more welcome than sensible advice about personal hygiene and the fundamental social and medical reforms that would be required for a sustained and successful public health crusade. Koch had found the microbe of anthrax and tuberculosis, but his French counterpart had developed preventive inoculations for anthrax and rabies. Thus Koch was under considerable pressure to follow up his identification of the tubercle bacillus with a cure. In 1890, the year that Germany served as host to the Tenth International Medical Congress, Koch announced that he had found a substance that could inhibit the growth of the tubercle bacillus in cultures and in guinea pigs; 1890 was also the year in which Emil von Behring (1854–1917) and Shibasaburo Kitasato (1852–1931) initiated the new era of serum therapy based on their experiments with diphtheria and tetanus antitoxins.

Although Koch may have been somewhat reluctant to discuss his studies in public, he apparently made his announcement under considerable pressure from government officials; some of his more cynical colleagues suggested that he had been bribed rather than coerced. News reports immediately referred to the mysterious remedy that Koch called "tuberculin" as "Koch's lymph," "Kochin," or "Koch's fluid." Three months later, after conducting some tests in humans, Koch published a "Progress Report on a Therapeutic Agent for Tuberculosis." Tuberculin was described as a stable, clear, brown liquid, but, despite German laws prohibiting secret medicines, its nature was kept secret and its production was entrusted to Eduard Pfuhl, Koch's son-in law. Hordes of consumptives and scientists flocked to Germany in hopes of gaining access to Koch's remedy. Joseph Lister, who brought his niece to Berlin for treatment, was impressed by German advances in medicine, which included new methods for the prevention or treatment of diphtheria and tetanus, as well as tuberculin. Before the safety and efficacy of tuberculin had been properly tested, the Grand Cross of the Red Eagle and other high honors were bestoyed on Koch by a grateful nation, but a wave of profound disappointment soon followed. In 1905 Koch was awarded the fifth Nobel Prize for Physiology or Medicine in recognition of his pioneering work on tuberculosis.

Tuberculin definitely discriminated between uninfected and infected subjects. While a healthy person had almost no reaction to the agent, tuberculous patients experienced severe reactions, including vomiting, fever, and chills. Human beings were much more sensitive to the agent than guinea pigs; when Koch tested tuberculin on himself he had a strong reaction indicative of his own latent infection. Immunologists later identified the tuberculin response as part of the complex immunological phenomenon called delayed-type hypersensitivity. As Koch predicted, tuberculin proved to be an indispensable diagnostic tool for detecting early and asymptomatic cases of tuberculosis. But Koch's claim that tuberculin would be more valuable in therapy than diagnosis proved tragically misleading. Patients from all over the world flocked to Berlin for the new wonder drug, but miraculous cures rarely occured, and some patients died from reactions to tuberculin. Public opinion rapidly turned against Koch when it became obvious that tuberculin was not a "magic bullet" for the cure of consumption. In 1891 Koch finally revealed the nature and means of preparation of tuberculin: it was simply a glycerine extract of tubercle bacilli. What came to be called the tuberculin fiasco could be regarded as a case study in the need for caution in evaluating potential remedies. One hundred years later, however, the AIDS crisis provided a powerful stimulus for calling into question the whole apparatus of controlled clinical trials. Like AIDS in the 1980s, tuberculosis in the 1880s was a mysterious, dreaded, and fatal disease. Clearly, withholding a drug that might cure a fatal illness would be a cruel and unethical act. The risks involved in offering futile remedies to victims of diseases as complex and uncertain in their course as tuberculosis or AIDS are more subtle. The threat of tuberculosis to society cannot, of course, be directly compared to AIDS, because everyone indulges in the major risk behavior for tuberculosis, which is breathing.

Up to the end of his life, Koch continued to hope that some improved form of tuberculin would provide a cure or a preventive vaccine, but this dream was never realized. While his work on tuberculosis was regarded as his greatest accomplishment, Koch and his students fought many other diseases, including cholera, malaria, rinderpest, and plague. Research on tropical diseases finally provided almost limitless opportunities for journeys to exotic locations. Eventually Koch's methods were exploited successfully in the search for the agents of typhus, diphtheria, tetanus, pneumonia, dysentery, relapsing fever, and other infectious diseases. Similarly, studies of the marked variations in the virulence of different varieties of tubercle bacilli led to attempts to create preventive vaccines. The most widely used vaccine, BCG (Bacille Calmette-Guérin), was derived from a live, attenuated strain produced by Albert Calmette (1863–1933) and coworkers. One hundred years later, with the development of antibiotic-resistant strains of tubercle bacilli, the World Health Organization recommended use of BCG despite continuing questions about the efficacy of the vaccine.

While colleagues generally considered Koch a patient, conscientious researcher and a master of bacteriological technique, after achieving his initial successes he seems to have become increasingly arrogant, dogmatic, and opinionated. Despite his professional success, his private life was apparently unfulfilling. Koch and his first wife Emmy had been estranged for some time when he met 17-year-old Hedwig Freiberg. Colleagues at the 1892 Congress of German Physicians admitted that there was more excitement about Koch's love affair with the young artist's model than the scientific papers. When he remarried in 1893, Koch was almost 50 and Hedwig was 20. Jealous of the success of his young colleagues, Koch lashed out at bacteriologists who disagreed with positions he had taken. For example, when Emil von Behring and the American bacteriologist Theobald Smith (1859–1934) warned that milk from cows with bovine tuberculosis was a danger to children, Koch insisted, quite incorrectly, that humans could not be infected with the bovine organism. Even at the 1908 International Tuberculosis Congress held in the United States, Koch continued to argue that the question of contaminated milk was irrelevant to human tuberculosis. When he returned to Berlin, Koch attempted to resume his research on tuberculin, but his health rapidly deteriorated and he died of heart disease in 1910. Two years after his death the Institute for Infectious Diseases was renamed the Robert Koch Institute. Eighty years later, according to the World Health Organization, tuberculosis, a disease that public health workers in the 1960s thought could be eradicated, was still claiming 3 million victims each year while causing an additional 8 million cases on a global basis. In the 1990s United States public health authorities found themselves dealing with significant increases in the incidence of tuberculosis, often in association with AIDS. The resurgence of tuberculosis in the 1990s, especially in virulent, multi–drug-resistant forms, highlights the vast chasm between what medical science in coooperation with systematic public health measures can do and the persistence of preventable infectious disease.

Robert Koch and his second wife Hedwig in Japanese costume, photographed in 1903

IMMUNOLOGY: THE ART OF SELF-DEFENSE

On the foundations built by Pasteur, Koch, and others, microbiology exploded and splintered into numerous subdisciplines. Physicians and scientists were fascinated with the search for the microbial agents of disease and the toxins that some of them produced, as well as the implications of these discoveries for prevention and therapy. As the nature of infectious diseases became more clearly understood, the ancient question of the body's own healing powers could also be investigated in the light of new theories. Given the ubiquity of the microbial hordes, it was obvious that the human body must be equipped with powerful weapons to fight off these microscopic invaders. The successful use of vaccinations for smallpox, anthrax, rabies, and other diseases provided the foundation for the new science of immunology. By the end of the nineteenth century, two schools of thought concerning the mechanism of immunity were competing for supremacy. According to the humoral theory, immunity was dependent on the induction of certain factors in the blood and fluids of the body. Other investigators contended that serum factors were insignificant compared to the role of certain cells that served as soldiers in the body's war against invaders. The humoral concept and studies of diphtheria and tetanus led to the discovery of bacterial toxins and the successful development of serum therapy. On the other hand, intriguing evidence for the doctrine of cellular immunity had already been established by the Russian zoologist Élie Metchnikoff (1845–1916).

In the twentieth century, immunology developed into a specialized discipline out of the convergence of concepts and methods representing major nineteenth- and twentieth-century fields, including microbiology and pathology, as well as embryology, evolutionary biology, and biochemistry. After a period dominated by biochemical approaches to immune function, by the end of the twentieth century, immunologists were once again considering cellular and integrative concepts close to those originally formulated by Metchnikoff. Thus, the roots of immunology, and other areas of modern biology, can be traced back to the commingling of traditional comparative embyology and evolutionary theory in the late 1860s.

Immunity is an ancient concept in law and in medicine. It was observed that during an epidemic, some people resisted the infection, and those who survived one episode of certain diseases were known to be safe from further attacks. Historians have argued that Metchnikoff should be considered the founder of modern immunology because he asked new questions about the nature of the active responses and defenses enlisted by an organism engaged in protecting itself against infection. Previous explanations and speculations about immunity dealt with the nature of the infective agent and its development but ignored the question of an active defensive mechanism in the host. The new principle that Metchnikoff formulated was that immunity to such infections was based on an active response by a specialized system of the body. Honored for his discovery of *phagocytes* (eating cells, from the Greek *phagos*, to eat, and *cyte*, cell) and the process of phagocytosis, Metchnikoff remained for many years the advocate of what seemed to be an eccentric and unproductive approach to immunology. While antibodies and antitoxins were creating new triumphs in therapeutics, Metchikoff continued to pursue his idiosyncratic studies of phagocytic cells. Indebted almost as much to the evolutionary theories of Charles Darwin as to the germ theories of Pasteur and Koch, Metchnikoff began his studies of inflammation with simple unicellular organisms and then pursued the mechanism of the response to injury and microbial invasion up the evolutionary ladder. Metchnikoff came to the concept of "active host immunity" from an interest not in medicine, but in embryology and evolution. Metchnikoff's interest in comparative embryology, the process of digestion, and his discovery of phagocytosis led him to a new approach to immunology. For Metchnikoff, phagocytosis was the key to functional integration of the activities of the organism throughout its life history; the phagocyte was responsible first for digestion in the embryo and for protecting the organism from infection.

The Russian-born zoologist began collecting specimens and natural history books as a child. He published his first article at the age of 16. While in Germany he bought Charles Darwin's *Origin of Species*, a work that exerted a profound influence on his scientific career. When Metchnikoff later reconstructed the development of his ideas, he claimed that he had decided to study zoology at the age of 18 because of his enthusiasm for Darwinian theory. Presumably, by then he had forgotten or had chosen to suppress the fact that after reading the *Origin* in 1863 he had written a very critical review, and he seems to have been especially skeptical of what he called Darwinian embryology. Later, in defending and explaining the phagocyte hypothesis, he apparently tried to reconstruct the steps that led to his discovery in a more logical manner.

While Metchnikoff was not the first to observe the presence of bacteria in blood cells, he was the first to see the link between intracellular digestion of microbes and the body's defense mechanisms. From the perspective of an embryologist and evolutionist, Metchnikoff drew attention to the nutritive, digestive role played by phagocytic cells in primitive organisms. In higher animals, which had specialized digestive organs, such cells could serve new functions. The white blood cells and their movements had previously been

Élie Metchnikoff

PROFESSEUR ROUX
Directeur de l'Institut Pasteur, Paris.

PROFESSEUR METCHNIKOFF
Sous-Directeur de l'Institut Pasteur, Paris.

PROFESSEUR CALMETTE
Directeur de l'Institut Pasteur, Lille.

Directors of the Pasteur Institute: Émile Roux, Élie Metchnikoff, and Albert Calmette

explained as accidental transport vehicles for pathogens; inflammation was thought to be a *reactive* phenomenon, rather than a *defensive* one. When Koch saw anthrax bacilli inside blood cells, he had concluded that the bacteria must have penetrated white blood cells in order to reproduce within them. In contast, Metchnikoff thought that the white blood cells must have devoured the parasites. Furthermore, contrary to prevailing opinion, Metchnikoff argued that the inflammatory response was not a purely passive and deleterious reaction to injury. Inflammation occurred when the white blood cells that were constantly wandering about the body seeking out foreign invaders and damaged, dying, or malignant cells were mobilized to defend the body. The heat, swelling, redness, and pain of the inflammatory reaction should be seen as part of a beneficial and protective but imperfect physiological process.

Metchnikoff did not react well to criticism or academic restrictions, and his relationships with mentors and colleagues were often stormy and acrimonious. Oscillating between extremes of enthusiasm and depression, his volatile nature led to several suicide attempts and to bitter and protracted battles with critics. Metchnikoff received his doctorate from Saint Petersburg and shared the prestigious Von Baer Prize for embryological research with his collaborator Alexander Kowalevsky. After a brief and turbulent appointment as an instructor at the new University of Odessa, he returned to Saint Petersburg in 1868, but this move did not significantly improve his status, and Metchnikoff succumbed to illness and depression. He was nursed back to health by Ludmilla Féderovitch, who became his first wife in 1869 despite the fact that she was so weak from tuberculosis that she had to be carried to the wedding in a chair. When Ludmilla died in 1873 Metchnikoff made two attempts at suicide. Thinking about a scientific riddle concerning the effect of natural selection on short-lived insects restored his will to live. He secured a faculty position at the University of Odessa and remained there from 1872 until 1882. In 1875 he married Olga Belokopitova, a young student, who became his devoted assistant and collaborator. Conditions at the university deteriorated rapidly after the assassination of Tsar Alexander II in 1881. Depressed by the growing power of reactionaries at the university and worried about Olga's battle with typhoid fever, Metchnikoff again attempted suicide. In order to make his death useful to science, he injected himself with the spirochete that causes relapsing fever. Instead of killing him, the inoculated illness restored his energy and optimism. During this troubled period he made his major discovery concerning the defensive role of the white blood cells.

As early as 1865, while studying planaria, simple ciliated worms which lack an intestinal cavity, Metchnikoff first observed the process of phagocytosis. Later studies of a disease of Daphnia, the common water flea, led Metchnikoff to conclude that the white blood cells played a role in protecting the organism from infectious materials. In this simple system the battle between the disease-causing fungus and the phagocytes of the transparent water flea could be studied for hours in simple, unstained preparations. These observations led to the idea that phagocytosis, or engulfment, might be a general biological defense mechanism as well as a way for cells to get their nourishment in accordance with the general rule "eat or be eaten." Disease in a system as simple as Daphnia revealed the battle between two living beings: the fungus and the phagocytes that acted as the destroyers of parasites. The phagocytes might then be regarded as the "bearers of nature's healing power." If this were the case, then an invasive injury might provoke migration of cells to the site. In 1882 while in Messina, Metchnikoff watched the mobile cells of a transparent starfish larva under the microscope. To test his phagocyte hypothesis, he introduced a thorn into a starfish larva; the next morning he found it surrounded by mobile cells like those that would gather around a splinter in a human finger. Metchnikoff presented his observations to Rudolf Virchow and Nicholaus Kleinenberg, who happened to be in Messina during the summer of 1883. When Metchnikoff published his study of disease in Daphnia, he called attention to Virchow's work and proposed that his studies of phagocytes should be seen as support for the basic theory of cellular pathology. Many critics of Metchnikoff's idea of the phagocyte as an active host defense were convinced that the all too common pus cells seen in wounds and postsurgical patients were dangerous traitor cells that spread the seeds of infection around the body—converting localized lesions into fatal systemic septic conditions—rather than valiant soldiers trying to fight off the enemy.

It is not surprising that Metchnikoff tried so hard to apply his theories to health and longevity considering the fate of his first wife, the near-fatal illness of his second wife, the early deaths of his three brothers, and his own medical problems. His studies of comparative anatomy and pathology led to the conclusion that senility was caused by the eventual transformation of friendly phagocytes into the agents of chronic degeneration. The damage wrought over time by misguided motile cells could be seen in every sign of aging from grey hair and wrinkles to weakness of muscle and memory. Intrigued by the possibility of reversing the chronic poisoning caused by intestinal microbes, Metchnikoff suggested disinfecting the large intestine with a hygienic diet that could neutralize the dangerous, toxin-producing bacteria harbored by this perfidious and useless organ. The yogurt industry was probably the chief benficiary of Metchnikoff's orthobiotic crusade, because frustrated by evidence that traditional purges and enemas did more damage to the host than to the noxious intestinal flora, he recommended the consumption of large quantities of fresh yogurt as a means of introducing beneficial lactic acid producing ferments into the digestive system.

Rudolf Virchow became one of Metchnikoff's major defenders. Like Metchnikoff, Virchow was concerned with disease processes: What are the interactions between host and pathogen. How does the pathogen cause disease. How does the host defend itself and become immune to further attacks? When Metchnikoff was attacked by the humoralists, Virchow defended the phagocyte theory by arguing that the factors in protective sera might be cell products. Metchnikoff's theory was actually proposed some 7 years before the humoralist school was established. Thus his immediate adversaries were not in a position to support an alternative theory of active immunity.

In 1888, Metchnikoff found a measure of peace when he joined the Pasteur Institute, bringing with him an unusual combination of experience and interests ranging from embryology and the evolution of digestive functions, to inflammation, immunity, and senescence. He collaborated on studies of syphilis with Émile Roux, but his main interest was applying his biological theories to issues of health and longevity and rebutting the humoralists who had demonstrated by the mid-1890s that bacterial toxins triggered an immune response. The growing influence of humoral pathology led to another period of severe depression and suicidal thoughts, but with the support of colleagues at the Pasteur Institute Metchnikoff was able to channel his energies into his 1891 lecture series on comparative pathology and a defense of an approach to health and longevity that he called *orthobiosis*. He later coined the term *gerontology* for a new scientific discipline dedicated to the study and control of senescence.

Unconventional to the point of eccentricity, throughout his career Metchnikoff nurtured a unique vision of the organism as intrinsically disharmonious. Physiological phenomena, therefore, represented not efforts to restore harmony, but a complex matter of integrating disharmonious elements. Many of Metchnikoff's publications, such as *The Nature of Man: Studies in Optimistic Philosophy*, relate to this concept and its applications in social and natural philosophy. Coming from a man who had made several suicide attempts, Metchnikoff's twentieth century philosophical writings are remarkably optimistic. Metchnikoff considered himself an atheist, but colleagues thought that his belief in the power of science to provide solutions to human suffering was a form of faith bordering on religious fanaticism. Recapturing the faith of his youth, Metchnikoff confidently predicted that scientific progress would improve the human condition; biomedical progress would not only save the young from infectious disease, but would benefit those confronted by the degenerative diseases of middle and old age.

When the humoralist school provided evidence of the protection provided by the passive transfer of immune serum, Metchnikoff rejected the posssibility that immunity involved both humoral factors and phagocytes. To account for the antibacterial activity in the serum demonstrated by Behring and Kitasato, Metchnikoff contended that it was the result of the breakup of white cells and release of their contents into the blood. In contrast, Behring and other humoralists argued that extracellular factors were the primary host defenses in the battle against invaders. Historians have argued that the bacteriologists and humoral immunologists were positivists and reductionists who attacked Metchnikoff for what appeared to them as an essentially descriptive and teleological approach. Despite the fact that Metchnikoff and Ehrlich shared the Nobel Prize in 1908 for their contributions to immunology, these scientists did not see their separate contributions as complementary aspects of a new integrated theory; both continued to maintain their own independent concepts concerning immunity, cancer, and senility. Perhaps to counter the exuberant claims of the humoral pathologists, in his Nobel Prize lecture Metchnikoff urged his audience to interpret his discoveries as an example of the "practical value of pure research." One year later he met with the great Russian writer Leo Tolstoy (1828–1910), who had told the story of Metchnikoff's eldest brother Ivan Ilyich in the tragic story *The Death of Ivan Ilyich*. The last years of his life were darkened by poor health and what he called the "criminal" return to "savagery" that had caused World War I and interrupted the vital war against disease that had been carried on at the Pasteur Institute.

Pasteur's work on preventive vaccines in the 1880s might be considered the beginning of the modern era of immunology; however, in the 1890s the discovery of antitoxins to diphtheria and tetanus by Behring and Kitasato raised great hope that a powerful and perhaps universal therapeutic approach to all infectious diseases was at hand. Like many other ambitious but impoverished men of his generation, Behring had obtained his medical degree at the Army Medical College in Berlin in exchange for a 10-year term in the Prussian Army. After serving as a member of Koch's staff at the Institute for Infectious Diseases, Behring obtained a professorship at Halle in 1894.

One year later he moved to the University at Marburg. In 1901 he was awarded the first Nobel Prize in Physiology or Medicine and, pleading ill health, gave up teaching in order to devote all his energies to the study of tuberculosis and the marketing of serum therapy. He founded the Behringwerke for the production of sera and vaccines in 1914. Behring's experience as a military doctor led him to speculate about the possibility of using "internal disinfectants" to kill the microbial agents of disease. By 1890 bacteriologists had established that both diphtheria and tetanus bacilli produced toxins that provoked the symptoms of the disease when injected into the body. At the Pasteur Institute, Émile Roux and Alexandre Yersin (1863–1943) found that although diphtheria bacteria generally remained localized in the throat, distant tissues were damaged by toxins released into the bloodstream. Behring and Kitasato showed that the blood of animals that had been challenged by a series of toxin injections produced *antitoxins*, that is, chemicals in the blood that neutralized the bacterial toxins. The antitoxins could be isolated from the serum of immunized animals, at least in crude form. Human tests of the antitoxins were carried out in 1891. Although horses and sheep could be used as antitoxin factories, Behring's antitoxin preparations were too variable and weak for commercial production or routine use. Afraid that further advances in serum therapy would be made by French scientists, much as Pasteur had used Koch's work on anthrax to develop protective vaccines, Behring sought help from Paul Ehrlich, who had used specific plant toxins to work out systematic methods of immunization. After an obscure series of transactions, it was Emil von Behring who reaped the benefits of commercial production of antitoxins by Hoechst, the German chemical company that had produced Koch's tuberculin.

When Behring gave his Nobel Prize lecture on "Serum Therapy in Therapeutics and Medical Science," he seemed to be justified in predicting that further developments in harnessing active and passive immunity would soon lead to a complete triumph over the infectious diseases. But before his death in 1917, profound disillusionment had set in and the promising science of immunology had entered a period immunologists would later recall as their Dark Ages. Even the originality of scientific studies of toxins seemed to be questionable, as indicated by reports about certain traditional healers in Africa and other parts of the world who told nineteenth century European observers that they could protect people against venomous snakes with a potion containing snake heads and ants or ant eggs. Presumably, formaldehyde-related chemicals in these preparations detoxified toxins; in 1928 Léon Ramon (1886–1963) demonstrated that formaldehyde could be used to modify diphtheria toxin. It should be noted that 100 years after Behring introduced serum therapy, genetic engineers were exploiting the "naturally engineered" properties of various bacterial toxins in order to create hybrid molecules linking toxins to specific antibodies. Such novel immunotoxins have been referred to as "poisoned arrows" or "smart bombs" which, at least in theory, could deliver more fire power than the "charmed bullets" first synthesized in the laboratory of Paul Ehrlich, the founder of chemotherapy.

Paul Ehrlich

Paul Ehrlich studied medicine at the Universities of Breslau, Strasbourg, and Freiburg-im-Breisgau, and finally graduated from Leipzig in 1878. His cousin Karl Weigert (1845–1904), a pioneer in histology, served as a model and inspiration as Ehrlich began his own researches. His doctoral thesis, "Contributions to the Theory and Practice of Histological Staining," reflected this influence and contained the germs of many of Ehrlich's later ideas. Soon after Koch demonstrated his discovery of the tuberculosis bacillus, Ehrlich developed a more efficient staining method, which took advantage of the acid-fast characteristics of the microbe. Using this method, Ehrlich was able to diagnose his own asymptomatic infection, for which he prescribed a successful regimen of travel. On his return to Germany he became a reader in medicine at the University of Berlin and joined Koch and Behring at the Institute for Infectious Diseases. His association with Behring and the development of efficient methods for producing diphtheria antitoxin proved to be a very bitter experience. Eventually, the success of the antitoxins led the Prussian minister of state, Friedrich Althoff, to establish a new institute for serology and serum testing; Ehrlich became director in 1896. Although he had only a small laboratory and a very small budget, Ehrlich and his associates carried out a prodigious amount of work. Between 1877 and 1914, Ehrlich published 232 papers and books. He often said that he could work in a barn as long as he had a water tap, a flame, and some blotting paper. He might have added cigars and mineral water to the list, for he practically lived on these two items. His thought processes were so dependent on smoking that he always carried an extra box of cigars under his arm. As director, Ehrlich managed his staff with a firm hand. Every day each person was given a series of index cards in various colors containing instructions for the day's experiments. Any assistant who dared to ignore the instructions was severely and publicly reprimanded.

As a student, Ehrlich had conducted systematic studies of the aniline dyes, which led to the discovery of different types of white blood cells and a means of classifying the leukemias in terms of the prevailing cell type. During a study of lead poisoning, Ehrlich noticed that the same organs that were most affected by the poison in the living being were the ones that most avidly accumulated lead when tissues were stained with lead solutions. Such observations led him to emphasize the importance of specific interactions between particular chemicals and cells, cell components, and microbes. Like Pasteur, Ehrlich did not separate his theoretical interests from the practical problems to be found in immunology, toxicology, pharmacology, and therapeutics. Indeed, Ehrlich suggested that there was a direct link between the

Paul Ehrlich

growth of the chemical industry and progress in histology that would lead to selective staining. On this foundation it would be possible to establish a better understanding of the body's natural immunological defenses, develop means of augmenting them, and create a new experimental pharmacology dedicated to the design of "magic bullets," that is, compounds that could specifically interact with pathogens without damage to normal cells.

While investigating serum-mediated immunity, Ehrlich discovered that the injection of poisons such as ricin led to the production of antibodies, nature's own magic bullets. To explain the development of antibodies, Ehrlich proposed his ingenious "side-chain" theory of immunity. Ehrlich's side-chain concept is generally considered the first theory of antibody formation to offer a chemical explanation for the specificity of antibody-antigen interactions. Antibody-producing cells, according to Ehrlich's theory, were studded with side chains, that is, groups that could combine with antigens, such as diphtheria toxin. When a specific antigen entered the body, it reacted with a specific side chain; this stimulated the antibody-producing cell to maximal production of the appropriate side chain. As the side chains increased in number, they were released from the surface of the cell and entered the circulation, where they were able to bind to their antigen. The fit between antigen and antibody was as specific as that between a lock and a key, although it was presumably the result of accident rather than design.

Whereas antibodies are nature's own magic bullets, chemotherapy was an attempt to imitate nature by creating synthetic magic bullets. In 1906, while expectations for serum therapy were still high, Ehrlich introduced the term "immunotherapy," but as the limitations of this approach became clearer, he shifted his emphasis to chemotherapy. If the body could not produce effective antibodies for every pathogen, then the goal of biomedical science must be to design and synthesize chemotherapeutic agents that would act to supplement and sustain the natural defense mechanisms. Tests of a vast array of arsenical compounds led to the discovery by Ehrlich and his associate Sahachiro Hata (1872–1938) of Preparation 606 (subsequently named salvarsan), which was effective in the treatment of syphilis. Despite charges that the drug was itself a dangerous poison, until penicillin became available after World War II, salvarsan, in conjunction with mercury and bismuth, remained the only effective remedy for syphilis.

Despite the initial success of serum therapy and salvarsan, the search for other magic bullets was largely unsuccessful until Gerhard Domagk's (1895–1964) work on dye substances led to the discovery of the sulfa drugs. While a pragmatic approach to therapeutics was to be expected from the director of research in experimental pathology and bacteriology at the Ger-

Gerhard Domagk

man chemical firm I. G. Farben, Domagk's research program was based on attempts to use dyes as a means of elucidating the vital functions of micro-organisms. Like many scientists of his generation, Domagk had been profoundly influenced by his experiences during World War I. As a member of the Sanitary Service, he had confronted the horrors of cholera, typhus fever, dysentery, battlefield surgery, and wound infection. Like Paul Ehrlich, Domagk systematically searched for chemical substances that could destroy bacteria without damaging the patient. When he began his studies in the early 1930s, few effective drugs were available for microbial infections, and none of them were useful for coccal infections. Using the methods developed by Ehrlich, Domagk tested many chemicals in infected animals. Systematic tests of the aniline dyes led to the discovery that prontosil, a red dye with no in vitro antibacterial activity, protected mice from streptococcal infections. In a typical experiment, using mice as experimental animals, the lethal dose of a particular strain of streptococci or staphylococci was determined. Mice were then inoculated with 10 times the lethal dose and half of the animals were given prontosil. By 1933 prontosil had been tested in humans with staphylococcal and streptococcal infections, but Domagk's results were not published until 1935, presumably because Farben was seeking patent protection for the new miracle drug. When Domagk's report was published, researchers in France discovered that prontosil was broken down in the animal body and its antibacterial activity was entirely due to the sulfonamide portion of the molecule. Not only did this finding explain prontosil's lack of antibacterial activity in the test tube, it gave researchers access to a new family of so-called sulfa drugs not covered by Farben's 1932 patent on prontosil. Sulfanilamide, which did not have the disadvantage of being a messy red dye, had been synthesized in 1908.

In 1939 Domagk won the Nobel Prize for Physiology or Medicine, but Nazi officials would not allow him to accept it. The perverted Nazi ideology and National Socialist policies, which effectively isolated Germany from the international research community and forced Jewish scientists to flee to England and America, also caused that nation to forfeit its leadership role in the development of chemotherapeutic agents. By 1947, when Domagk finally received the medal and presented his acceptance speech, at least 5000 sulfanilamide derivatives had been laboriously synthesized and tested. Unfortunately, only about 20 of these compounds were safe and effective enough to be therapeutically useful. Disillusioned chemists came to realize that finding a magic bullet was no more probable than winning a lottery. Moreover, the sulfa drugs had been so widely, indiscriminately, and inappropriately used that resistant strains of bacteria seemed to appear even more rapidly than new wonder drugs. By the end of World War II, the former miracle drugs were considered largely obsolete. In his Nobel Lecture, Domagk suggested that the declining efficacy of the sulfa drugs might have been accelerated by the stress and upheavals caused by the war. His warning that future miracle drugs would also lose their therapeutic efficacy if physicians failed to confront the factors that led to the development and spread of drug-resistant microorganisms was, of course, fully vindicated by the 1980s.

A magic bullet more powerful than sulfanilamide had been discovered in 1928, but it was not isolated and tested in humans until the 1940s. According to popular mythology, the antibiotic era began with Alexander Fleming's (1881–1955) chance discovery that the mold *Penicillium notatum* produced a substance inimical to the growth of certain bacteria. Of course the path that led to the development of powerful antibiotics from obscure molds was much more convoluted. Actually, several scientists had called attention to the implications of "antibiosis," the struggle for existence among different microorganisms, as early as the 1870s. When Fleming delivered his Nobel Prize lecture in 1945, he suggested that the discovery of antibiotics had taken so long because bacteriologists of his generation had simply taken the existence of antibiosis for granted rather than as a riddle worth solving. It is true that Fleming discovered the effect of the mold *Penicillium notatum* on bacteria in 1928 and that within a year he had demonstrated that crude preparations of penicillin that seemed to be harmless to higher animals could kill certain bacteria. However, Fleming did not seem to anticipate the therapeutic potential of penicillin. Perhaps the failure of lysozyme, his earlier discovery, to be therapeutically useful had discouraged him almost as much as his inability to purify the seemingly unstable antibacterial activity of his mold juice.

Serving in the Royal Army Medical Corps during World War I stimulated Fleming's interest in antibacterial agents, but experience in fighting wound infection suggested that antiseptics and disinfectants posed more of a threat to human tissues than to pathogenic bacteria. After the war, Fleming returned to the Inoculation Department at St. Mary's Hospital in London, which was directed by the eminent but highly eccentric bacteriologist Sir Almroth Wright (1861–1947). Wright was immortalized by George Bernard Shaw in *The Doctor's Dilemma*, but the failure of his experimental inoculations and vaccines gave him the title "Sir Almost Wright." Proceeding in his own idiosyncratic way, in 1922 Fleming discovered lysozyme, a powerful antibacterial agent found in nasal secretions, tears, saliva, and egg white. While lysozyme plays a role in the body's natural defenses against disease, Fleming and his contemporaries could not find a way to press it into service

Alexander Fleming

as a magic bullet. Six years later, Fleming noted that a mold growing on a petri dish seeded with staphylococci had destroyed the bacteria in its immediate vicinity. Fleming decided that this was a case of antibiosis worth pursuing. In 1928, the year that Fleming discovered penicillin, textbooks of therapeutics and pharmacology were still recommending the heavy metals and aromatics that had so long been used in the treatment of wounds, along with relatively new chemicals such as carbolic acid, hydrogen peroxide, iodoform, and the hypochlorites. But medical authorities generally agreed that while it might be possible to disinfect an external lesion, it would be impossible to disinfect the tissues of the body as a whole, because any drug circulating in the bloodstream in a concentration high enough to kill bacteria would inevitably destroy the tissues that harbored them.

Contamination of bacteriological materials with molds is a not uncommon accident, generally regarded as a sign of poor laboratory technique, but Fleming liked to argue that the discovery of both lysozyme and penicillin could not have been made in an overly meticulous setting. Indeed the plate on which the penicillin effect was discovered had been left in the laboratory when Fleming went away on vacation. An obsessively neat researcher would presumably have cleaned the laboratory before leaving. Attempts to recreate the discovery suggest that a spore of the relatively rare mold *Penicillium notatum* must have fallen on a plate seeded with staphylococci and proceeded to grow and secrete penicillin. The bacteria, however, were unable to start multiplying until warm weather triggered their growth. Penicillin, which only kills growing bacteria, caused inhibition of growth in the vicinity of the mold. Fleming and his students tested crude preparations of penicillin against white blood cells and various species of bacteria with some success, but he did not perform the animal tests that could have revealed penicillin's remarkable therapeutic power. Unfortunately, the bacteriacidal agent appeared to be unstable and difficult to purify. Not until the sulfa drugs had been successfully exploited and World War II made chemotherapy a high priority would researchers isolate the active ingredient in Fleming's old "mold juice."

In 1938, Howard Walter Florey (1898–1968), director of the Sir William Dunn School of Pathology at Oxford University, and Ernst Boris Chain (1906–1979), a refugee from Nazi Germany, began a systematic evaluation of naturally occurring antibacterial agents. Their partially purified preparations of penicillin protected mice against streptococci, staphylococci, and several other pathogenic bacteria. Finally, in the 1940s Florey and Chain, rather than Fleming, took penicillin out of the test tube and put it into patients. The first patient treated, a 43-year old man with a mixed infection of staphylococci and streptococci, improved when treatment began, but died because supplies of the drug were rapidly exhausted. Clearly, penicillin production was a wartime priority, but Britain's resources were too strained by the war to support extensive research and development. With American support, industrial production of penicillin was accomplished within 2 years. Clincial trials demonstrated the effectiveness of the drug against infections caused by staphylococci, streptococci, and pneumococci. Because penicillin was also effective in the treatment of syphilis and gonorrhea, the miracle drug could treat soldiers wounded in battle as well as those wounded in the brothels. Self-righteous moralists warned that penicillin would stimulate immorality. When data on mortality rates for military personnel were studied, some authorities suggested that penicillin and the sulfonimides had had a major impact on the mortality rate among both the wounded and the sick. By 1948, penicillin, which had been in limited supply during the war, was widely available thoughout the world. Because of the problems associated with the care and feeding of the mold and the difficulties in isolating pure penicillin, some attempts were made to achieve the chemical synthesis of the drug. It soon became clear, however, that this was an impractical approach. The chemical synthesis of penicillin was finally accomplished by John C. Sheehan in 1957. Battles over patent rights and commercial considerations proved to be more convoluted than the path from Fleming's "chance discovery" to Sheehan's total synthesis. The success of penicillin stimulated the growth of the pharmaceutical industry and the search for other "miracle molds" in samples of dirt and putrefying materials from every corner of the earth. During the postwar golden age of antibiotics, streptomycin, neomycin, chloramphenicol, aureomycin, erythromycin, nystatin, and other valuable drugs were discovered. Despite optimistic predictions that more active and less toxic agents would continue to appear, the golden age of discovery seemed to have ended by about 1960. Moreover, little or no progress was made in developing antiviral agents. Worse yet, the careless use of essential antibiotics revealed dangerous side effects and allowed the proliferation of drug-resistant bacteria.

INVISIBLE MICROBES AND THE SCIENCE OF VIROLOGY

As bacteriologists successfully exploited the methods of Pasteur and Koch to identify the specific causative agents of epidemic infectious diseases, it seemed reasonable to accept Pasteur's assertion that every virus is a microbe. But the meaning of the word *virus* as used by Pasteur was quite different from the modern definition of a virus as a minute entity composed of a core of nucleic acid and an envelope of protein that can only reproduce within living cells. Originally, the Latin word *virus* meant something like *slime,* which was rather unpleasant but not necessarily harmful. Medical writers, however, tended to use the term in reference to obscure infectious agents, or things that were dangerous to health, such as a poison, venom, or contagion. In this sense, both the first-century Roman encyclopedist Celsus and the nineteenth-century chemist Louis Pasteur could allude to the virus of rabies.

Long before the nature of viruses was understood, traditional practitioners, at least in Africa, Asia, India, and the Mediterranean region, had discovered means of combating certain viral diseases. In the 1720s, England and New England learned about the ancient use of inoculation from Lady Mary Wortley Montagu (1689–1762) and the Reverend Cotton Mather (1663–1728), respectively. Lady Mary, wife of the British Ambassador to the Turkish court, had been introduced to the Turkish method of inoculation against smallpox while living in Constantinople. Cotton Mather learned about smallpox inoculation from a young African slave given to him by members of his congregation. Inoculation, or ingrafting, involved taking fresh material from smallpox pustules and inserting it into cuts in the skin of a healthy individual. Generally, inoculated smallpox was less dangerous than the naturally acquired infection, and those who survived were rewarded with lifelong immunity. In 1798 Edward Jenner (1749–1823) investigated the folk belief that the mild disease known as cowpox, or *variolae vaccinae*, provided protection against smallpox. Experiments proved that inoculations of cowpox did indeed confer immunity to smallpox. Jenner named the new procedure *vaccination* in honor of the cow. Thus, when Pasteur discovered that attenuated cultures of the microbe that causes chicken cholera could immunize experimental animals against the disease, he assumed that the phenomenon was analogous to the use of cowpox vaccine to protect against the virus of smallpox.

During the nineteenth century, therefore, the term virus acquired a more specific meaning within the germ theory of disease as an agent with infectious properties. By the beginning of the twentieth century, the techniques of microbiology were sufficiently advanced for scientists to state with some degree of confidence that certain diseases were caused by specific pathogenic microbes. Still, the pathogens for some diseases refused to be isolated by the most sophisticated bacteriological techniques available. When such diseases seemed to be associated with submicroscopic infectious agents, the term virus became increasingly restricted to this category of disease. Working on many exotic tropical diseases, Koch himself came to the conclusion that not all infectious diseases could be attributed to conventional bacteria. Some infectious agents seemed to have complex life cycles and growth requirements that could not be met by conventional laboratory techniques. Clearly some diseases were caused by agents like protozoa that were larger than bacteria, but it was also possible that some unconventional pathogens might be smaller than bacteria.

Providing unequivocal proof that unconventional agents, whether larger or smaller than bacteria, caused disease was difficult because it was virtually impossible to grow these agents in laboratory cultures. Nevertheless, eventually the rickettsias, chlamydias, mycoplasmas, and brucellas were added to the ranks of the pathogenic bacteria, fungi, and protozoa. This still left the infectious agents that caused a large number of important diseases of humans and animals in the unsatisfactory category of "filterable viruses." The criterion of filterability was the outcome of work conducted by one of Pasteur's associates, Charles Chamberland (1851–1908). In 1884 Chamberland published a paper describing the use of a porous porcelain vase to prepare what he called "physiologically pure water." Chamberland's filter was used to separate visible microorganisms from their culture medium and to prepare bacteria-free liquids. Perhaps sharing Pasteur's obsession with bacterial contamination, Chamberland suggested using filtration to prepare pure drinking water. Chamberland was also instrumental in the development of the autoclave, a device for sterilizing materials by means of steam heat under pressure. The availability of these techniques made it possible to establish, by means of operational criteria, a new category of infectious agents, the *invisible-filterable viruses*. Many important diseases of humans and animals were thought to be caused by such agents, but their nature remained obscure; they were defined operationally as "filterable" and "invisible" microbes. But technique-based negative criteria provided little insight into the genetic and biochemical nature of viruses. It was a disease of the tobacco plant that established Adolf Eduard Mayer, Martinus Beijerinck, and Dimitri Ivanovski as the founders of virology.

In 1892 Dimitri Ivanovski (1864–1920) read a paper to the Academy of Sciences of St. Petersburg on tobacco mosaic disease. Using the filter method of Chamberland, Ivanovski found that a filtered extract of infected tobacco leaves produced the same disease in healthy plants as an unfiltered extract. He thought that the simplest explanation of his observations was the presence of a toxin in the filtered sap or that an unusual bacteria had passed through the pores of the filter. Apparently unaware of Ivanovski's work, Martinus Willem Beijerinck (1851–1931) also reported that a filterable agent caused tobacco mosaic disease. Beijerinck concluded that a *contagium vivum fluidum,* which could pass through the finest filter and reproduce only in living plant tissue, was the cause of tobacco mosaic disease. Based on reports in the botanical literature, Beijerinck thought many other plant diseases were caused by similar soluble germs. According to his Russian biographers, Ivanovski was the true founder of virology, but because he was too shy and self-effacing to promote his innovative concepts, the rest of the scientific world tended to think of him as "Ivanovski the obscure." Unable to isolate or culture the invisible tobacco microbe, Ivanovski abandoned this line of research and turned to studies of alcoholic fermentation.

Beijerinck first attempted to isolate tobacco mosaic virus while working on his doctoral degree under Adolf Mayer (1843–1942) at the University of Leiden. As professor of microbiology at the Polytechnical School in Delft, Beijerinck initiated studies of various problems in botany, chemistry, and genetics before returning to the puzzle of tobacco mosaic disease. In thinking about the way in which filtered plant sap could be used to transmit the disease to a large series of plants, he concluded that the disease must be caused by an agent that needed the living tissues of the plant in order to reproduce itself. In other words, the virus could be thought of as some kind of *contagium vivum fluidum*, or contagious living fluid. The crucial difference between a virus and other microbes might not, therefore, be a matter of size; some microbes might be obligate parasites of living organisms that could not be cultured in vitro on any cell-free culture medium.

Similar observations were made by Friedrich Loeffler (1862–1915) and Paul Frosch (1860–1928) in their studies of foot-and-mouth disease, the first example of a filterable virus disease of animals. Attempts to culture bacteria from the fluid in lesions in the mouths and udders of sick animals were unsuccessful. The apparently bacteria-free filtered fluid could, however, be used to transmit the disease to experimental animals even after passage through a Chamberland filter. Experimentally infected animals were capable of transmitting the disease to other animals. These experiments suggested that a living agent, rather than a toxin, must have been in the filtrate. Loeffler and Frosch suggested that other infectious diseases, such as smallpox, cowpox, and cattle plague, were also caused by very minute organisms. Nevertheless, they continued to think of the infectious agent as a very small and unusual microbe that had not yet been cultured in vitro rather than a novel entity that could *only* multiply within the cells of its host.

Within a few years of these studies, filterable viruses were suspected of being the cause of various plant, animal, and human diseases. In 1915 Frederick William Twort (1877–1950) reported that even bacteria could be the victims of diseases caused by invisible viruses. As Jonathan Swift (1667–1745) had predicted in a satirical poem on the use of the microscope, or flea-glass, scientists had found that the "smaller fleas" that preyed on fleas had "smaller still" to bite them. While trying to grow viruses in artificial medium, Twort noted that colonies of certain contaminating bacteria growing on agar became glassy and transparent. If pure colonies of this micrococcus were touched by a tiny portion of material from the glassy colonies, they too became transparent. Twort later became obsessed with speculative work on the possibility that bacteria evolved from viruses, which had developed from even more primitive forms. Working independently, Félix d'Hérelle (1873–1949) also called attention to the existence of bacterial viruses. In 1917 he published his observations on "An invisible microbe that is antagonistic to the dysentery bacillus." Because the invisible microbe could not grow on laboratory media or heat-killed bacilli, but grew well in a suspension of washed bacteria in a simple salt solution, d'Hérelle concluded that the antidysentery microbe was an *obligate bacteriophage*, that is, an eater of bacteria. Bacterial viruses were sometimes called Twort-d'Hérelle particles. D'Hérelle had obtained the invisible microbe from the stools of patients recovering from bacillary dysentery. When an active filtrate was added to a culture of Shiga bacilli, bacterial growth soon ceased and bacterial death and lysis (dissolution) followed. A trace of the lysate produced the same effect on a fresh Shiga culture. More than 50 such transfers gave the same results, indicating that a living agent was responsible for bacterial lysis.

Speculating on the general implications of the phenomenon he had discovered, d'Hérelle predicted that bacteriophages would be found for other pathogenic bacteria. Although the natural parasitism of the invisible microbe seemed species specific, d'Hérelle believed that laboratory manipulations could transform bacteriophages into microbes of immunity with activity against various germs. The hope that bacteriophages could be trained as weapons in the war on bacteria was never realized. Until pressed into service as the experimental animal of molecular genetics, bacterial viruses remained a laboratory curiosity. Viruses became the "living molecules" that served as a kind of scientific Rosetta stone and made it possible to decipher the instructions for cell structure and function.

Wendell M. Stanley (1904–1971) reported in 1935 that he had isolated and crystallized a protein having the infectious properties of tobacco mosaic virus. Starting with a ton of infected tobacco leaves and utilizing the tedious conventional purification techniques then available, Stanley obtained a spoonful of crystalline material which essentially consisted of protein, but still possessed the attributes of the tobacco mosaic virus found in cell sap. Stanley concluded that tobacco mosaic virus was an autocatalytic protein that seemed to require the presence of living cells to accomplish its own multiplication. Further work showed that the virus was not a pure protein but rather a combination of protein and nucleic acid. Stanley won the Nobel Prize in Chemistry in 1946 for this provocative demonstration that an apparently pure chemical substance could behave as if alive. Generally ignoring metaphysical debates about the nature of viruses and life, biochemists and geneticists described viruses as particles composed of an inner core of nucleic acid enclosed in a protein overcoat. Perhaps the most appropriate commentary on the great philosophical debates was Nobel Laureate André Lwoff's paraphrase of a famous line by Gertrude Stein: "Viruses should be considered viruses because viruses are viruses." While viruses may still be considered "living molecules" on the borderline between macromolecules, genes, and cells, research on diseases that have been attributed to slow viruses, viroids, and prions suggest that the submicroscopic world may well be populated by other obscure, intriguing, and perhaps dangerous creatures.

SUGGESTED READINGS

Ackerknecht, E. H. (1953). *Rudolf Virchow: Doctor, Statesman, Anthropologist*. Madison, WI: University of Wisconsin Press.

Adelberg, E. A. (1960). *Papers on Bacterial Genetics*. Boston: Little, Brown.

Ainsworth, G. C. (1981). *Introduction to the History of Plant Pathology*. New York: Cambridge University Press.

Ainsworth, G. C. (1986). *Introduction to the History of Medical and Veterinary Mycology*. New York: Cambridge University Press.

Andrews, M. L. A. (1977). *The Life That Lives on Man*. New York: Taplinger.

Applequist, J. (1987). Optical activity: Biot's bequest. *American Scientist 75:* 59–68.

Baer, George M., ed. (1975). *The Natural History of Rabies*. New York: Academic Press.

Barlow, C., and Barlow, P. (1971). *Robert Koch*. Geneva: Heron.

Bäumler, E. (1983). *Paul Ehrlich: Scientist for Life*. Trans. by G. Edwards. New York: Holmes & Meier.

Beauchamp, D. E. (1989). *The Health of the Republic. Epidemics, Medicine, and Moralism as Challenges to Democracy*. Philadelphia, PA: Temple University Press.

Bloch, M. (1973). *The Royal Touch: Sacred Monarchy and Scrofula in England and France*. Trans. by J. E. Anderson. London: Routledge & Kegan Paul.

Brock, T. D., ed. (1961). *Milestones in Microbiology*. Englewood Cliffs, NJ: Prentice-Hall.

Brock, T. D. (1988). *Robert Koch. A Life in Medicine and Bacteriology*. Madison, WI: Science Tech.

Brock, W. H., McMillan, N. D., and Mollan, R. C., eds. (1981). *John Tyndall. Essays on a Natural Philosopher*. Dublin: Royal Dublin Society.

Bulloch, W. (1938). *The History of Bacteriology*. New York: Dover (reprinted 1979).

Burnet, Sir MacFarlane, and White, D. O. (1972). *Natural History of Infectious Diseases*. 4th ed. New York: Cambridge University Press.

Calmette, A. (1923). *Tubercle Bacillus Infection and Tuberculosis in Man and Animals*. Baltimore, MD: Williams and Wilkins.

Chanock, Robert M. and Lerner, Richard A. eds. (1984). *Modern Approaches to Vaccines: Molecular and Chemical Basis of Virus Virulence and Immunogenicity*. New York: Cold Spring Harbor Laboratory.

Clark, P. F. (1938). Alice in virusland. *Journal of Bacteriology 36:* 223–1241.

Cole, Leonard A. (1988). *Clouds of Secrecy. The Army's Germ Warfare Tests over Populated Areas*. Totowa, NJ: Rowan and Littlefield.

Collard, P. (1976). *The Development of Microbiology*. Cambridge: Cambridge University Press.

Conant, J. B., ed. (1952). *Pasteur's Study on Fermentation*. Cambridge, MA: Harvard University Press.

Conant, J. B., ed. (1953). *Pasteur's and Tyndall's Study of Spontaneous Generation*. Cambridge, MA: Harvard University Press.

Diener, T. O., ed. (1987). *The Viroids*. New York: Plenum.

Doetsch, R. N., ed. (1960). *Microbiology: Historical Contributions from 1776–1908*. New Brunswick, NJ: Rutgers University Press.

Douglass, Joseph D., and Livingstone, Neil C. (1987). *America the Vulnerable. The Threat of Chemical and Biological Warfare*. Lexington, MA: Lexington.

Dowling, H. F. (1977). *Fighting Infection: Conquests of the Twentieth Century*. Cambridge, MA: Harvard University Press.

Dubos, R. (1960). *Louis Pasteur: Free Lance of Science*. New York: Da Capo Series in Science.

Dubos, R. (1988) *Pasteur and Modern Science*. Madison, WI: Science Tech Publishers.

Dubos, R., and Dubos, J. (1952). *The White Plague: Tuberculosis, Man and Society*. Boston, MA: Little, Brown, & Co.

Duclaux, E. (1920). *Pasteur, The History of a Mind*. Trans. by E. F. Smith and F. Nedges. Philadelphia, PA: Saunders.

Ehrlich, P. (1956-60). *The Collected Papers of Paul Ehrlich*. Ed. by F. Himmelweit. 3 volumes. London: Pergamon.

Farley, J. (1977). *The Spontaneous Generation Controversy from Descartes to Oparin*. Baltimore, MD: Johns Hopkins University Press.

Fenner, F., and Gibbs, A., eds. (1988). *A History of Virology*. Basel: Karger.

Foster, W. D. (1979). *History of Medical Bacteriology and Immunology*. London: William Heineman.

Fox, S. W., ed. (1965). *The Origins of Prebiological Systems*. New York: Academic Press.

Geison, G. L. and Secord, J. A. (1988). Pasteur and the process of discovery. The case of optical isomerism. *ISIS 79:* 6–36.

Grafe, A. (1991). *A History of Experimental Virology*. Trans. by E. Reckendof. Berlin: Springer-Verlag.

Hamlin, C. (1990). *A Science of Impurity: Water Analysis in Nineteenth Century Britain*. Berkeley, CA: University of California Press.

Hare, R. (1970). *The Birth of Penicillin*. London: George Allen & Unwin.

Henle, J. (1938). *Jacob Henle: On Miasmata and Contagia*. Trans. by George Rosen. Baltimore, MD: Johns Hopkins University Press.

Himmelweit, F., and Marquardt, M. (1957). *The Collected Papers of Paul Ehrlich*. New York: Pergamon.

Hitchens, A. P. and Keilind, M. C. (1939). The introduction of agar-agar into bacteriology. *Journal of Bacteriology 37:* 485–493.

Hughs, S. S. (1977). *The Virus: A History of the Concept*. New York: Science History Publications.

Iterson, C. Van, Den Dooren de Jong, L. E., and Kluyver, A. J. (1983). *Martinus Willem Biejerinck: His Life and His Work*. Madison, WI: Science Tech Publishers.

Kaprowski, H. and Plotkin, S., eds. (1985). *World's Debt to Pasteur*. Wistar Sympoosium Series, Vol. 3. New York: Alan R. Liss.

Kay, Lily E. (1986). W. M. Stanley's Crystallization of the Tobacco Mosaic Virus, 1930–1940. *ISIS 77:* 450–472.

Knapp, Vincent J. (1989). *Disease and Its Impact on Modern European History*. New York: Mellen.

Koch, R. (1932). *The Aetiology of Tuberculosis*. Trans. of the 1882 paper by Dr. and Mrs. Max Pinner. Intro. by A. K. Krause. New York: National Tuberculosis Association.

Koch, R. (1890). *Investigations into the Etiology of Traumatic Infective Diseases*. Translated by W. W. Cheyne. London: New Sydenham Society.

Latour, B. (1988). *The Pasteurization of France*. Trans. by A. Sheridan and J. Law. Cambridge, MA: Harvard University Press.

Lechavalier, H. (1972). Dimitri Iosifovich Ivanovski (1864–1920). *Bacteriological Reviews 36:* 135–146.

Lechevalier, H. A., and Solotorovsky, M. (1974). *Three Centuries of Microbiology*. New York: Dover.

Marquardt, M. (1949). *Paul Ehrlich*. New York: Schuman.

Metchnikoff, É. (1905). *Immunity in Infective Diseases*. Trans. by F. G. Binnie. London: Cambridge University Press.

Metchnikoff, É. (1939). *The Founders of Modern Medicine. Pasteur. Koch. Lister*. New York: Walden Publications.

Metchnikoff, É. (1968). *Lectures on the Comparative Pathology of Inflammation*. New York: Dover.

Metchnikoff, O. (1921). *Life of Élie Metchnikoff, 1845–1916*. Trans. by R. Lankester. Boston: Houghton Mifflin.

Moschcowitz, E. (1948). Louis Pasteur's credo of science: his address when he was inducted into the French Academy. *Bulletin of the History of Medicine 22:* 451–466.

National Academy of Sciences (1966). *Extraterrestrial Life: An Anthology and Bibliography*. Washington, DC: National Academy of Sciences-National Research Council.

Nutton, Vivian (1990). The reception of Fracastoro's theory of contagion. *Osiris 6*: 196–234.

Oparin, A. I. (1964). *The Chemical Origin of Life*. Springfield, IL: Charles C. Thomas.

Palladino, P. (1990). Stereochemistry and the nature of life. Mechanist, vitalist, and evolutionary perspectives. *ISIS 81*: 44–67.

Parascandola, John, ed. (1980). *The History of Antibiotics: A Symposium*. Madison, WI: American Inst. Hist. Pharmacy.

Pasteur, Louis (1968). *Correspondence of Pasteur and Thullier Concerning Anthrax and Swine Fever Vaccinations*. Trans. and ed. by R. M. Frank and D. Wrolnowska. Tuscaloosa, AL: University of Alabama Press.

Porter, J. R. (1972). Louis Pasteur Sesquicentennial (1822–1972). *Science 178*: 1249–1254.

Robinson, Victor (1921). *The Life of Jacob Henle*. New York: Medical Life.

Root-Bernstein, R. S. (1989). *Discovering*. Cambridge, MA: Harvard University Press.

Sagan, D., and Margulis, L. (1988). *The Garden of Microbial Delights. A Practical Guide to the Subvisible World*. New York: Harcourt.

Scott, Andrew (1985). *Pirates of the Cell. The Story of Viruses from Molecule to Microbe*. New York: Basil Blackwell.

Sheehan, J. C. (1982). *The Enchanted Ring: The Untold Story of Penicillin*. Cambridge, MA: MIT Press.

Silverstein, A. M. (1989). *A History of Immunology*. New York: Academic Press.

Stanley, W. M. (1935). Isolation of a crystalline protein possessing the properties of tobacco mosaic virus. *Science 81*: 644–645]

Swazey, J. P., and Reeds, K. (1978). *Today's Medicine, Tomorrow's Science: Essays on Paths of Discovery in the Biomedical Sciences*. Washington, DC: U.S. Dept. of HEW, PHS, NIH.

Tauber, A. I., and Chernyak, L. (1991). *Mechnikoff and the Origins of Immunology. From Metaphor to Theory*. New York: Oxford University Press.

Tauber, A. I., ed. (1991). *Organism and the Origins of Self*. Boston, MA: Kluwer Academic Publishers.

Topley, W. W. C., and Wilson, G. S. (1941). *The Principles of Bacteriology and Immunity*. Baltimore, MD: Williams & Wilkins.

Twort, Frederick W. (1915). An investigation on the nature of the ultramicroscopic viruses. *Lancet 2*: 1241–1243.

Tyndall, John (1881). *Essays on the Floating-Matter of the Air in Relation to Putrefaction and Infection*. London: Longmans, Green & Co.

Vallery-Radot, R. (1923). *The Life of Pasteur*. Trans. by R. L. Devonshire. New York: Doubleday.

Vandervliet, G. (1971). *Microbiology and the Spontaneous Generation Debate During the 1870s*. Lawrence, KS: Coronado University Press.

Virchow, R. (1971). *Cellular Pathology*. New York: Dover.

Virchow, R. (1990). *Rudolf Virchow: Letters to His Parents, 1839–1864*. Ed. by M. Rabl. Trans. and ed. by L. J. Rather. Canton, MA: Science History Publications.

Waterson, A. P. and Wilkinson, L. (1978). *An Introduction to the History of Virology*. New York: Cambridge University Press.

Wilson, G. S. (1967). *The Hazards of Immunization*. London: The Athlone Press.

Winslow, C. E. A. (1980). *The Conquest of Epidemic Disease. A Chapter in the History of Ideas*. Madison, WI: Univeristy of Wisconsin Press.

Yates, F. E., Garfinkel, A., Walter, D. O., and Yates, G. B., eds. (1987). *Self Organizing Systems. The Emergence of Order*. New York: Plenum Press.

8

EVOLUTION

The publication of Charles Darwin's *Origin of Species* in 1859 and the controversy that it ignited have overshadowed other aspects of the development of the life sciences during the nineteenth century, a period in which the new science of biology replaced traditional natural history and the study of nature became the domain of the professional scientist. The term *biology*, which was first introduced at the beginning of the nineteenth century, was popularized in the writings of the French zoologist Jean Baptiste Lamarck (1744–1829) and the German naturalist Gottfried Treviranus (1776–1837). According to Lamarck, biology encompassed the study of all that pertained to living bodies, their organization, development, special organs, and vital movements. Auguste Comte (1798–1857), the French social philosopher, considered biology one of the principal sciences of positive philosophy. The idea of progress was central to Comte's system, and his law of the three stages called for the intellectual evolution of humankind from theological to metaphysical and finally to scientific knowledge; his classification of the sciences was based on their increasing complexity from mathematics, astronomy, physics, and chemistry to biology, and sociology. Biologists were urged to turn away from work that was merely descriptive towards the study of the vital functions of plants and animals. Eighteenth-century naturalists had exemplified that era's passion for the accumulation and classification of masses of facts about plants, animals, and peoples from all over the world, but the new task of biologists was to apply physical theories to the rational solution of physiological problems.

All human societies have wrestled with the problem of how to group individual objects into useful general categories. Simple classification systems are usually based on easily comprehensible, useful properties, such as moist and dry, edible and inedible. Plato and Aristotle were among the first to attempt to create systematic classifications based on *form* and *idea*, *genus* and *species*. The search for systematic schemes of classification dominated the life sciences during the seventeenth and eighteenth centuries. Several factors contributed to this new preoccupation with organization, but the sheer bulk of new materials certainly provided a powerful stimulus. Millions of different species may exist, but until 1600 only about 6000 species of plants were known. By 1700 botanists had 12,000 new ones to assimilate. Zoologists also faced the problem of information overload. It was clear, moreover, that eighteenth-century natural history lagged far behind the physical sciences. Physicists had created an orderly universe, bound by natural laws. Certainly, naturalists argued, there must be laws that would create order from the chaotic mass of information about plant and animal species as well as some means of purging the bestiaries and herbals of exotic and dubious entities. Even Conrad Gesner (1516–1565), the greatest of the sixteenth-century encyclopedic naturalists, included creatures like the tree goose and the basilisk in his *Historia Animalium*. Many of the herbals of this period merely listed plants in alphabetical order, or by relative importance, but some botanists attempted to organize plants according to overall affinities or by the properties of particular parts.

The first naturalist to develop a taxonomic system applicable to plants and animals, using species as a unit, was John Ray (1627–1705). His mother's local success as an herbalist probably stimulated Ray's early interest in the medicinal virtues of plants. A man of strong convictions, Ray was willing to lose his livelihood rather than sign the Act of Uniformity adopted during the reign of Charles II. Reduced to barely genteel poverty, Ray spent the rest of his life studying natural history and composing sermons, religious essays, and treatises on folklore and natural science. Whatever the dangers and uncertainties of the political and religious milieu of the seventeenth century, in the study of plants and animals Ray was able to find his bedrock of order and purpose in accord with God's own design. Dedicating his *General History of Plants* to furthering Adam's task of naming all of God's creatures, Ray also undertook the more complex labor of describing, classifying, and interpreting the meaning of form and function in living beings. His ideas were given further development in his most popular and influential work, *The Wisdom of God as Manifested in the Works of Creation* (1691).

The period between the publication of *The Wisdom of God* and *Origin of Species* might well be considered the golden age of natural theology, especially in England. Scientists and clergymen shared the comfortable conviction that, when properly interpreted, studies of the works of nature, God's universal and public manuscript, would always be in harmony with the Bible, God's written word, because both revealed the attributes of God to human reason. In nature Ray found a stable and perfect world created by God, but in his descriptive work religious convictions rarely interfered with his attempts to achieve a sound measure of objectivity. Because he treated the species as the fundamental unit of classification, Ray analyzed the concept of species carefully; he suggested that a breeding test could be used to determine whether plant specimens were members of the same species. Given the conviction that God in his plenitude had filled the entire Scale of Nature, Ray argued there could be no great gaps between living forms. Eventually naturalists would discover all the species that had been established by God at the Creation. Nevertheless, Ray admitted that new varieties of plants seemed to arise by means of degeneration. Natural theology also incorporated the belief that God preserved the species, despite the death of individuals. Fossilized specimens with no known living counterparts, therefore, presented a problem for this doctrine. Despite the doubts raised by fossilized plant and animal remains, Ray comforted himself with the belief that man was God's most recent creation. It is possible that Ray's ideas, if further developed, could have led to natural systems of classification that might have yielded a form of developmental taxonomy more compatible with evolutionary ideas than the well-known Linnaean system.

Carl Linnaeus (1707–1778), taxonomist, botanist, entomologist, zoologist, and physician, defined the fundamentals of biology in terms of nomenclature and classification. For Linnaeus the task of the naturalist was to complete the assignment given to Adam: to name the plants and animals, contemplate the handiwork of God, and marvel at His creations. Primarily interested in botany and natural science, Linnaeus studied medicine at Lund and Uppsala. In 1732 he was able to make an extensive field trip to Lapland, then a virtually unexplored region. During the trip he met his future wife, the patient Sara-Lisa Moraeus, to whom he was engaged for 8 years because her father would not allow the marriage to take place until Linnaeus obtained his doctor's degree.

While Linnaeus had little interest in medical practice, he earned recognition and honors for his *Systema Naturae, or The Three Kingdoms of Nature Systematically Proposed in Classes, Orders, Genera, and Species* (1735). This text contained the basic principles of the Linnaean binomial system, although the work was often revised in the light of new ideas and information. Well aware that his classification system was not natural, in the sense called for by Aristotle, Linnaeus emphasized the virtues it undeniably had. It was practical and easy to use; plants and animals were grouped into species, genus, order, class, and kingdom. The binary nomenclature standardized by Linnaeus became the universally accepted method of naming plants and animals. Although Sweden rewarded his efforts with a university appointment and many honors, Linnaeus often complained that his fame prevented him from undertaking new expeditions. Practically a national monument, he was buried in 1778 with the honors proper to royalty. Quarrels over the disposition of his collections soon followed, and, unfortunately for Sweden, his widow sold his collections and books to an Englishman. Legend has it that Sweden sent out a warship to recapture this national treasure, but the British ship escaped. In 1829 the Linnaean Society of London acquired these materials.

For Linnaeus the act of naming and classifying objects was the very foundation of science. But in addition to taking up the task of Adam, Linnaeus also assumed the labor begun by Aristotle in his *Historia animalium*, that is, knowing living things as they exist in nature. While Linnaeus praised animals as the "highest and most perfect works of the Creator," he thought that it was in his studies of procreation in plants that he had discovered "the very footsteps of the Creator." Censors today would find this subject somewhat less than inflammatory, but Linnaeus must have shocked many readers with his descriptions of the "celebration of love" in the plant world, including scandalous episodes of adultery as well as pure marriage. Classification required careful attention to the reproductive organs and sex life of the plants because plants were divided into *classes* according to the number and character of the stamens; classes were divided into *orders* by the number and character of the pistils. Physician, scientist, and poet Erasmus Darwin (1731–1802), grandfather of Charles Darwin, made the Linnaean sexual system seem almost pornographic in his satirical treatise "The Loves of the Plants" (1789). Erasmus Darwin also accorded Linnaeus very serious and respectful treatment in his *System of Vegetables* (1783).

Critics called Linnaeus an arrogant man who wrote as if personally present at the Creation, but in private letters he expressed doubts about the fixity of species and complained about the conservative religious authorities who could punish those who expressed unorthodox opinions. Probably Linnaeus was troubled by the difficulty of distinguishing between "true species," unchanged since the Creation, and the varieties and sports (unusual variations or mutants) known to gardeners and animal breeders. In his published work, however, he generally accepted the doctrine that all species had been created by God and had remained fixed in form.

EVOLUTIONARY SPECULATIONS

Searching for the ideas that were subsumed by the Darwinian Revolution, historians have traced the tenuous threads of evolutionary thought through the writings of numerous naturalists and philosophers. Lists of Darwin's "precursors" include Anaximander and Empedocles, among the ancient Greeks; Benoit de Maillet, Étienne Geoffroy St-Hilaire, Georges Buffon, Jean de Lamarck, and Pierre Maupertuis, among French naturalists; and Erasmus Darwin, William Charles Wells, Patrick Matthew, Robert Chambers, and Alfred Russel Wallace, among the English. Some of these individuals are fairly well known, while others have been largely forgotten despite their previous notoriety. For example, a once scandalous book entitled *Telliamed: Conversation Between an Indian Philosopher and a French Missionary on the Diminution of the Sea, the Formation of the Earth, the Origin of Men and Animals, etc.*, first appeared in 1748, 10 years after the death of the author, a diplomat and traveler named Benoit de Maillet (1656–1738). Even though he had suppressed the book during his lifetime, de Maillet disguised his heretical ideas by ascribing them to a mythical foreign sage named Telliamed (the author's name spelled backwards). This exotic spokesman could safely express unorthodox ideas that his interrogator, a respectable but curious French missionary, knew to be contrary to religious dogma. Popular in France and England, *Telliamed* brought theories of cosmic and organic evolution to the attention of readers who had little interest in purely scientific texts.

De Maillet attempted to link cosmic and biological evolution in an all-embracing cosmology reminiscent of the pre-Socratic Greeks. As in those ancient cosmologies, Telliamed's world was one of eternal cycles of creation and destruction. Incorporated into the text were discourses on the inheritance of acquired characteristics, the successive deposition of geological strata, the nature of fossils, and the great age of the earth. The all-important seas were the agents responsible for changes in the earth's crust as well as the site of the origin of life forms. Living beings developed only in response to favorable conditions and adapted to their changing circumstances. Thus, when the seas began to recede, living beings were able to change and make their way onto the land. Finding it difficult to reconcile the apparent fixity of species with his conviction that species were always changing, de Maillet suggested that changes in species are never directly observed because they take place over long periods of time. Since the time of Galileo, it had become irrefutable that the allegedly immutable heavens were indeed subject to change; profound changes in the earth and its living beings were not, therefore, totally implausible. De Maillet suggested that microscopic living entities might be the primeval seeds of life that evolved into ever more complex creatures. Improvement and progress were not inevitable in the long run, because the continued diminution of the seas would eventually lead to a period of universal volcanic action that would cause the earth to catch fire and become another sun.

In a remarkable little book entitled *The Earthly Venus*, Pierre Louis Moreau de Maupertuis (1698–1759) promised to explore the great riddle of life as an anatomist, rather than a metaphysician. Best known as a mathematician, the ingenious Maupertuis was also interested in the problem of heredity and the origin of species. After evaluating ovist and spermist theories of generation, Maupertuis proclaimed that both theories were inadequate and incorrect. According to preformationist theories only one parent could contribute heritable traits to the offspring, but there was considerable evidence to the contrary, including mules and mulattos. Maupertuis suggested that mating the same types for several generations would result in the production of new species from varieties. New varieties presumably arose by chance, but climate and food might have some influence as well. For example, the heat of the tropics might be more favorable to the particles that compose pigmented skin. Maupertuis used this argument to explain the development of the various "races" of mankind from what he thought of as the original human type. Not surprisingly, Maupertuis believed that the "European" was the original human type and that other "races" had formed by degeneration under the influence of the unfavorable conditions found in other regions of the earth. These and other speculative ideas were further developed in his *System of Nature*, which was published under a pseudonym in 1751. Many of Maupertuis's ideas were explored and extended by Georges Louis Leclerc, Comte de Buffon (1707–1788).

Born into a family enjoying all the advantages of wealth and power, Buffon acquired a reputation as a charming and elegant man who believed that style was the essence of all things. A youthful love affair led to a duel that forced Buffon to leave France. Escaping from the rigidities of the ancien régime, Buffon encountered the excitement of the new experimental sciences in England. He demonstrated his interest in Newtonian physics and English science by translating Isaac Newton's *Fluxions* and Stephen Hales's *Vegetable Staticks* into French. Although his first scientific interest was physics, Buffon was drawn to biology. A large inheritance freed him from the nuisance of having to earn a living and allowed him to devote his life to science. He died in 1788, 10 years after the death of Linnaeus and one year before the French Revolution claimed the life of his only son.

In 1739 Buffon became an associate of the French Academy of Sciences and was appointed Keeper of the Jardin du Roi, which he turned into a valuable research center. His work resulted in the publication of a popular multivolume compendium of natural history. The 44 volumes of Buffon's popular *Natural History* (1749–1804) presented a complete picture of the cosmos, from the stars and solar system to the earth and its organic and inorganic constituents. Known as an elegant writer, when he was admitted to the French Academy in 1753, Buffon said, "Style is the man himself." Buffon managed to touch on many dangerous and forbidden topics so delicately that he nearly concealed his unorthodox views by pointing out how the facts might lead one to support an evolutionary interpretation, if one did not know what Genesis and the True Church taught. Nevertheless, Buffon aroused the suspicion of orthodox scientists and theologians. He was called before the Syndic of the Sorbonne, the Faculty of Theology, in 1751 and warned that certain parts of the *Natural History* were in conflict with religious dogma. The offending passages dealt with the age of the earth, the derivation of the planets from the sun, and the idea that truth can only be derived from science. Acknowledging the infallible authority of the church, Buffon promised to subdue his heterodox speculations and express himself in a more circumspect manner, at least in public.

Completely won over by Newtonian science, Buffon struggled to extend the concept of the rule of natural law to the living world. Seeing the great complexity and diversity of nature, he ridiculed the oversimplifications and arbitrary arrangements found in Linnaean taxonomy. At one stage of his thinking Buffon argued that the conceptual basis of Linnaean taxonomy was artificial, arbitrary, and mischievous because genus and species were no more real than the lines of longitude on maps, for in nature only individuals exist. Further reflections concerning animal breeding forced Buffon to revise his ideas and define species operationally as groups of animals that could interbreed to produce fertile offspring rather than sterile mules. Species appeared to be nature's only "objective and fundamental realities," but the fixity of species was called into question by evidence of imperfect, useless, and rudimentary organs. Why should such structures exist if all existing species had been perfectly designed at the Creation?

Exploring the ways in which the earth had developed into a place suitable for living creatures, Buffon offered an alternative to biblically based calculations that had fixed the Creation at 4004 B.C. Scriptural authority could be maintained by treating the 7 days of creation as seven epochs of indeterminate length. During the first epoch, the earth and the other planets were formed as the result of a collision between the sun and a comet. Rotating and cooling for some 3000 years, the incandescent gases torn from the sun turned into spheroids. During the second epoch the earth hardened and congealed into a solid body. In the third epoch the earth was covered by a universal ocean. As the waters began to subside during the fourth epoch, volcanic activity was the major influence. Once the land surface was exposed, plants and land animals began to appear. During the sixth epoch the original land mass began to break up and the continents separated from each other. Finally, in the seventh epoch human beings appeared. Simple experiments provided a tenuous basis for this speculative time frame. Spheres made of a variety of materials, such as iron, granite, and limestone, were heated in a blacksmith's furnace and placed on wire racks so that Buffon could measure the time they took to cool. He then estimated the length of time it would have taken before the red-hot earth became cool enough to support life.

With his provocative, elegant writings, Buffon helped to spread new ideas about comparative anatomy, geology, physics, and natural history. For example, Buffon's denigration of the native animals and peoples of the Americas provoked Thomas Jefferson to a rebuttal in his *Notes from Virginia*. Special comparative tables were included to demonstrate American superiority in the size of various mammals. Jefferson went to great trouble and expense to have the skeleton of a 7-foot moose and the bones of American caribou, elk, and deer sent to Paris to prove the "immensity of many things in America."

By the turn of the century, scientists were ready to explore the idea of species transformation more consistently and openly than Buffon. One of the best remembered and probably most misunderstood and maligned of the advocates of evolutionary theory was a man who had enjoyed Buffon's patronage, Jean Baptiste Pierre Antoine de Monet, Chevalier de Lamarck (1744–1829). Although he was the youngest of 11 children, the family title eventually devolved on him because of a series of deaths among his brothers. Unfortunately for Lamarck there was no fortune to go with the title. As

Jean Baptiste Pierre Antoine de Monet, Chevalier de Lamarck

a young man with no prospect of an inheritance, Lamarck was sent to a Jesuit school to become a priest. When he was 16 and the Seven Years' War was drawing to an end, his father's death brought him just enough money to buy a horse and join the army. According to Lamarckian family tradition, all the officers of his company were killed in battle and Lamarck assumed command. His bravery was so outstanding that he was immediately promoted, but at age 22 he was forced to leave the army because of ill health. While convalescing in Paris, Lamarck studied medicine, botany, and music.

In 1778 Lamarck published a popular three-volume guide to the flowers of France. Impressed by this study, Buffon helped to establish Lamarck's scientific career. The French Revolution destroyed the family of his patron, but in 1793 when the King's Garden became the National Museum of Natural History, Lamarck and Étienne Geoffroy Saint-Hilaire (1772–1844) were appointed to new professorships in zoology created by the National Convention. Dividing up the animal kingdom, Geoffroy Saint-Hilaire took the vertebrates and Lamarck became professor of insects, worms, and microscopic animals. Despite this advance in his career, Lamarck remained modest, retiring, and poor. Lamarck survived four wives and several of his children. When he became blind in 1817 he had to give up his professorship. He died in Paris in 1829 and was buried in a pauper's grave; his bones were thrown in a trench to mingle with the skeletons of other nameless unfortunates.

As a self-taught zoologist, Lamarck indulged in interests so broad his researches in any particular area were likely to be lacking in depth. Little is known about the sources of Lamarck's ideas because few of his papers survived and his published works rarely contained references to other authors. His ideas can be interpreted, however, as refutations of the natural theological traditions of eighteenth-century France. His *Memoirs on Physics and Natural History, Based on Reason Independent of Any Theory* was an attempt to formulate a general theory of existence based on physics, chemistry, and biology, a term he had coined. Considered primarily a taxonomist by his contemporaries, Lamarck was more excited by considerations of comparative anatomy and the origin of species. His views on the transformation of species were generally ignored or ridiculed until the geologist Charles Lyell (1797–1875) introduced Lamarck's ideas to the English speaking world in the 1830s. Taking his cue from Lyell, Charles Darwin gave Lamarck credit for being the first writer to create significant interest in the study of the origin of species.

The concept of the inheritance of acquired characteristics is so firmly associated with Lamarck that it is often assumed that he invented it. Actually what Lamarck did was to apply a very old assumption about heredity in an original way to the problem of evolution. According to Lamarck's theory, as conditions on earth changed, geographic and climatic alterations affected plant and animal life. Given time without limit, transmutation of species occurred in accordance with the laws of evolution. In *Zoological Philosophy* (1809), evolution was attributed to a fundamental innate tendency to evolve towards increasing complexity and the transmission of increasingly complex structures to each generation. Simple creatures like worms and infusorians, produced by spontaneous generation, were imbued with the innate tendency to evolve into higher life forms. If the environment was stable, this innate drive would itself lead to more complex life forms. When the environment changed, living beings were forced into new kinds of exertions, which eventually produced modifications; for example, increased usage of particular parts would lead to an increase in the size of the organs involved. Changes in size, strength, or structure brought on by increased usage would then be passed on to future generations.

In *A Natural History of Invertebrate Animals* (1815), Lamarck elaborated four laws pertaining to evolution. The most important and misunderstood was the law that essentially stated that under new conditions, animals experienced a strong physiological need for new organs. The French word *besoin* was often mistranslated into English as *desire* rather than *need*. Some romantics and transcendentalists found the idea that *desires* drove evolution quite inspiring, as demonstrated by Ralph Waldo Emerson's poetic explanation of this putative tendency towards greater perfection:

> And striving to be man, the worm
> Mounts through all the spires of form.

When all living animals were known, Lamarck believed they could be arranged in a linear series with some small collateral branches. Such an arrangement would reflect the natural order of animals in nature, encompassing all forms from the lowest monad to the highest creature on the scale, the human being. If all species were included, the gradations between neighboring species would be nearly imperceptible; apparent gaps in the scale would represent undiscovered species. The relationship between humans and other animals was somewhat obscure, although Lamarck seems to have suspected a common origin for humans and the great apes. Perhaps in the future, Lamarck suggested, the apes could be brought to a higher level of intelligence through training and practice. After all, Lamarck pointed out, few individuals in all of history had achieved real intelligence, and the rest remained in a state approximating bestial ignorance.

Lamarck, old and blind

Lamarck's colleague Geoffroy Saint-Hilaire was the author of equally speculative and imaginative theories. Before his appointment as a professor of zoology in 1792, Geoffroy had been a mineralogist, but his interests included chemistry and natural history, as well as crystallography. Nature philosophy shaped Geoffroy's philosophy of biology. Basically, he was convinced that all animal species could be thought of as variants on a common fundamental archetype. Although he tended to exaggerate possible affinities, his theory of the archetype served many comparative anatomists as a workable alternative to either a strict creationist or evolutionary framework. According to his son Isidore, who adopted and extended many of his ideas, Geoffroy began to suspect as early as 1795 that modern species were degenerations of a primeval organism. It was not until 1828, however, that he publicly discussed this idea.

Erasmus Darwin was a physician, author, poet, inventor, member of the remarkable band of prodigies comprising the Lunar Society, and a friend of Josiah Wedgewood, Charles Darwin's uncle and father-in-law. Known as a man with an insatiable appetite for food and ideas, Erasmus Darwin explored a great range of botanical, zoological, physiological, and medical questions. More radical, if less systematic, a thinker than his grandson Charles Darwin, Erasmus Darwin proposed a theory of the transmutation of species in which all organisms were descendants of some primordial filament. His speculations about the nature of life were presented in *Zoonomia*, *The Botanic Garden*, and *The Temple of Nature*. Although his writings might not excite much attention today if not for his grandson, at the time of their publication these books were considered threatening enough to earn a place on the *Index of Prohibited Books*. Translated into several languages, *Zoonomia* was especially popular among German natural philosophers. It was read twice by Charles Darwin and probably had more of an impact on him than he cared to admit. Nevertheless, while condescendingly expressing his disappointment at the disproportion between the amount of speculation and the number of facts in his grandfather's writings, Charles Darwin came close to accusing Lamarck of stealing ideas from Erasmus Darwin. It is more likely that both Erasmus Darwin and Lamarck had been inspired by the work of Georges Buffon and others, as well as their own observations and inferences drawn from traditional knowledge about plant and animal breeding. Always rather self-conscious and even defensive about his originality, Charles Darwin seems to have underestimated his grandfather's achievements.

Unlike Charles Darwin, who focused his attention on the origin of species from preexisting but still complex antecedents, Erasmus Darwin envi-

Erasmus Darwin

sioned the history of the earth in terms of millions of ages during which "one living filament" was transformed into all the myriad forms of animals that have ever existed. All such transformations were the result of "THE GREAT FIRST CAUSE" which endowed a primordial living filament with the ability to acquire new parts and properties. Acted upon by various stimuli, these new entities acquired further improvements and passed these changes on to their posterity, "world without end." Although the "whole warring world" might now appear to be "one great slaughter-house," Erasmus Darwin suggested that sympathy should exist among all life forms because of their relatedness through descent. Rudimentary structures and vestigial organs revealed the "wounds of evolution." Ultimately, we would realize that "the whole is one family of one parent." Even "imperious man," had arisen from "rudiments of form and sense, an embryon point, or microscopic ens!" (In philosophy "ens" refers to that which has being in its most general sense.)

Naturalists like Lamarck and Erasmus Darwin were intrigued by the eighteenth-century idea that unlimited progress and organic change were possible, but the fears generated by the excesses and terrors of the French Revolution did much to eclipse the hope of progress. Stability in society and nature might well be perceived as more desirable than limitless, unpredictable change. Indeed, evolutionary theories and their advocates were rejected and ridiculed by one of France's most eminent scientists, Georges Leopold Chrétien Frédéric Dagobert, Baron Cuvier (1769–1832). Georges Cuvier, preeminent comparative anatomist and founder of paleontology, was an implacable opponent of Lamarckian ideas in general and evolutionary theory in particular. A veritable prodigy who had been reading since the age of 4, Cuvier entered the Academy of Stuttgart at 14 where he distinguished himself for his prodigious memory, hard work, and self-discipline. His biology teacher, Karl Friedrich Kielmayer, taught that taxonomic order would be achieved when comparative anatomists discovered the archetype of all living creatures. Cuvier's attempts to bring order to the study of marine animals came to the attention of Geoffroy Saint-Hilaire and led to Cuvier's appointment as assistant in comparative anatomy at the Museum of Natural History in Paris in 1795.

Despite Cuvier's lack of formal preparation for his duties, which included training medical students, he soon surpassed Geoffroy Saint-Hilaire and Lamarck in prestige, wealth, and fame. To his contemporaries he was no less than a second Aristotle. In addition to his research and teaching, he took on the reform of the French educational system as Inspector General in the Department of Education, and also served as Chancellor of the Imperial University and Councillor of State. While Cuvier disentangled the taxonomic relationships of many different groups of animals, his most exciting studies were of fossil mammals and reptiles. The term *fossil* had been used by Georg Bauer Agricola (1494–1555) in *De re metallica* (1555) to refer to anything dug out of the earth. Those things that resembled animal or plant forms were generally regarded as sports of nature, a kind of art form that nature practiced rather capriciously to amuse and mystify human beings.

Understanding fossil forms was quite difficult because these entities were rare, often incomplete and usually in poor condition. By the end of the eighteenth century, however, large quantities of fossil teeth and bones had been collected, especially from caves in the Harz Mountains and French Alps. Caves appeared to be primordial features of the earth that gave access to the remote bowels of the planet, the archival vault of its antediluvian history. Some naturalists thought that fossilized bones had been washed into caves during great floods; others argued that peculiar bones had been deposited in caves by humans. Goethe suggested that the close study of caves and fossils could provide profound insights into the distant chapters in the still unfolding history of the earth. Johann Friedrich Blumenbach (1752–1840), whose work in paleontology was eventually overshadowed by that of Cuvier, attempted to use fossils as a means of classifying the geological epochs of earth history. In his popular *Textbook of Natural History* (1779) and his *Contributions to Natural History* (1790, 1806) Blumenbach called attention to the uses of comparative anatomy for paleontology and geology. He noted that certain species such as the dodo had become extinct in historical times and suggested that other extinctions had probably occurred during the most ancient epochs.

Most collectors of fossils knew too little about the structure of living beings and their geographic distribution to say whether or not particular bones might be those of known species. Indeed, many of those who had the opportunity to observe fossils within rock formations and caves were primarily interested in practical aspects of mineralogy and mining. In contrast to the antiquity of mineralogy and the ancient lore of mining, modern paleontology essentially began in the late eighteenth century. Paleontology eventually became part of the domain of historical geography, but it was originally considered a part of anatomy and was, therefore, associated with the medical sciences. To understand fossil animals, Cuvier analyzed the bones of living creatures for comparative purposes. Most importantly, Cuvier used his knowledge of the functional relationships between form and function in living animals to analyze fossils; he called this approach his *correlation theory*. Comparative anatomy provided the clues that allowed Cuvier to reconstruct the form and function of unknown animals from surviving bits and pieces.

For example, Cuvier proved that fossil bones that had been identified as the remains of a man who had died before the great flood of Noah were really the remains of a gigantic extinct salamander.

Recognizing discontinuities between extinct and living species, Cuvier explained changes in the animal world in terms of catastrophes or revolutions that had destroyed whole populations of living things in prehistoric times. Many cultures have legends about global catastrophes, particularly floods, which supposedly took place in the distant past. Both the physical evidence of local geological strata and recent events in France seemed to demonstrate the power of sudden, violent transitions. The sharp lines of demarcation between various geological strata, the inversion and disorder of strata at some sites, and the great numbers of fossils in some layers could readily be interpreted as evidence of mass extinctions and rapid changes in the surface of the earth. Thus, the discovery of frozen mammoths in Siberia was taken as evidence for sudden and violent changes. To account for such observations, Cuvier proposed a series of catastrophes in the history of the earth, culminating with one about 6000 years ago that had established the geology of the present epoch.

Despite much authoritative condemnation, evolutionary theories continued to attract the attention of imaginative amateurs such as Robert Chambers (1802–1871), the author of a much maligned book called *Vestiges of the Natural History of Creation*. Chambers admitted that his goal as a writer was to be "the essayist of the middle class," but he was fascinated by the idea of anonymity that was commonplace in nineteenth century fiction, journalism, and political writing. Special precautions were taken to assure the anonymity of "Mr. Vestiges" until 1884, long after the publication of Charles Darwin's *Origin of Species*, when the posthumous twelfth edition of *Vestiges* was published. Robert and his brother William published a very successful series of journals and textbooks. Presumably Chambers thought that publicly supporting radical ideas like evolution might have been detrimental to the family business.

The Chambers brothers had a personal interest in the problem of variety in nature; both brothers were hexadactyls, that is, they were born with six digits on their hands and feet. *Vestiges*, a layman's guide to scientific theories concerning the development of nature as a whole from the evolution of the solar system to that of human society, created a sensation when it appeared in 1844. Clergymen attacked *Vestiges* as a work of "black materialism" that threatened the foundations of morality and religion, while scientists dismissed it as "bad science." *Vestiges* was read by hordes of middle- and working-class readers in Europe and America, as well as by outstanding individuals such as Abraham Lincoln, Frances Cobbe, Benjamin Disraeli, Charles Darwin, Alfred Russel Wallace, and Thomas Henry Huxley. Disraeli satirized the text in his *Tancred* (1846), where *Vestiges of Creation* appears as *The Revelations of Chaos*. The English geologist Adam Sedgwick (1785–1873) said that he loathed and detested *Vestiges*, and that the book was so stupid he knew the author must be a woman. Others suspected that Prince Albert was "Mr. Vestiges." Charles Lyell, no mean stylist himself, credited Mr. Vestiges with familiarizing the English reading public with Lamarckian views of transmutation and progression in a clear and attractive style. Huxley was very much on target when he said that the author of *Vestiges* obviously knew no more about science than could be picked up in *Chambers' Journals*. While scientists and clergymen competed in vilifying "Mr. Vestiges," his book went through 10 editions in as many years and was translated into several languages.

Conventional references to divine providence were incorporated into Chambers' general writings, but religion seems to have been a minor influence in his own life. Presumably, the writing of *Vestiges* was not a response to a personal religious crisis, but a product of Chambers' interest in physiology, phrenology, and physics. By extrapolating from the nebular hypothesis concerning the formation of the universe, Chambers attempted to apply the "law of progress" to the whole realm of nature, from geology and biology, to anthropology and archaeology. Seeing a fundamental unity in the development of the universe, the earth, and living beings, Chambers contended that his theories did more to glorify the Creator than the idea that God was directly involved in fashioning every kind of creature from the elephant to the dung beetle. Special Creation entailed the embarrassing juxtaposition of a perfect and efficient Creator with incontrovertible evidence of created species burdened by the "blemishes or blunders" manifested in vestigial and rudimentary organs.

God was the First Cause and the laws of nature were His mandates for the working out of the divine will. The processes of cosmogony were not, however, necessarily fixed at one moment in creation; thus, the universe was always developing according to laws for which physicists could establish a sound mathematical basis. There was no reason, Chambers wrote, to deny the fact that God had created living beings, but this did not mean that God was constantly involved in the creation of the myriad forms of life. This could be done more economically through the operation of natural laws which were expressions of the will of God.

While Chambers seemed quite confident that his evidence proved that evolution had occurred, he was unable to invent a new mechanism or driving force. Generally, Chambers ascribed improvement or degeneration of individuals and species to environmental influences. Citing the well-known fact that bees could produce a queen or a worker from the same starting material by changing the conditions of development, he argued that similar influences must be at work in other cases. Such changes might even lead to the evolution of human beings. Assuming that the laws of evolution had produced the various "races," Chambers arranged them hierarchically and argued that just as the embryo passed through stages corresponding to the lower animals during development, so too did the human brain develop through the stages represented by human "racial types." He had, however, argued that speech was the only real novelty attending the creation of human beings. Because studies of language had shown that some so-called primitive peoples had more complex language systems than Europeans, Chambers placed himself in the awkward position of denigrating the difficulty of acquiring the one truly unique human trait. Complexity of language, he rationalized, is not directly correlated with intelligence, because little children learn whatever language their parents speak. Man was fundamentally, Chambers concluded, a "piece of mechanism, which can never act so as to satisfy his own ideas of what he might be."

THE HISTORY OF THE EARTH

Geology, the science that led many biologists to think unorthodox thoughts about the history of the earth and its inhabitants, was also a source of inspiration for advocates of Special Creation. Indeed, many geologists considered their science the true handmaiden of theology. Although applied knowledge of the earth is very old, involving mining, metallurgy, and surveying, a new scientific discipline grew out of eighteenth century controversies about the nature of the forces that had shaped the earth. The emerging science of geology eventually developed into two major schools of thought, which have been called *Neptunist* and *Vulcanist*. The disciples of Abraham Gottlob Werner (1750–1817), the founder of geognosy and Neptunist geology, believed that the formation of the crust of the earth could best be explained by the action of water, such as the great flood described in the Bible. Despite the limited extent of his own field work, Werner exerted a tremendous influence on the study of geology through his success as a lecturer at the new Mining Academy of Freiberg, where he taught for 40 years. Satisfied that his geognosy brought order and system to the chaotic state of classification of rocks and mineral, Werner proceeded to more sweeping pronouncements about minerals, mining, art, industry, and human affairs. Werner suggested that about one million years ago the earth had been covered by a universal ocean. The oldest crystalline rocks had, therefore, originated by chemical precipitation out of the water. Eventually the ocean subsided and more new rock originated mechanically as older formations were exposed to the air and the process of erosion. Exactly how the ocean had held the great mass of minerals in solution and where the great waters had gone were problems the Neptunists had difficulty explaining.

When Charles Darwin was a student, the most popular English geologists, William Buckland, Adam Sedgwick, and William Daniel Conybeare, often lectured about the marvelous harmony between Neptunist geology and biblical accounts of the Creation and the great flood. But such ideas had already been challenged by James Hutton (1726–1797), the guiding spirit of the Vulcanists, or Plutonists, who emphasized the action of heat in shaping the earth. Although Hutton had studied medicine at Edinburgh, Paris, and Leiden, and received his Doctor of Medicine in 1749, he never practiced the healing art. Hutton preferred to supervise his family farm and study agriculture, mineralogy, and natural history. In 1785 Hutton presented his novel ideas about the history of the earth to the Royal Society of Edinburgh. In 1795 Hutton published his massive two-volume treatise *Theory of the Earth, with Proofs and Illustrations*. Except for attacks by Werner's followers, Hutton's book attracted little attention; fewer than 500 copies were printed.

Hutton formulated the two major tenets of geology: the geological cycle and the principle of uniformitarianism, which states that scientists must explain the history of the earth in terms of the natural processes acting in the present. Observing that heat, pressure, and water were constantly but gradually shaping and reshaping the earth, Hutton assumed that its present form must be the outcome of similar changes occurring over limitless stretches of time. Most rocks, Hutton argued, formed by the erosion of dry land; this material was deposited on the ocean floor, turned into rocks by heat from deep within the earth, and eventually uplifted to become dry land once again. In other words, the earth as we know it today is the product of an endless cycle of erosion, deposition, lithification, and uplift. When Hutton investigated the history of the earth, he found "no vestige of a beginning,—no prospect of an end."

Uniformitarian geologists, who modeled their work on the concepts of Lyell and Darwin, tended to portray Hutton as the hero and Werner as the villain in a battle between scientific, secular geology and an archaic biblically based Neptunist cosmogony. But Hutton's work was also deeply teleological in outlook. While he ridiculed those who used the eruption of volcanoes as a way to frighten superstitious people into piety and devotion, when he investigated the structure of the earth, he too saw evidence of a grand design reflecting the wisdom of the power that had produced it.

Few people became acquainted with Hutton's theory of the earth until his friend John Playfair (1748–1819), professor of mathematics and natural philosophy at the University of Edinburgh, published his popular *Illustrations of the Huttonian Theory of the Earth* in 1802. It was Playfair who enthused that the mind "seemed to grow giddy by looking into the abyss of time." While defending Hutton's theory, Playfair attacked the hostile champions of the Neptunian system for misusing the weapons of both theology and science and thus doing as much damage to the dignity of religion as to the freedom of scientific inquiry. By pretending to carry their own theory of the earth to a period prior to the present world of causes and effects, such critics had committed crimes against true empirical science. Claiming that they had been guided by a superior light, they cast their geological speculations in the form of a commentary on the books of Moses. Creation itself, Playfair insisted, would always be hidden from scientific, philosophical, empirical enquiry.

Playfair's explication of Huttonian theory was considered a major landmark in the history of British geology, but it was William Smith (1769–1839), a largely self-educated drainage engineer and surveyor, who was awarded the title "father of English geology" and the first Wollaston Medal of the Geological Society (1831). Smith's great insight into the arrangement of rock strata was that beds of the same kind of stone, at different levels of succession, could be recognized by means of the fossils they contained. By 1799 Smith had noted the relationship between fossils and geological strata and had explained his ideas to some of his friends. But Smith regarded himself as a practical man rather than a scientist and he had a profound aversion to writing. Smith did not publish his famous map of the strata of England and Wales until 1815. However, his friend the Reverend Joseph Townsend used Smith's ideas as the foundation for his own treatise *The Character of Moses Established for Veracity as an Historian, Recording Events from the Creation to the Deluge* (1813). Avoiding speculation, Smith contended that science, especially geology, should be useful as a means of promoting industry and manufacturing. His geology was essentially descriptive and claimed no higher, broader vision or challenge to biblical creation.

Many early structural geologists, such as the Reverend Adam Sedgwick (1785–1873), professor of geology at Trinity College, Cambridge, were clergymen eager to study the great book of nature in order to illuminate the attributes of the Creator. When Sedgwick was appointed Woodwardian Professor of Science in 1818, he admitted he knew nothing about geology. Another candidate had been nominated for the position, but Sedgwick knew his rival had no chance because he was burdened by a great deal of geological knowledge, all of which was wrong. Energetically immersing himself in research and fieldwork, Sedgwick found evidence to support the biblical account of the creation and the great flood everywhere he looked. This was to be expected, he explained, because truth must always be consistent with itself; thus, when science was properly interpreted, it must be consistent with the scriptures. Sometimes Sedgwick's enthusiastic teaching had unintended consequences. It was a summer the young Darwin spent exploring the geology of North Wales with Sedgwick's guidance that quickened his interest in the history of the earth.

Another important influence on the young Darwin was the work of Charles Lyell (1797–1875), the geologist who became known as the great high priest of uniformitarianism, an ironic title considering it was Lyell's work that severed the chains that bound geology to theology. As a young man Lyell was not a particularly good student, but he was zealous in pursuit of his hobbies, such as lepidoptery and geology. His plans for a career in law were abandoned because of an eye disease, but this apparent misfortune left Lyell free to pursue his real interests. In his elegant writings, Lyell reasserted Huttonian principles so persuasively that the concept of the historical and physical continuity of nature became inextricably linked with his name. Generous to his predecessors, Lyell suggested that the revolutionary events of the eighteenth century had promoted a religious and political climate in which the work of Hutton and Playfair was viewed with suspicion. In contrast, Neptunianism and Wernerianism appealed to orthodox religious principles and assumptions. During his own era, Lyell noted, the intellectual climate had become more propitious for uniformitarian geology. Lyell's three-volume *Principles of Geology* was kept up to date with constant revisions; the twelfth edition was published posthumously in 1875.

Like Hutton, Lyell adopted a strict uniformitarian attitude towards geology, insisting that geological data must be interpreted in accordance with forces known to be acting on the earth today. He did not, however, extend this concept to the natural and historical succession of plant and animal life forms. Although Charles Darwin considered Lyell one of his most important friends and guides, for most of his life Lyell resisted a uniformitarian system of organic evolution. Lyell believed that each species had been created for its proper niche and could neither change nor adapt to altered conditions; if a species was not already suitable for the new environment, it would be replaced with other, more suitable species. The creative force that originally produced species was, Lyell believed, outside the proper scope of scientific inquiry.

Convincing Lyell of the validity of the transmutation of species by means of natural selection was one of Darwin's proudest achievements. Lyell disagreed with Darwin's theory of human evolution, but he had no great difficulty accepting evidence for the "antiquity of man" and the vast distance of time separating Stone Age humans from civilized Europeans. In the first stage of his existence, Lyell argued, man would have been "just removed from the brutes" in terms of intellectual powers. It was impossible to set specific dates for the beginning and end of the first Stone Age, but Lyell argued that the original stock of mankind could not have been endowed with superior intellectual powers, inspired knowledge, or the same "improvable nature" enjoyed by their modern counterparts. Once civilization had been achieved, the rate of progress in the arts and sciences would have increased geometrically as knowledge increased. In analyzing various theories of human progression and transmutation, Lyell was not unwilling to refer to Lamarck. He was also willing to accept the idea that all the major human races were the descendants of a single pair. Rather than using this concept of a common human heritage to argue for a strict biblical chronology and a young earth, Lyell saw it as evidence that a great amount of time had been necessary to form the modern human races.

Tracing the lines of progress in biology from Lamarck to Darwin forced Lyell to face the question of whether the general hypothesis of transmutation also applied to the human race: Were human beings the product of a continuous line of descent from the lower animals? This conclusion seemed inescapable, at least for physical characteristics, but might not apply to specifically human moral and intellectual attributes. While not abandoning the hypothesis of variation and natural selection, Lyell objected that it was not necessary to assume an absolutely insensible and unbroken line of improvement in intelligence from the inferior animals to uniquely improvable human reason. Lyell called attention to the occasional birth of an individual of "transcendent genius" to parents of average intellectual capacity. Perhaps, Lyell reasoned, similar rare and anomalous events had established the great chasm between the unprogressive intelligence of the lower animals and the first stage of the improvable reason found in humans. Citing the American botanist Asa Gray (1810–1888), Darwin's chief advocate in America, Lyell concluded that the Darwinian doctrine of variation and natural selection did not weaken the foundations of Natural Theology. The Darwinian hypothesis of the origin of species could be held along with the belief that the deity might have initiated a chain of events that could proceed without any subsequent interference, or with some occasional direct intervention. Thus, Lyell was sure that the idea of change and progress in the history of the earth need not foster a materialistic tendency, but could be seen as "the ever-increasing dominion of mind over matter."

THE DARWINIAN REVOLUTION

In formidable waves and faint ripples, the Darwinian revolution inundated virtually every domain of human thought and belief: scientific, religious, political, philosophical, historical, social, and literary. Like Hutton's vision of the history of the earth, there seems to be "no prospect of an end" of books, articles, and new interpretations where Charles Darwin and the Darwinian revolution are concerned. Modern historians of science tend to believe that all ideas and theories are largely socially determined, but this hardly explains how and why Darwin formulated a theory with such power and notoriety. Although his family tree was a veritable Victorian *Who's Who*, Charles Robert Darwin (1809–1882), the sixth of eight children born to Susannah (Wedgewood) Darwin and Dr. Robert Waring Darwin, was not a very promising young man. But Darwin's life and work remain the subjects of intense interest more than 100 years after his entombment in Westminster Abbey, directly under the monument to Sir Isaac Newton.

Darwin's son Francis speculated that Charles Darwin had inherited his sweetness of disposition from the Wedgewood side of the family and his genius from grandfather Erasmus Darwin. When Darwin reflected on his education, he complained that his early schooling had been detrimental to the development of his mind. Moreover, he was well aware of the fact that his teachers and his formidable father considered him a very ordinary boy, below commonly accepted academic standards, who would probably be a disgrace to the family. Nevertheless, Charles was dispatched to Edinburgh to study medicine in the tradition of his father and grandfather. It was soon obvious that he was temperamentally unsuited for a career in medicine; anatomy and Latin merely bored him, but the sight of blood made him ill. Seeing operations performed without anesthesia convinced him that medicine was a "beastly profession." Moreover, Darwin did not feel the need to exert himself in preparing for any profession because he expected to inherit enough money to live well without working. He did, however, pursue his interest in natural science. While Darwin was at Edinburgh, an anonymous paper vigorously supporting the theory of the transmutation of species appeared in the *New Philosophical Journal*. The author was Darwin's friend Dr. Robert Grant, who had astonished Darwin by his enthusiastic support for Lamarck's evolutionary theory. Despite his association with Grant and his reading of *Zoonomia*, Darwin later claimed that in his youth he had not entertained any doubts about the strict and literal truth of every word in the Bible, augmented by close readings of William Paley's *Natural Theology: or, Evidences of the Existence and Attributes of the Deity, Collected from the Appearances of Nature* (1802).

Accepting the fact that Charles could not be turned into a respectable physician, his father decided that the next best profession was that of clergyman, although Robert Darwin was himself a skeptic where religion was concerned. Dutifully, Charles began to prepare himself to take holy orders, but his attitude towards education and the need for career planning was unchanged. The event that transformed Darwin's life was an offer to serve as unpaid naturalist on the H.M.S. *Beagle*. The voyage of the *Beagle*, which lasted from December 1831 to October 1836, stands out as the great adventure and determining influence in Darwin's life. Indeed, the first sentence of the introduction to *Origin of Species* reflects back on this experience and how the distribution of species in South America and the geological relationship between the region's living beings and its fossil forms seemed to throw light on the great "mystery of mysteries"—the origin of species.

It was only through a peculiar chain of accidents and compromises that Darwin managed to secure a place on the ship. The offer was extended to this untested young man because more suitable candidates had turned down the position. The geologist John Henslow recommended Darwin despite his youth and inexperience, but Darwin's father objected to this wasteful enterprise. Dr. Darwin said he would change his mind only if a man of common sense would support his son's strange idea. Fortunately, Uncle Josiah Wedgewood, unquestionably a man of common sense—and uncommon means—gave Charles his invaluable support. The ship and its captain posed the next set of problems. Captain Robert Fitz Roy, a follower of phrenology, thought that the shape of Darwin's nose suggested a certain lack of energy. The *Beagle*, a 10-gun brig with rotting timbers and decks, refitted for the surveying mission as a three-masted bark, needed thorough reconstruction. The mission of the *Beagle* was to complete the survey of Patagonia and Tierra del Fuego. While circumnavigating the earth, the ship was to make stops at the Cape Verde Islands, South America, the Galapagos Islands, Tahiti, New Zealand, Australia, Mauritius, and South Africa. After two unsuccessful efforts to set sail, on December 27, 1831, the *Beagle* finally departed with 74 persons crammed into every available nook and cranny. Accommodations, even for the captain and his naturalist, were far from luxurious. Darwin, who slept in a hammock slung over the chart table in the poop cabin, soon discovered that he was very susceptible to seasickness and that his quarters were in the roughest-riding section of the ship.

As the *Beagle* proceeded southwards along the coast of South America, Darwin noticed how plant and animal forms differed with geographical location. Like Chambers, he must have worried about the uneconomical way the direct creation of all these slightly different forms would have taken up God's time and energy. These questions became most striking when he investigated the Galapagos Islands, a group of about 20 barren islands of volcanic origin on the equator, about 600 miles off the west coast of South America. In general, the animals on these desolate, geologically young islands were similar to those on the mainland of South American, but many species were unique. The Galapagos Islands served as a living laboratory of evolution where reproductive isolation produced peculiar species living under essentially identical environmental conditions; different species of birds, tortoises, and so forth could clearly be associated with specific islands. Studying the creatures of these bleak islands was probably the single most decisive experience in Darwin's conversion from a naively orthodox future clergyman into a scientist of infinite curiosity and endless patience.

Darwin always regarded the voyage as his first real education; he discovered that the pleasure he found in "observing and reasoning" could be attained only by hard work and concentrated attention. All former interests gave way to a love of science. By the time he returned to England, his mind and habits had been so thoroughly transformed that when Robert Darwin first saw his son again, he resorted to a diagnosis rooted in phrenology and exclaimed: "Why, the shape of his head is quite altered." The change in Darwin's formerly rugged and indefatigable constitution was even more profound; the bold adventurer of the *Beagle* was transformed into a frail, nervous, reclusive invalid. Darwin paid the *Beagle* one last visit in May 1837 just before she left for Australia and wrote that he would gladly sail around the world again, if not for seasickness. However, he never left England again.

The nature of Darwin's mysterious illnesses and chronic ill health has long fascinated scholars, especially for the insights such studies might offer into the nature of the creative process. Many diagnoses and theories have been advanced, including the suggestion that Darwin's recurrent bouts of debilitating illness were psychological in origin. Some of the evidence suggests that he enjoyed the iron constitution of the true hypochondriac, but hereditary weakness, anxiety attacks, depression, arsenic poisoning, and a debilitating malady peculiar to South America known as Chagas' disease have been blamed for the sufferings he recorded in his meticulously detailed diary of ill health. Unusually severe episodes of illness were likely to occur when Darwin finally completed a special phase of his work, when he was overwhelmed by concern about the health of his wife and children and the deaths of family members. Attempts to find relief included rest cures, visits to spas, hydrotherapy, electrotherapy, and the usual and unusual panoply of Victorian drugs.

Like his father, Darwin married a Wedgewood; cousin Emma was the daughter of the uncle who had convinced Robert Darwin to allow Charles to serve as naturalist on the *Beagle*. The marriage that took place in 1839 provided a handsome dowry and an annual allowance. With a comfortable guaranteed income, Darwin was able to devote himself to research while Emma devoted herself to worrying about his health and her endless series of pregnancies. Because of poor health, Darwin chose to leave London and moved his family to the village of Down, where the Darwins lived in almost total seclusion. Despite chronic ill health, punctuated by episodes of mysterious, debilitating maladies, Darwin produced a prodigious amount of systematic scientific work, a revolutionary theory, and 10 children, 7 of whom survived to adulthood.

Using the diaries and notebooks accumulated during the voyage, Darwin published several rather turgid volumes on geology and the *Voyage of the Beagle*, which became a popular success. His scientific reputation was first built on his geological work, especially his explanation of the formation of coral atolls. During his years in London, Darwin served as secretary of the Geological Society and met many important leaders of the scientific and intellectual community. Charles Lyell, the geologist, and Joseph Dalton Hooker (1817–1911), the botanist, became his closest friends. Lyell and Hooker served as confidants, critics, and as Darwin's special agents in a narrowly averted priority dispute with Alfred Russel Wallace (1823–1913).

Although Darwin later stated that his conversion from special creationism to evolution occurred during the voyage of the *Beagle*, 20 years elapsed before he published his theory of the origin of species. While Darwin may have convinced himself of the fact of evolution as early as 1837, he did not write out his preliminary sketch of the idea until 1842. Alluding to the persecution of Galileo, Darwin later recalled his reluctance to make public views that his contemporaries, especially his wife, would find unorthodox and offensive. But surely Darwin knew that the last burning for scientific heresy was that of Giordano Bruno in 1600. Moreover, being independently wealthy, Darwin did not have to worry about losing a job. Probably, Darwin was less concerned with being labeled a heretic than he was afraid that he might not be able to convince his peers of the validity of his ideas.

Of course Darwin did not claim that the theory of the transmutation of species was in itself new and original. But despite much speculation about the nature of the transmutation of species, no convincing mechanism of evolution had been proposed. Darwin recognized that the question of *whether* evolution had occurred could be separated from the question of the *mechanism* by which it might occur. Just how Darwin worked out the details of his theory is not completely clear; perhaps the most important moments in the creative process are never recorded, but there are surprising gaps in the written record, despite the fact that Darwin was a compulsive hoarder. Almost all of Darwin's voluminous correspondence was preserved, along with journals of his research, annotated books and articles sent to him by fellow scientists, minutely detailed financial and family records, and bits of paper containing scribbled notes.

By the time Darwin published *Origin of Species*, he was finally beginning to believe that he had enough evidence to build a persuasive case for his theory, if not a compelling one. His contemporaries acknowledged that for every fact adduced by Lamarck, Darwin presented 100. He had, moreover, provided a mechanism for evolution quite different from that of Lamarck. In reflecting on the major literary sources of his inspiration, Darwin called attention to Charles Lyell's *Principles of Geology* and the *Essay on Population* by that cheerful prophet of doom Robert Malthus (1766–1834). In his *Autobiography*, Darwin states that while reading the *Essay on Population* for diversion in September of 1838, he was struck by the way the Malthusian struggle for existence could lead to positive as well as negative results. It is rather difficult today to imagine anyone reading a work so ponderous and gloomy for amusement, but the work of Malthus was apparently a great stimulus for evolutionary thought; Alfred Russel Wallace also discovered the theory of natural selection after reading the *Essay on Population*. While Darwin's *Autobiography* gives the impression that the idea of natural selection came to him in a flash of insight, his letters and notebooks indicate that he was already thinking in terms of "selection" months before he turned to Malthus.

When critics wanted to denigrate Charles Darwin's achievements, they contemptuously charged that only a Victorian gentleman could have produced a book as unoriginal as *Origin of Species*, a mere application of the doctrine of Malthus to the world of plants and animals. Robert Malthus had written the *Essay on Population* after a dispute with his father concerning the nature of man and the prospects for human society. Daniel Malthus had been a friend of the philosopher Jean Jacques Rousseau (1712–1778) and had attempted to instill in his son a belief in the perfectibility of human beings. But Robert rejected such hopeful visions of the future and developed

Charles Darwin

an awesome and apparently inexorable calculus of human misery. He would have been scandalized to find out that his work had been used in the construction of the Darwinian theory of evolution. The thesis propounded in the *Essay on Population* was that all utopian visions of future human societies were impossible, because in any society the innate and immutable defects in human nature were the source of all major problems. Thus, the natural state of a large part of humankind was misery and vice, sickness, war, perversion, and promiscuity. All of this misery was the inevitable outcome of two basic needs: food and sex. Misery was inevitable because population always tended to overwhelm the available resources; the reproductive capacity of human beings universally tended to exceed the food supply and led to violent competition for the resources necessary to life. As a consequence of the fixed relationship between land and production, food supplies increased arithmetically in the series 1, 2, 3, 4, 5, and so forth. But, Malthus explained, populations increased geometrically so that the human population, if left unchecked, would tend to increase every 25 years in the series 1, 2, 4, 8, 16, 32, 64, and so forth. As precise and unarguable as a multiplication table came the proof that in 200 years the ratio of population to food would be 256/9. In 300 years this would reach 4096/13. It was obvious that the poor would be more severely affected by the rigors of competition than the rich; the misery of the poor classes, therefore, could only tend to increase. An unsettling paradox is embedded in Malthus's seemingly definitive exposition of the future: the ratios he calculated were those that would obtain *if population growth continued unchecked*. Yet Malthus also argued that misery and vice provided constant and unavoidable checks on population growth. In later editions of the *Essay,* Malthus compounded this inconsistency by conjuring up another check on population growth which he called *moral restraint*, a very fallible form of birth control.

The Malthusian vision of the unceasing competition for resources and the ensuing struggle for survival could be translated into the theory of evolution by natural selection. Since variation and fertility clearly exist in abundance, and different varieties have different degrees of fitness in the struggle for existence, it must follow that selection will operate on all varieties and allow only the most fit to survive. The varieties that survive will leave behind the most offspring; if their descendants inherit favorable variations, they and their descendants will in turn be favored in the battle for survival. Over a fairly short period of time, Darwin reasoned, human beings had domesticated plants and animals with useful traits by acting as agents of *artificial selection*. When selection was carried out by nature over unlimited periods of time, the effects must be superior to those produced by humans. *Natural selection* had two effects on populations: it led to divergence of characters, or speciation, and to the extinction of less fit varieties and species.

As early as 1844 Darwin was hinting in rather melodramatic tones to his friend Joseph Hooker that he was working on the species problem from a very unorthodox viewpoint. Comparing his heresy to a murder confession, Darwin admitted to Hooker that he was almost convinced that species were not immutable. For Linnaeus and other taxonomists, the problem of defining the nature of *species* was central to biology; but once organic life was seen as the product of long eons of change and diversification *species* became a rather arbitrary term used primarily for the sake of convenience. While it might be necessary for cataloging and descriptive purposes, from an evolutionary viewpoint the species was merely a group of organisms that resembled each other at a given point in time. *Varieties*, which Linnaeus had dismissed as fluctuations of no real importance, now took on an important role as incipient species. Viewing living entities in this dynamic fashion rather than as static creations suggested a truly natural system of classification based on lines of descent from common ancestors. Nevertheless, the practical problems of taxonomy were not easily resolved, because in most cases all the information available to taxonomists was morphological. Descent, therefore, had to be traced by inference from morphology. Many enthusiastic evolutionists were, however, eager to turn morphological and embryological relationships into lines of descent.

The "origin" in Darwin's *Origin of Species* clearly referred to the transmutation of one species into another and not to a process in which all living creatures were derived from some primordial fluid, fiber, or living molecule. Speculation about the ultimate origin of life was outside of the scientific problem that Darwin chose to explicate. The starting point for his argument was the great abundance of variation in nature, a finding that should be obvious, he wrote, to even the most casual observer. The existence of evolution as change was, Darwin suggested, essentially self-evident. The more difficult problem was to explain the mechanism of change. This was where Darwin believed he had made his greatest contribution: natural selection, he contended, was the engine of evolution.

Having convinced himself that modern species were the products of evolution, Darwin wrote out two trial essays for his own use in 1842 and 1844. It was not, however, until about 1854 that he attempted to bring order to his collection of evidence. His friend Lyell, although still skeptical about evolution, urged Darwin to publish his theory before someone else anticipated him. Slowly and methodically, Darwin began to write out his ideas in great detail. About halfway into this project, Darwin received a letter from Alfred

Russel Wallace, a naturalist working in Malay, along with an essay entitled "On the Tendency of Varieties to Depart Indefinitely from the Original Type." Wallace wanted Darwin's opinion of his theory. Should it have any merit, Wallace requested that Darwin send the essay to Lyell for review and publication.

Reduced to a state of profound shock and disappointment, Darwin realized that Lyell's prediction had been definitively confirmed. Not only had Wallace developed a theory of transmutation by natural selection, but his clear and concise essay seemed to be an excellent abstract of Darwin's 1842 manuscript on the theory of evolution. Moreover, unlike Darwin's preliminary essay, Wallace's account of the theory was well written. Left to himself in this crisis, Darwin might have withdrawn from the potential priority battle. Luckily for Darwin's reputation, his loyal friends Charles Lyell and Joseph Hooker took over as mediators and champions of fair play. A joint presentation of the papers written by Wallace and Darwin was quickly arranged, as was simultaneous publication in the August 1858 *Proceedings of the Linnaean Society.* With this unsettling experience to focus his attention, Darwin polished off the *Origin of Species* in little more than 13 months. In his autobiography Darwin declared that he did not really care whether others attributed originality to him or to Wallace, but in his letters to Lyell and Hooker he admitted how hard it was to think of seeing 20 years of priority lost. Later, when critics attacked the *Origin,* he complained that he had been rushed into premature publication.

With the help of his friends, Darwin was awarded the lion's share of the honors for revolutionizing ideas about evolutionary theory. Wallace came close to becoming the forgotten man, but his role in the discovery of natural selection was often used to belittle Darwin's achievements or to prove that the basic idea of survival of the fittest was already commonplace in the intellectual milieu shared by men like Darwin and Wallace. Whereas Thomas Henry Huxley has been called "Darwin's bulldog," Wallace has been called "Darwin's moon." Reacting with stoic grace, Wallace accepted his role as minor codiscoverer as just another disappointment in a life full of bad luck. Even though Wallace apparently believed that his essay alone would not have created the revolution associated with the *Origin,* others charged Darwin and his friends with the greatest sins that scientists can commit, i.e., the misappropriation of another scientist's ideas and a conspiracy to cover up the crime. Suggestive evidence of opportune gaps in the surviving collections of pertinent letters and manuscripts, as well as odd discrepancies in Darwin's records of important dates, need not indicate a full-fledged conspiracy, but it does demonstrate the need for a fuller appreciation of Alfred Russel Wallace as a scientist and highly original thinker in his own right.

One of the nineteenth century's most colorful and unlucky scientists, Wallace endured misadventures in the jungles of the Amazon and the Malay archipelago that made the adventures of Charles Darwin on the *Beagle* look like a picnic in the park. Unlike Darwin, who came from a background of wealth, privilege, and high expectations, Wallace had a rather meager formal education and spent his entire adult life in pursuit of a variety of modest employments. By the time Darwin had written his preliminary essays on evolution, Wallace had become a teacher at the Collegiate School at Leicester. In 1848 Wallace was able to persuade the entomologist Henry Walter Bates (1825–1892), who is remembered for his discovery of mimicry in animal coloration, to accompany him on an expedition to the Amazon. The two planned to study natural history in the tropics and then pay for their trip by selling their collections. While planning this expedition, Wallace was already thinking about the question of the origin of species, having been inspired by the writings of the great traveler and naturalist Alexander von Humboldt (1769–1850), who was often called the second discoverer of the Americas, and those of Charles Darwin. Wallace was sure that he would find the answer to the "question of questions" in the lush and exotic world of the tropics. While the venture added immeasurably to his knowledge of natural history, his collections, obtained with such difficulty, were lost when his ship caught fire and sank in the middle of the Atlantic on the return voyage in 1852. After spending 10 days in a leaky lifeboat, Wallace and his shipmates were rescued by another decrepit and unseaworthy ship. Recovering from this disaster, Wallace began a new phase of his research in the Malay archipelago (now Indonesia), observing and collecting. During an attack of a tropical fever, Wallace experienced a flash of insight in which he realized that natural selection could serve as the mechanism of evolution. Within a few days he had completed his essay and sent it to Darwin.

Having ceded priority to Darwin, Wallace continued to publish works on natural history and travel, such as *The Malay Archipelago* (1869), *Contributions to the Theory of Natural Selection* (1870), *Geographical Distribution of Animals* (1876), and *Island Life* (1880), while helping to spread the new gospel of evolutionary theory as a popular scientific lecturer. He bombarded the journals with articles about geology, geography, taxonomy, astronomy, anthropology, education, social and economic problems, land ownership, the occult, and the evils of vaccination. Despite major differences in their approach to scientific questions, Darwin and Wallace remained good friends and collaborators. It was Wallace who called evolution by means of natural selection "Darwinism" in order to distinguish this theory from its predecessors. With excessive modesty, Wallace said that his own discovery had a ratio to that of Darwin's work of "one week to 20 years." In many ways Wallace remained more "Darwinian" than Darwin; their differences in emphasis and interpretation involved issues still unresolved 100 years later.

In his later years Wallace supported a variety of unorthodox and radical causes, such as women's suffrage and socialism. These activities, along with his studies of spiritualism led critics to describe him as half genius and half crank. Gilbert Keith Chesterton (1774–1936), poet, polemicist, and journalist, called Wallace one of the two most important figures of the nineteenth century, having the unique distinction of being a leader of a major revolution in thought for his advocacy of materialistic evolutionary theory and its own counterrevolution as a champion of antimaterialistic social theories and psychical research.

At one point an admirer of Herbert Spencer's doctrine of individualism, Wallace eventually rejected Spencerian social Darwinism. Wallace had been impressed by Spencer's assertion in *Social Statics* (1851) that improving adaptation of human beings to the social state would in the future establish a society with each citizen possessed of a strong moral sense so that no governmental restraints on individual liberties would be needed. The essence of progress, according to Spencer, was individuation with differentiation and mutual adaptation; the process was the same ultimately for cells in the animal body and human beings in society. But Wallace became an enthusiastic follower of Robert Owen's philosophy and his cooperative commercial ventures. Wallace came to believe that natural selection could explain the existing diversity among human races; it could also lead to a prediction that in the distant future all human beings would reach a new and exalted level. In speaking to the Anthropological Society in 1864, Wallace merged Darwinism and Spencerian philosophy and grandly predicted that natural selection would eventually eliminate all but the most "intellectual and moral" race; the remaining men (sic) would then be so well adapted to the social state that it would be a virtual paradise in which the moral faculties of all would be so well balanced that compulsory government would be unnecessary. Later, in reviewing Darwin's *Descent of Man*, Wallace came to the unhappy realization that there might not be an evolutionary guarantee of such a relationship between fitness and happiness. Indeed, Darwin's work suggested that selection might increase the same tribe's competitive survival without increasing its moral sense. Invariably described as a self-effacing man, unable to secure full-time employment, the aging and financially distressed Wallace was enormously pleased when he was notified in 1881 that, thanks to the efforts of

Alfred Russel Wallace

Charles Darwin and other eminent scientists, he would receive a Civil List Pension for £200 a year.

As the Darwin/Wallace priority question indicates, the core of modern evolutionary theory appears to be simple enough to be contained within the pages of Wallace's 1858 essay. While the *Origin* is a challenging and complex book, Darwin always thought of it as merely an abstract of the comprehensive treatise he intended to write. It can also be seen as simply the first installment of the long argument with nature that consumed the rest of Darwin's life. After publishing the first edition of the *Origin*, Darwin returned to the book many times, struggling in each new edition to answer his critics. The sixth and last edition was published in 1872; Darwin had reached the point of exhaustion and could not face further revisions. His audience, however, continued to grow; seven reprints of *Origin* were released between 1872 and the year of Darwin's death.

Probably the most difficult criticism came from the eminent physicist William Thomson (1824–1907), better known as Lord Kelvin, and his friend and collaborator Fleeming Jenkin (1833–1885), professor of engineering at Glasgow University. The former challenged Darwin's concept that evolution proceeded through continuous small steps by declaring that the earth was not old enough to allow for such gradual transmutations. Jenkin accepted Kelvin's estimate of the age of the earth and used the prevailing theory of inheritance to challenge Darwinism. Given a "blending theory" of inheritance, it was reasonable to assume that a rare new mutant would be swamped out of existence because of matings between mutants and normal individuals.

Although Darwin did not see blending as a fatal flaw, he was distressed by Jenkin's review of the *Origin*. But as Wallace noted, variations were probably more common than Jenkin or even Darwin supposed. Eventually, Darwin acknowledged Jenkin's argument and confessed that he saw no simple way around it, except to argue that given enough time and slight variations, modifications would accumulate and new species would appear. Still, Darwin, Lyell, and their supporters refused to succumb to "physics envy." Geologists and Darwinians believed that their vision of virtually unlimited time periods was true and hoped that some way would be found to refute the apparently devastating logic of the physicists.

Only 3 years after the publication of *Origin of Species*, Kelvin wrote an article for a popular magazine entitled "On the Age of the Sun's Heat." Citing established physical laws and apparently impeccable calculations, Kelvin argued that the sun could not have illuminated the earth for more than 100 to 500 million years. Worse yet, the inhabitants of the earth could not anticipate further solar light and heat for many more millions of years "unless sources now unknown to us are prepared in the great storehouse of creation." Convinced that physics was the superior science, Kelvin contended that, given a choice between the estimates of time made by a physicist and those based on geology, any rational person must accept the work of the physicist. Kelvin contemptuously compared Lyell's steady-state model of the earth to a perpetual motion machine operating in violation of the laws of thermodynamics. In later work dealing with the relationship between the distribution of heat within the earth and the age of the earth, Kelvin challenged biologists and geologists to accommodate themselves to his best estimate of the age of the earth, i.e., about 100 million years. Of course Darwin never entered into a direct battle with Kelvin, but Huxley insisted that physics need not be privileged over geology in estimating the age of the earth. Both biology and geology, Huxley argued, provided independent evidence of processes operating over immense time periods. Essentially what Huxley said in his elegant nineteenth-century prose was "garbage in, garbage out." Ultimately, the discovery of radioactivity by Henri Becquerel (1852–1908) established a powerful new rationale for rejecting Kelvin's time scale. Radioactive decay not only provided a source of huge quantities of energy unknown to nineteenth-century physicists, it also revealed the existence of radioactive "clocks" that have been of immeasurable benefit to biologists, paleontologists, and archaeologists.

If Darwin was reluctant to publish *Origin of Species* because of the attacks that were sure to follow, he must have expected the level of hostility to increase when he published *Descent of Man* (1871). In the *Origin* Darwin merely hinted that his ideas might throw light on the origins of human beings. Probably he hoped that someone else would assume the burden of writing the book that would elaborate on this theme. But he discovered that while other evolutionists were willing to write about man's place in nature and the survival of the fittest in human society, they were not willing to undertake the study of human evolution as a purely biological process. Many supporters of evolutionary theory, Darwin realized, reserved some role for divine intervention where human intelligence and the soul were concerned. Allowing no such exceptions, Darwin's analysis of the evidence led to the conclusion that "man is descended from a hairy, tailed, quadruped, probably arboreal in its habits."

Through the analysis of qualities shown by animals, such as curiosity, memory, imagination, reflection, loyalty, and the tendency to imitate, Darwin proceeded to the evolution of human qualities. Because he was very cautious and vague about the time and place, as well as the racial identity of the first humans, it is difficult to be sure where Darwin stood on many issues. He suggested so many possibilities and ranged so widely that adherents of all kinds of ideologies were able to cite Darwin in support of their particular cause. As indicated by the complete title, a major portion of the book was devoted to the subject of sexual selection rather than human evolution.

The original theory of natural selection, Darwin reasoned, had failed to explain the evolution of the secondary sexual characteristics because such variations were not directly involved in the struggle for existence. According to the concept of sexual selection, traits involved in the competition between males for attracting the females of their own species were not, strictly speaking, useful in the struggle for existence carried out between species. Some secondary sexual traits might even be dangerous or cumbersome, such as the bright plumage of male birds or the large antlers of certain deer. Like Aristotle, Darwin assumed that strength, bravery, and intelligence were "male traits." Females, in contrast, were described as passive, weak in body, and deficient in brains. Even here Darwin was sufficiently ambiguous to leave room for various interpretations. For example, Darwin noted that the male brain was larger, but he admitted that he did not know whether or not the absolute size difference was lost when the differences in body size between males and females was taken into account. Since the theory of sexual selection depended on intrasexual competition for mates, Darwin overemphasized male traits and generally neglected the very demanding female functions of the care, feeding, and training of the young. Needless to say, Darwin's assertions of scientific evidence for male superiority won him many supporters. Even the centennial tribute to Darwin entitled *Sexual Selection and the Descent of Man 1871–1891* presented essentially the same perspective and emphasis. Animals in which the average female is larger than the average male tend to be ignored, although that is the case in about 30 of the 122 families of living mammals, including some rabbits, hares, bats, whales, seals, antelopes, and the hairy-nosed wombat. No single theory has so far been able to account for sexual dimorphism with respect to size, the diversity of mating systems, and intrasexual competition. These questions are obviously quite complicated since similar species may differ as to the relative size of the two sexes despite similarities in habitat, food preferences, and behavior.

Darwin considered the idea of sexual selection essential to his theory of evolution as a means of explaining elaborate and apparently nonadaptive sexual characteristics, but Wallace did not find the idea very persuasive. Indeed, the theory was largely ignored until a resurgence of interest in the

1960s. Generally, scientists found the idea of "choice" by animals, especially female animals, incompatible with their view of behavior. Renewed interest in sexual selection might be explained in terms of interest in finding Darwinian explanations for characteristics that might initially seems anomalous, from sexual ornamentation to cooperation.

Given the fact that human paleontology and cultural anthropology were still in their infancy at the time Darwin was working on *Descent of Man*, his views of human evolution were remarkably prescient. From the fragmentary evidence of comparative anatomy and psychology, Darwin surmised that the ancestor of modern beings was related to the gorilla and the chimpanzee; that the first humans originated in a hot climate, probably in Africa; and that the early human ancestor appeared between the Eocene and Miocene periods.

Before the publication of *Origin*, naturalists were divided as to whether all human "races" had descended from the same ancestors or different races were the products of separate creations. Despite such differences, European scientists were sure that there was a hierarchy of race, with the Hottentot at the bottom and the white man at the top. After the triumph of Darwinian evolution, most scientists accepted a common origin for all human races but assumed that fixed and ineradicable differences had accumulated during evolution. Despite the lack of any significant scientific evidence of the existence of a biological hierarchy based on race, such specious assumptions were widely used to provide a presumably natural basis for a hierarchy of power and privilege.

An examination of the question of where humans evolved reveals the remarkable reluctance of scientists to accept the evidence of an African origin. A century after publication of the *Descent of Man*, a combination of fossil evidence and new biochemical techniques suggested a rather recent origin for all the human races. Efforts to reconstruct the family tree of modern humans by the use of mitochondrial DNA have led to popular accounts of the alleged discovery of "Mitochondrial Eve." While suggestive and intriguing, the mitochondrial DNA evidence remains too ambiguous to support any firm conclusions as to human origins and racial diversity.

By the time *Descent* was published, several other naturalists had independently concluded that humans beings were the descendants of some ancient, lower, and extinct form. But many others still rejected evidence of the antiquity of the human race. Discouraged by the unceasing attacks on his work and reputation, Darwin began to doubt "whether humanity is a natural or innate quality." When scientists, philosophers, and clergymen asserted that human origins could never be discovered through science, Darwin admonished

them to remember that such confidence was more often the product of ignorance than knowledge. "It is those who know little, and not those who
know much," he wrote, "who so positively assert that this or that problem
will never be solved by science."

Always seeking new evidence and arguments to support the theory of
evolution, Darwin was obsessed by his one great subject, even though he
also worked on more conventional topics before publishing the *Origin*. For
8 long years he studied the taxonomy and physiology of living and fossil
barnacles; later he wondered whether this respectable but extremely tedious
study had been worthwhile. His friend Huxley called it an excellent form of
self-education and discipline. After the publication of the *Origin*, Darwin
produced a stream of volumes on a variety of subjects sharing the evolutionary theme. *The Expression of the Emotions in Man and Animals* provided
the fundamentals of modern research in ethology: studies of infants, the insane, paintings and sculptures, photographs of expression evaluated by different judges, and comparative studies of expression among different
peoples. Darwin argued that the evolution of behavior is like the evolution
of organs. Thus, because behavioral patterns were subject to the laws of
inheritance and selection, some behavior patterns might outlive their original
function and become vestigial, just like certain physical characteristics inherited from distant ancestors.

Observations of animal behavior had long been part of the domain of
natural history and amateur nature studies; Darwin's work made it a key
aspect of evolutionary theory. That is, Darwinian evolution suggested that an
animal's behavioral adaptation to its environment was as important as its
anatomy and physiology. The establishment of ethology as a new science
occurred in Germany, but it soon found affinities with the British traditions
of natural history. Konrad Lorenz, who shared the 1973 Nobel Prize with
Nikolaas Tinbergen and Karl von Frisch for studies of behavioral patterns,
can be seen as the key figure in the establishment of this field, especially for
his discovery of imprinting in Greylag geese. Unfortunately, such studies
have often generated facile comparisons between animal and human behaviors and questionable conclusions about human evolution and human nature.

Perhaps because of the intensity with which Darwin approached complex,
controversial scientific ideas, he has often been portrayed as an obsessive,
lonely, and even neurotic man. But more detailed and careful explorations of
the way in which Darwin wrestled with "the question of questions" suggest
a strong, tenacious mind and generous, warm personality. Focusing on evolutionary theory to the exclusion of almost everything else, Darwin complained to his friend Hooker that "it is an accursed evil to a man to become

so absorbed in any subject as I am in mine." Ostensibly a modest, reticent, and private man, Charles Darwin provided many surprisingly candid insights into his life and thought in his letters, journals, and autobiography. Despite some apparent discrepancies between his memories and the records, Darwin's autobiography has been called one of the "most sincere" ever written. In 1876, Darwin began to compose an autobiography for his children and their descendants. Salient passages on religion, as well as various overly candid references to friends and critics, were deleted when the *Autobiography* was published in 1887 as part of the *Life and Letters of Charles Darwin*. The normally loving Darwin clan was brought to the point of litigation when Francis Darwin, who had been entrusted with the manuscript of his father's *Autobiography*, attempted to publish the unexpurgated text. Within the Darwin family religion was strictly relegated to the woman's sphere. Thus, Francis was surprised by the violent objections his mother and sisters raised to the section headed "Religious Belief." A compromise was achieved by deleting this section and placing parts of it in a separate chapter of the *Life and Letters*. As part of the celebration of the centenary of the publication of the *Origin*, Darwin's granddaughter Nora Barlow published an unexpurgated edition of the *Autobiography*; some 6000 words were restored.

In his discussion of his religious beliefs, Darwin explained that he had lost his belief in Christianity and, furthermore, he could not see how anyone would want to believe in a "damnable doctrine" that purported to consign eminently good, thoughtful, and moral people such as his grandfather, father, and many of his best and brightest friends to everlasting punishment. Perhaps part of Charles Darwin's belief in the inferiority of the female mind was the result of the conviction, handed down from Erasmus and Robert Darwin, that the female part of the family maintained religious beliefs that the more enlightened males had discarded.

Given the profound psychological and intellectual difficulties involved in constructing a complex scientific theory, the scope and quantity of the work carried out by Charles Darwin, a gentle, modest, and chronically sick man, was truly extraordinary. He answered his own intriguing questions in a manner so direct that both his critics and his allies tended to underestimate the quality of his work. For example, Darwin's studies of earthworms, including observations of a pot of worms kept on Emma's beloved piano, called attention to the great effects that resulted from "small causes often repeated." Indeed, when Thomas Henry Huxley first read the *Origin,* he felt forced to voice the reaction of many eminent thinkers: "How extremely stupid not to have thought of that oneself." Disciples and critics understood that Darwin had provoked a revolution in thought by challenging the long-cherished picture of nature as the result of God's design, intention, and direct intervention and the earth as a place created especially for human life. The *Origin* was cited by many who had never read the work, but an amorphous concept called Darwinism was invoked in support of many causes and quite contradictory ideas from atheism to new forms of Christian faith, from socialism to fascism, from robber-baron capitalism to Marxist economic theory. Only those with tremendous self-discipline were able to think of evolution by means of natural selection as a purely scientific hypothesis.

It has been said that so much has been written for and against Darwinian evolutionary theory because it appears to be so simple that almost anyone can misunderstand it. Familiar arguments from natural theology had been turned inside out to create a new world in which the fit between creatures and their environment was not proof of the wise design of the Creator, but the result of natural selection acting blindly and impersonally on accidental variations over the course of vast periods of time. Nevertheless, some theologians accepted the idea of evolution, or at least abandoned the idea that all plants and animals were the unchanged and unchangeable descendants of species fabricated during the first 6 days of the Creation. As Chambers and others had suggested, it was not impossible to see evolution as consistent with a more glorious and efficient, if more distant and abstract, concept of God. Many others were, however, so deeply committed to arguments from design and the literal truth of the scriptures that they could not accept any compromise or alternative.

Because his chronic ill health was exacerbated by excitement, discord, or arguments, Darwin could not enter into direct confrontations with opponents of evolutionary science. Fortunately for Darwin, the battle was taken up by some extremely pugnacious, determined, and ingenious naturalists. Chief among them were Thomas Henry Huxley and Ernst Haeckel. The amazing effect of Haeckel's writings on a young mind is well illustrated by Richard Goldschmidt's (1878–1958) reminiscence in *Portraits from Memory*. On reading Haeckel's history of creation as a young man, Goldschmidt immediately felt that all the problems of heaven and earth had been solved simply and convincingly. Every question that had troubled him had been answered. All previous beliefs and creeds could be discarded, thought Goldschmidt, because evolution was "the key to everything."

For his role as leader of the evolutionists' army and debating team, the irrepressible Thomas Henry Huxley was awarded the title "Darwin's bulldog." Huxley was, however, an eminent scientist in his own right who was definitely not a slavish follower of Darwinian hypotheses. Indeed, Darwin considered converting Huxley to evolutionary theory one of his greatest accomplishments. Long after Huxley accepted descent with modification, he continued to think of natural selection as a working hypothesis. Unlike Darwin, whose wealth gave him access to the best available education (even if he considered it dull and worthless), Huxley had only a few years of formal schooling. He was the seventh of eight children born to Rachel and George Huxley, an impoverished schoolmaster. When only 15 years old, Huxley designed his own rigorous and highly successful program of self-instruction. Later, Huxley had to struggle to find a position with a salary sufficient to bring his fiancée Henrietta Anne Heathorn to England from her home in Australia. Physicians told Huxley that his ailing bride was unlikely to live another 6 months; this estimate proved to be wrong by about 50 years. Even as Huxley anxiously awaited the birth of his first child on the last night of 1856, he was outlining a plan of study to provide a "new and healthier direction to all Biological Science." Like Darwin, Huxley experienced a period of profound grief at the death of a child, which may have triggered his questioning of orthodox religion. Known for his love of controversy, Huxley dedicated himself to the campaign to "smite all humbugs," large or small.

Like Darwin and Wallace, Huxley developed his talents as a naturalist through a long voyage to exotic places. A prodigious worker, Huxley turned out over 150 research papers covering fields as diverse as zoology, paleontology, geology, anthropology, and botany, as well as many books and essays. While Darwin was wholly consumed by his work on evolutionary theory, Huxley became a key figure in debates about virtually every aspect of English civilization, including education, metaphysics, politics, and theology. His research, writings, and position as president of the Royal Society made him one of the most powerful scientists in Britain. He was called the fiery prophet of Victorian science, as well as the gladiator-general with the razor-sharp claws and beak who was always ready to do battle with all those who attempted to obstruct the sacred cause of science. A master stylist and one of the most influential writers of his generation, Huxley coined words such as *biogenesis*, *agnostic*, and *bishopophagous* (to describe his ability to eat up conservative clergymen) when the English language failed to provide a term precise enough to suit his needs. Part of Huxley's success derived from his gift for language, spoken or written. The gift was apparently handed down through three generations of Huxleys, most notably in the case of his grandsons Aldous Huxley, Sir Julian Huxley, and Sir Andrew Fielding Huxley, who independently earned their places in what has been called the world's "reigning dynasty of the mind." By the time Huxley began what might be called his second career, that of cultural critic of English society, his professional reputation as anatomist, physiologist, and defender of science was already well established. Perhaps the greatest warrior in the battle to convert the world to evolutionary thought, "Darwin's bulldog" was also given the ironic title "Pope Huxley" in recognition of the influence he wielded as a scientific humanist and educator.

One of Huxley's most famous debates involved Samuel Wilberforce, the Bishop of Oxford, whom Lyell called "that Jesuit in disguise." The occasion for the "Battle of Oxford" was the June 1860 meeting of the British Association. Huxley did not like the medieval atmosphere of Oxford, but he was talked into attending the meeting by Robert Chambers, the still anonymous author of the infamous *Vestiges*. Darwin, of course, was conspicuously absent. After reexamining accounts of the verbal duel between Huxley and Wilberforce, some scholars have declared it a legend and a myth. There is, however, general agreement that hundreds of people packed into the lecture room to hear the Bishop denounce the dreadful "monkey theory" presumed

Thomas Henry Huxley

to be at the heart of *Origin of Species*. In conclusion the Bishop turned to Huxley and asked whether he claimed his descent from a monkey through his paternal or maternal lineage. Relishing the chance to respond to this gratuitous insult, Huxley explained Darwin's ideas and closed the discussion by saying that he would "rather have a miserable ape for a grandfather" than a man who used his great gifts and influence "to introduce ridicule into a grave scientific discussion."

In this episode, as well as others less distorted by the shadows and shards of myth and memory, Huxley gave a spectacular performance in his role as the preeminent champion of the Darwinian worldview. Devoted to the task of establishing a new morality founded on natural knowledge, Huxley the agnostic preached the gospel of a moralizing naturalism. Instead of a pulpit, Huxley saw himself standing at the lectern of "Nature's university," the world institution in which all "mankind" was inescapably enrolled. Of course, Huxley and his contemporaries assumed that the curriculum and opportunities available to "womankind" were very different from those appropriate to mankind. Despite Huxley's reputation as a champion of enlightened thought and education for women, his actions in late nineteenth-century disputes concerning the professionalization of science, the control of Victorian anthropology, and the role of women reveal the depth of his very conventional belief in female inferiority.

While Huxley and his beloved wife Henrietta believed in educating their daughters in science so that they would be "fit companions of men," Huxley was sure that the vast majority of women would remain in the "doll stage" of evolution. Huxley and his contemporaries took it for granted that anthropology, the science of man, along with any discussion of sex, reproduction, anatomy and physiology, heterodox religious ideas and agnosticism were not fit subjects for women. In 1871, the year in which Darwin published *Descent of Man*, the Ethnological and Anthropological Societies merged, thanks to Huxley's efforts, and the Darwinians, safe from the intrusions of frivolous females, gained control of the "science of man."

Like the physical anthropologist Karl Vogt (1817–1895), Huxley was convinced that scientific studies of the structure of the brain had provided proof of the inferiority of women and the "lower races." Vogt claimed that the crania of men and women were so different that it was scientifically appropriate to classify them as different species. According to Vogt, the crania of adult women were invariably more childlike than those of men of the same race; this difference increased with racial evolution and the progress of civilization so that the difference was most marked in the advanced European white race and less well established in the black race. The arguments made by Darwin in *Descent of Man* and Huxley in "Emancipation—Black and White" (1865) concerning female inferiority may have been couched in more chivalrous language than that of the openly racist Vogt, but they were not fundamentally different.

In general, Huxley asserted, the cerebral convolutions of women and those of males and females of the "lower races" were less complex than those of white European males. Given the central role of the brain in all of that was truly human, the average woman was, therefore, invariably inferior to the average man in every important mental and physical characteristic. Nature's irrevocable laws had, Huxley assured his contemporaries, given the human male all the advantage of physical size and strength, as well as a more massive brain. With such a guarantee of superiority, men should not deliberately add to the biological burden already placed on women; after all, Huxley insisted, no amount of education or opportunity would allow women to compete successfully with men.

Although it could be argued that other aspects of nineteenth-century biology, such as bacteriology and physiology, have had a greater effect on life and health than evolution, no other science has had more impact on human thought. Evolutionary theory was extended into many fields far removed from the biological limits to which Darwin generally confined himself. In a very general way the assumption that Darwinian thought could be extended to human history and the evolution of culture and civilization became known as *social Darwinism*. Many admirers of Darwin saw a political and economic message in evolutionary science: if the status quo is the result of eons of natural evolution, things *must be* the way they are. Present hardships must, therefore, be accepted as the result of the workings of inexorable natural laws. Supporters of the status quo warned that any attempts to improve the conditions of the poor or mitigate harsh aspects of modern industrial society should be regarded as futile, misguided, and dangerous interference with natural processes and natural laws. On the other hand, Karl Marx and Friedrich Engels tried to use the very same evolutionary laws to justify the inevitability of communism. Ambiguities, contradictions, and revisions in the Darwinian corpus could be used to rationalize a variety of positions, but it was obvious that many who claimed to have an infallible interpretation of the Darwinian canon had never directly confronted the pages of *Origin of Species* or *Descent of Man*.

The elaboration of the doctrine of social Darwinism, i.e., the application of evolutionary laws to human society, was most closely associated with Herbert Spencer (1820–1903), engineer, inventor, writer, and social philosopher. Among Spencer's many books on philosophy, sociology, education, and science, *Principles of Biology* was particularly important as a link between the general reader and the scientist. Almost forgotten today, Spencer was one of the most widely read and influential philosophers of his era. Enthusiastic admirers of Spencer considered him a second Aristotle and an intellectual giant besides whom Darwin was a dwarf.

The Spencerian Law of Evolution, which asserts that the entire universe is characterized by the inexorable development of complexity, richness, and excellence, is not found in *Origin of Species*. According to Spencer's law of universal history, gradual, but inexorable progressive development encompassed all cosmic, evolutionary, biological, social, economic, and cultural processes. Progress, according to Spencer, was not an accident, nor something within human control; it was instead a "beneficent necessity" deeply embedded in the universe itself. Within Spencerian philosophy, society was analogous to an organism; changes in the social organism must therefore be measured on the scale of evolutionary progress. Inevitably, without human interference, in the *very remote future* society would evolve into a better state. Followers of Spencer considered the great natural law of evolution to be as fundamental to the workings of human society as Newton's law of gravity was to the orderly movements of the heavenly bodies; the survival and advancement of the fittest was, therefore, an axiomatic corollary.

While many social Darwinists adopted a pessimistic view of social evolution as a Malthusian struggle, others were quite optimistic about the inevitability of progress. Despite the difficulties inherent in defining and delimiting the many doctrines that have come to be known as social Darwinism, virtually all of those who applied evolutionary theory to human society agreed on one thing: nature selected the best and the fittest, and to them she gave her rewards. The best were the ones who were most fit as demonstrated by their riches and their position in society. Spencer's followers identified the unfit as those who were weak, foolish, lazy, inefficient, reckless, and irresponsible. The workings of evolutionary law, they asserted, were immutable; the fit would be rewarded and the unfit would be destroyed. Such arguments led many skeptics to remark that the whole supposedly scientific package was nothing but biological Calvinism.

The fit were urged to reproduce while preventing the multiplication of the unfit. It is in this concern with the differential fertility of rich and poor that the analogy between biological and social Darwinism most obviously breaks down. In biology, fitness and successful competition are measured in terms of survival and reproductive success, that is, leaving a larger number of offspring. In society, the measure of success was (*and is*) the accumulation of wealth, not the number of offspring. Indeed, the well-to-do feared that advanced nations were committing "race suicide" because the fit, so worn out in the struggle for wealth, were not doing their share in the struggle to produce more offspring. The unfit, in contrast, were allegedly doing nothing but reproducing their own unfit kind.

Opponents of pessimistic and Spencerian versions of social Darwinism proposed alternative visions of the nature of human society and the operations of evolutionary law. Still committed to a belief in Darwinism, thinkers such as John Dewey, Lester Frank Ward, Charlotte Perkins Gilman, and others challenged the view of society founded on struggle, competition, and the inevitability of human misery. The idea that cooperation and altruism were more powerful factors in evolutionary change than competition and the struggle for existence had been developed by Peter Kropotkin (1842–1921), author of *Mutual Aid* (1902). Kropotkin, a geographer and naturalist, was a Russian prince who became one of Europe's most prominent anarchist-nihilist revolutionaries. Many of Kropotkin's biological ideas developed as a result of surveys of the flora and fauna of Manchuria and Siberia. In areas thought to be the very antithesis of the Garden of Eden, it seemed logical to expect the struggle for existence to be especially severe. Surprisingly, Kropotkin found much competition between species but little evidence of intraspecific struggle. Thus, he proposed that in addition to the law of struggle, a primary factor in the evolution of many species must be a law of mutual aid and support. He argued that this law could be found, at least implicitly, in Darwin's *Descent of Man*. Consciousness of species solidarity produced an instinct that drove the social insects and led wolves to hunt in packs. Nature was, Kropotkin asserted, neither all pitiless struggle nor all peace and harmony. In Russia, biologists had generally accepted Lamarckian evolutionary theory by the time that *Origin* appeared. Darwinism was received warmly and led to discussions of how natural selection could complement Lamarckian mechanisms. Eventually, these disparate elements were drawn together in the form of Lysenkoism, a doctrine that had particularly destructive effects on Soviet biology, genetics, and agriculture.

Long after Darwin called attention to the different reception his work had received in France and Germany, scholars began to analyze the reception of Darwinism in great detail. While many interesting international differences have been found, no simple correlation seems to exist between the reception of Darwin's theory around the world and the special characteristics of specific countries. Perhaps the major general conclusion of the scholarship on the international and comparative reception of Darwinism has been that in many areas where the biological sciences had not yet been firmly established, the discussion of Darwinism was essentially confined to arguments concerning social, political, and philosophical questions. It is ironic that, in virtually every part of the world, Darwin, who saw a world without goals, purpose, or progress, was generally misread and misinterpreted as having proved the inevitability of progress. In terms of non-Western philosophical systems, China and the Islamic world provide the most striking examples of reinterpretations of Darwinism. In the Islamic world interest in Darwin was primarily social, political, and philosophical, rather than scientific. By the 1880s Arab secularists were using Darwinism as a symbol of modernization.

A version of Darwinism more closely allied to the ideas of Herbert Spencer than those of Charles Darwin reached China by the end of the nineteenth century. By the 1880s Christian missionaries were providing translations of the works of Charles Lyell, Charles Darwin, and other scientists, but so-called Darwinian concepts were more forcefully introduced to China by men like British Consul Rutherford Alcock. An unshakable belief in the idea that the British had discovered and demonstrated the truth of the natural and moral law that governed individuals, nations, and races and invariably led to the triumph of the strong over the weak was not uncommon among British diplomats and businessmen. After 1895, the year of China's defeat in the Sino-Japanese War, Spencer's slogan "the survival of the fittest" entered Chinese and Japanese writings as "the superior win, the inferior lose." Concerned with evolutionary theory in terms of the survival of China rather than the origin of species, Chinese scholars saw the issue as a complex problem involving the evolution of institutions and ideas, especially the belief in progress. Many adaptations of Darwinism evolved in China, including varieties that might be called Taoist Darwinism, Confucian Darwinism, Legalist Darwinism, and Buddhist Darwinism. Eventually China absorbed, transformed and was transformed, by a world of ideas including those of Charles Darwin, Karl Marx, and Mao Zedong.

Japanese translations of *Descent of Man* and *Origin of Species* appeared in the 1880s and 1890s. According to Japanese scholars, traditional Japanese culture was not congenial to Western science because the Japanese view of the relationship between the human world and the divine world was totally different from that of Western philosophers. Nevertheless, Darwinian ideas were introduced to Japan by American zoologist Edward S. Morse, who arrived in Japan in 1877. Although Morse had come to Japan to further his own study of brachiopods, he also gave lectures about evolutionary theory. Whereas European and American scientists and theologians became embroiled in debates about the evolutionary relationship between humans and other animals, Japanese debates about the meaning of Darwinism primarily dealt with the national and international implications of natural selection and the struggle for survival. Late nineteenth century Japanese commentators were likely to see Darwinism as an "eternal and unchangeable natural law" that justified militaristic nationalism directed by a supposedly superior elite.

In German scientific circles Darwinian evolutionary theory was most enthusiastically welcomed and publicized by Ernst Haeckel. In the popular press, Darwinism became synonymous with progress. Generally, political conservatives assumed that Darwin's primary message was competition and the struggle for survival, while radicals and reformers found a message conducive to social evolution. When geneticists rejected the Lamarckian version of heredity and environmentalism, moderate versions of social Darwinism lost their base of scientific support, and strict new hereditarian versions of Darwinism were promoted by right wing radicals. Arguments about the nature of "superior" germ plasm were, as might be expected, based on political factors rather than scientific inquiry. However, the relationship between Nazi ideology and Darwinism is not a simple one; an evolutionary framework might have been seen as contrary to Nazi racist assumptions about the "immutable superiority" of the German people.

Surprisingly, during 1982, the centennial of Darwin's death, Italy held the largest number of commemorative symposia and produced the most centennial publications. The explanation for this phenomenon appears to be that by 1982 interest in Darwinism in the United States was generally confined to scientists and academics (and Creationists), whereas in Italy and Spain Darwin was still of general interest as a political symbol. Darwin had corresponded with several Italian naturalists and incorporated their criticisms and ideas into his writings.

In America, Darwinism found a sympathetic reception among scientists, intellectuals, businessmen, and politicians promoting the goals of competition, capitalism, and manifest destiny, even if evolution was not granted a place in American high school biology texts. Asa Gray (1810–1888), the eminent American botanist and friend of Sir Joseph Hooker, was the first and foremost of Darwin's supporters in America. Indeed, Darwin had outlined his theory of evolution in a letter to Gray in 1857, 2 years before he published the *Origin*. For his attempts to reconcile Darwinism and theology, Gray has been called a Darwinian theist. During the 1860s and 1870s, Gray published numerous essays on natural selection and theology in leading scientific and literary journals; a collection of his essays was published in 1876 under the title *Darwiniana*.

As Gray's work indicates, despite the myth of perpetual warfare between science and religion, the way in which Darwin was interpreted by religious thinkers was not a simple matter. Darwin's theory resonated with many aspects of natural theology as well as economic science, which is not surprising given the role that Malthus allegedly played in the development of the idea of natural selection. In Darwin's work, nature serves as a substitute for God as a means of explaining design, or the adaptation of organisms for their place in nature. By reasserting God's role as First Cause, orthodox Christian evolutionists, especially those with a Calvinist worldview, were able to accept a Darwinian view of the struggle of one against all in nature more readily than the so-called modernizing liberals who preferred a more hopeful Lamarckian approach to progressive, directed evolution.

Clearly, those who praised Darwin and those who condemned him often failed to understand him. Even those who spoke glibly of the struggle for existence and manifest destiny might join the opponents of Darwinism to condemn the idea that human beings, apes, and monkeys shared a common ancestry. Whatever scientists might think about the nature of Darwin's work, a large majority of Americans knew that they did not want evolutionary theory discussed in their schools. The campaign to suppress the teaching of evolution led to the infamous Scopes Trial in Dayton, Tennessee, in 1925. John Thomas Scopes, then a 25-year-old high school teacher, was tried and convicted of the crime of teaching "the theory of the simian descent of man" in violation of state law. A graduate of the University of Kentucky, Scopes expected to teach just long enough to obtain enough money to attend law school when he accepted a coaching and teaching position at Central High School in Dayton.

During the 1920s, most scientists were sure that it was impossible to teach biology without references to evolution, but Christian fundamentalists saw evolution as a threat to religious belief. Fundamentalist texts, with provocative titles like *Hell and the High Schools: Christ or Evolution—Which?* and *God or Gorilla*, were widely disseminated, and William Jennings Bryan (1860–1925) was energetically campaigning against the teaching of the anti-Bible heresy known as evolution. Bryan was a man who had suffered many defeats, including three unsuccessful campaigns for the presidency, but he had established a solid reputation as a great orator and crusader for various causes, including women's suffrage, free silver, independence for the Philippines, Prohibition, and the income tax.

A bill prohibiting the teaching of "Darwinism, Atheism, Agnosticism, or the theory of Evolution as it pertains to man" was narrowly defeated in the Kentucky legislature. There was little opposition to the Fundamentalists in Tennessee, and in 1925 an antievolution bill written by State Representative John Washington Butler passed by a wide margin. Butler was a farmer, part-time schoolteacher, and clerk of the Round Lick Association of Primitive Baptists. When Governor Austin Peay signed the bill, he said that in all probability the law would never be actively enforced, but opponents warned that this "pot shot at science will prove a boomerang." The Butler law made it unlawful for any teacher in any school in the state that was supported in whole or in part by public funds to teach "any theory that denies the story of the Divine Creation of man as taught in the Bible, and to teach instead that man has descended from a lower order of animals." The penalty for the crime of teaching about evolution was a fine of not less than $100 dollars or more than $500.

An advertisement placed in the Chattanooga *News* by the American Civil Liberties Union (ACLU) offering to pay the expenses of anyone willing to test the constitutionality of the Butler law precipitated a spirited discussion among some of Dayton's leading citizens. The town's foremost lawyer argued in favor of the Butler law, despite the fact that George William Hunter's *Civic Biology*, the textbook in use in Dayton since 1919, contained an evolutionary chart and a discussion of evolution. When the regular biology teacher refused to be the subject of a test case, Scopes was asked to volunteer to test the Butler law and win some publicity for Dayton. A few days after the Dayton test case was announced in a Chattanooga newspaper the story was picked up by the Associated Press. National interest in the case that H. L. Mencken of the Baltimore *Sun* labeled the Monkey Trial exploded after Bryan agreed to assist the prosecution on behalf of the World's Christian Fundamentalist Association.

John Randolph Neal, a constitutional expert who ran a private law school in Knoxville, became one of the ACLU lawyers for the defense. After the state accepted Bryan as a special prosecutor, Neal received a telegram from Clarence Darrow and Dudley Field Malone expressing their desire to assist the defense without fees or expenses. Expecting a defeat in Dayton, the defense team planned to appeal the case to the federal courts in order to test the constitutionality of the Butler Act. The Scopes Trial was heard by Judge John T. Raulston in a special term of the Eighteenth Judicial Circuit in July 1925. With all 700 seats taken and hundreds trying to force their way into the courtroom, many in the audience passed out from the heat. Eventually the heat became so unbearable that the judge moved the trial outside. A radio microphone was set up and a movie camera recorded the scene for the newsreels.

Experts on science, biology, evolution, and religion came to Dayton to testify against the Butler Act, but the prosecution successfully moved to exclude expert testimony. Denied the testimony of their expert witnesses, the defense lawyers moved to call Bryan to the stand as an expert witness on the Bible, thus creating the confrontation between Bryan and Darrow that has appeared in various fictionalized versions of the trial. Bryan not only rejected the theory that man evolved from lower animals, he objected to diagrams that portrayed man among the mammals. As expected, Scopes was found guilty; Judge Raulston imposed a fine of $100. Shortly after the trial Bryan died of a stroke; his followers portrayed him as a martyr who had been killed by Darrow. A Bryan Memorial University Association was founded in Dayton to raise money for William Jennings Bryan University, dedicated to Fundamentalist Christian principles. Classes began in September 1930 in the old Rhea County High School building where Scopes had taught. Despite the guilty verdict, Scopes was asked to stay in Dayton, but he never returned to teaching. Instead of applying to law school, he decided to do graduate work in geology.

On January 14, 1927, the decision made in Dayton was reversed on a technicality; Judge Raulston had imposed a $100 fine, but the size of the fine should have been decided and set by the jury. The Tennessee Supreme Court denied motions for a new hearing, and the Butler Act remained in effect. Although the publicity generated by the Scopes Trial may have stopped some states from enacting similar laws, antievolution laws were passed in several others.

The Scopes Trial became the model for a play by Jerome Lawrence and Robert E. Lee called *Inherit the Wind,* which became a popular movie after its premiere in Dayton on the thirty-fifth anniversary of the trial. Still seeking publicity, Dayton proclaimed a new Scopes Trial Day. Although Scopes was awarded the key to the city, he was convinced that if the trial had been held again the verdict would have been the same as in 1925. Indeed, in 1960 teachers in Tennessee were still required to sign a pledge that they would not teach evolution. In 1972 a survey of Dayton high school students revealed that 75% of them still believed the King James Bible's version of the creation and rejected Charles Darwin. A plaque erected by the Tennessee Historical Society on the courthouse grounds in Dayton tells the whole story with admirable brevity:

> Here, from July 10 to 21, John Thomas Scopes, a County High School teacher, was tried for expounding the theory of the simian descent of man, in violation of a lately passed state law. William Jennings Bryan assisted the prosecution: Clarence Darrow, Arthur Garfield Hayes [sic] and Dudley Field Malone the defense. Scopes was convicted.

While the infamous Monkey Trial created a national sensation, the outcome was far from clear. Evolutionists viewed the publicity generated by the case as a triumph for modern science, but the teaching of evolution actually declined. Most high school biology textbooks written after 1925 did not include the word *evolution* in the index or glossary. One exception was a book written by Alfred C. Kinsey, professor of zoology at Indiana University. His *Introduction to Biology* (1926) deliberately incorporated an evolutionary framework. But, for the most part, professional zoologists had little interest in high school texts and publishers willingly censored out potentially offensive materials. Profound anxiety about the state of American scientific literacy was triggered by the launch of the Soviet satellite Sputnik in 1957. Driven by the need to compete with the Russians, the federal government provided support through the National Science Foundation for the development of new curriculum materials. The resulting Biological Sciences Curriculum Study (BSCS) series set a new standard for high school biology texts. Within a decade, the success of the Russian space program accomplished what legions of lawyers could not do in the 1920s: the Tennessee antievolution law was repealed in 1967.

Americans may prefer to remember the 1960s and 1970s as the glorious Space Age, but these decades also saw the resurgence of fundamentalism and the appearance of this movement's most zealous televangelists and textbook watchers. Attempts to improve science education galvanized the old antievolutionary forces and led fundamentalists to develop new tactics. The success of the antievolution movement was obvious in 1964 when Texas rejected the BSCS biology books and nervous American textbook publishers rushed to water down or omit provocative references to evolution. When the United States Supreme Court invalidated all antievolution education laws in 1968, the antievolution movement responded with both its traditional zeal and the high-tech tactics of the New Right. The most successful strategy, based on the so-called Fairness Doctrine of the Federal Communications Commission, is known as the equal time or balanced treatment approach. Rather than call for a total ban on teaching evolution, the new fundamentalists demanded that if evolution was part of the curriculum, then Special Creationism must be taught as an alternative scientific model and that all references to evolution must mention that it is "just a theory."

The Creation Research Society, one of the organizations leading the campaign for balanced treatment, was founded in 1963 in order to "publish research evidence supporting the thesis that the material universe, including plants, animals, and man are the result of direct creative acts by a personal God." Membership in this organization was granted only to Christian men with a postgraduate degree in science who openly professed a belief in the literal truth of the Bible. Various disputes led the organization to split into several factions by the 1970s. One of the most active offshoots of the Creation Research Society was the Creation Science Research Center, formed in 1970 in San Diego, California, in order to campaign for the introduction of "scientific biblical creationism" in the public schools. Scientific creationists lobbied for state laws mandating that if biology courses included evolution, they must provide equal time for other accounts "including, but not limited to, the Genesis account in the Bible." In 1969 the California Board of Education succumbed to demands by the Creation Research Society that evolutionary theory and creationist theory be included in the state's public school biology curriculum as alternatives.

Creationists managed to expropriate the FCC's Fairness Doctrine and convince many legislators, parents, politicians, and students that their Equal Time Doctrine is a principle justified by American democratic principles of fairness and justice. Most scientists, however, find the idea that scientific theories should be evaluated by public debate and popular preference contrary to the dictates of reason and the most fundamental principles of science. Scientists argue that to teach creationism rather than evolution is like teaching Aristotelian physics and Ptolemaic astronomy as if the Scientific Revolution and the Enlightenment had never happened.

Many scientists find it impossible to believe that the theory of evolution needs defending more than a century after Darwin's death, but public opinion polls indicated that the majority of Americans did not accept evolutionary theory, especially with respect to human origins. Creationists demand that students must be told that evolution is *only* a theory. Scientists argue that the evolutionary process is a fact; the statement that natural selection is the mechanism of evolution is a theory. Moreover, as used by scientists, the term *theory* is not to be denigrated or trivialized. Theory, as in atomic theory or evolutionary theory, refers to a powerful set of explanations and not a mere hypothesis or guess.

Evolutionary theory, especially as it pertains to human origins and development, depends on evidence from geologists, paleontologists, physical anthropologists, the human fossil record, archaeology, etc. Such records are admittedly imperfect and fragmentary, and when cautiously interpreted give rise to useful, but always tentative theories. Creationism is based on a belief in the inerrancy of the Bible and takes its explanation of human origins from the Book of Genesis. Creationists believe creation occurred just once and that God created all possible species, which creationists call "kinds," for the duration of the world. Based on an essentially literal interpretation of biblical chronologies, creationists assume that the earth is only a few thousand years old and that human beings and all other living things were directly created in the not too distant past essentially in their present form. Fossils of extinct forms are explained as the remains of creatures that were destroyed by the Great Flood in the time of Noah.

Rejecting the methods of conventional, or "establishment" science, creationists and cult archaeologists have developed their own methods of ascribing specific dates to various events, artifacts, and sites. Creationists not only claim that their beliefs are justified by the word of God, as dictated and preserved in the King James English Bible, but that they can provide scientific evidence for this biblical version of creation. Scientific creationists argue that there are no discrepancies between their interpretation of geology, paleontology, biology, and the Bible. Leaders of the creation movement attempt to redefine science as "knowledge" and the search for "truth." Given this archaic, essentially medieval definition of science, with its rejection of what they call "naturalism," creationists then argue that the facts of science cannot contradict the Bible because biblical revelation is absolutely true and authoritative. Modern scientists, when exhibiting appropriate humility, acknowledge that they do not traffic in "truth" but in attempts to test hypotheses and construct theories. The true believers of the "creation movement," in contrast see themselves as committed evangelists who must go out into the world and preach their version of truth. Unlike conventional modern scientists, creationists are not trying to test a hypothesis but are involved in a grim battle between the forces of good and the forces of evil. Seeing the world in terms of absolutes, creationists argue that there can only be two possible ultimate worldviews: evolutionism or creationism. One hundred years after the death of Charles Darwin, creationists still contend that the true author of the concept of evolution is Satan himself.

At a trial popularly labeled "Scopes II," Judge William R. Overton in the U.S. District Court of the Eastern District of Arkansas, Western Division, ruled on the "Balanced Treatment for Creation-Science and Evolution-Science Act," which had been signed into law in 1981 by Frank White, then governor of Arkansas. The purpose of the law was to enforce balanced treatment of "creation-science" and "evolution-science" in the public schools of Arkansas. In a suit filed on May 27, 1981, challenging the constitutionality of the Balanced Treatment Law, the plaintiffs charged that the Act constituted an establishment of religion prohibited by the First Amendment to the Constitution, which was made applicable to the states by the Fourteenth Amendment. Second, the Act violated the right to academic freedom guaranteed to students and teachers by the Free Speech Clause of the First Amendment. Third, the plaintiffs argued that the Act was impermissibly vague and thus a violation of the Due Process Clause of the Fourteenth

A portrait of the elderly Charles Darwin

Amendment. Individuals and organizations listed as plaintiffs included the resident Arkansas Bishops of the United Methodist, Episcopal, Roman Catholic and African Methodist Episcopal Churches, the American Jewish Committee, parents of children attending Arkansas public schools, the National Association of Biology Teachers, and the National Coalition for Public Education and Religious Liberty. The plaintiffs charged that the Arkansas law threatened rather than enhanced religion by making the biblical literalism of a particular religious faction the basis of public education. The defendants included the Arkansas Board of Education and its members, the Director of the Department of Education, and the State Textbooks and Instructional Materials Selecting Committee.

Although the plaintiffs and the scientific community were not surprised that the Judge ruled against the Arkansas Balanced Treatment Act, the clear, decisive, strong ruling was unexpected. Judge Overton stated without qualification or equivocation that so-called creation science "is simply not science." The two-model approach, he concluded, is "fallacious pedagogy." Students are supposed to conclude that any evidence alleged to call evolutionary theory into question constitutes evidence for "creation science." In practice, when scientists disagree about the age of various fossils or the rate of mutation, creationists use the dispute as evidence for creationism; such "logic" is obviously specious. No reasonable person should accept disagreements about "neutral mutations" or "punctuated evolution" as evidence for a young earth and a worldwide flood. Judge Overton ruled against the Balanced Treatment Law on the following grounds: 1) Creation Science is a religion; 2) it is illegal to teach religion in the public schools; 3) Creation Science is not science.

Creationists complained that their case had not been well presented and insisted that the judge was biased. Televangelists and leaders of the Moral Majority openly accused the Arkansas Attorney General, who had defended the Balanced Treatment Law for the state, of being in collusion with the ACLU. A similar Louisiana Balanced Treatment law was brought to the U.S. Supreme Court in 1987. Although the court ruled against the Louisiana law, the decision was not as clear and forceful. Justice Brennan, who wrote the majority report, said that Creation Science as presented in the case under consideration was *religion*, but the teaching of alternative scientific views on origins would be acceptable. Special Creationists therefore claimed that the decision made it possible to bring creationism into the biology classroom as a new science. Certainly, creationists were able to take comfort and hope from the fact that despite these setbacks in the courts most public opinion polls continued to show that Americans basically supported the position that the public schools should "teach both sides." Theologians from main-stream religious denominations tend to be among the most persuasive opponents of this proposition, because they regard Special Creation as religion, not science, and it is definitely not their religion. Not all of those who revere and respect the Bible sympathize with the creationists. Many view the Bible as a set of human documents, blessed by divine inspiration, which need not be accepted literally, and find inspiration in the Bible without rejecting Darwin, Copernicus, or a spherical earth.

After winning the Arkansas decision, academics seemed ready to assume that the war was over. In contrast, creationists continued to work tirelessly on the legal, popular, and political fronts, winning considerable support from politicians, journalists, parents, judges, and educators who wish to be "fair" and open-minded. Judge Overton's decision against creationism and equal time and the Louisiana case led to renewed fundamentalist efforts at the local and state level, and more pressure on school boards, curriculum adoption committees, and the highly competitive world of textbook publishing.

Several surveys conducted in the 1980s among American college students indicated that more than 80% of the California respondents and about 30% of those in Texas and Connecticut had been taught creationism along with evolution in high school. Less than 40% of the students in Texas and Connecticut and about 8% of those in California had been taught evolution without creation; about 20% had not been taught evolution at all. Moreover, related studies of college students indicated that they did not understand what the modern theory of evolution is about. Fewer than 10% of the students polled knew that evolution refers to change in populations due to differential reproduction by differing organisms; the answer chosen most frequently was that evolution refers to a progressive Lamarckian advance from microbes to man. Significant numbers of students responded that they did not accept the scientific validity of evolutionary theory because it contradicted their own convictions. Many seemed to think that "theory" is equivalent to "guess" and that only "facts" are "certain" or "true." More than half of all students polled said that dinosaurs and humans were contemporaries, and 30–40% said there was plenty of evidence against evolution and for creationism. Significant numbers agreed that man was created about 10,000 years ago. The idea that "the Bible's account of creation" should be taught in public school was accepted by a majority of respondents in Texas and by about 40% in Connecticut and California. If a 1985 survey of 132 members of the Illinois Science Teachers' Association is representative of American science teachers, the results found among college students should not be surprising. Twenty percent of the Illinois science teachers agreed that "the Bible is an authoritative and reliable source of information" about the age of the earth and the origin of life. Many of the teachers thought that the earth is relatively young (probably less than 20,000 years old). Such surveys should give the leaders of the "balanced treatment" movement a great deal of satisfaction; they seem to have convinced many Americans that presenting both the "evolutionary model" and the "creation model" is appropriate in the interest of fair play.

The campaign for teaching Special Creationism as an acceptable alternative to Darwinian evolution raised many questions about the separation of church and state, academic freedom, and the ability of scientists to communicate with the public. The controversy over evolution proves more clearly than any other episode in the history of science that science is not limited to the laboratory, nor is it a catalog of facts; science is clearly an integral part of the history of ideas and culture.

SUGGESTED READINGS

Appleman, P., ed. (1979). *Darwin*. New York: Norton.

Bannister, R. C. (1979). *Social Darwinism: Science and Myth in Anglo-American Social Thought*. Philadelphia, PA: Temple University Press.

Barthelemy-Madaule, M. (1984). *Lamarck the Mythical Precursor*. Cambridge, MA: MIT Press.

Bowler, P. J. (1989). *Evolution. The History of an Idea*. Berkeley, CA: University of California Press.

Brooks, J. L. (1984). *Just Before the Origin: Alfred Russel Wallace's Theory of Evolution*. New York: Columbia University Press.

Burkhardt, R. W., Jr. (1977). *The Spirit of System. Lamarck and Evolutionary Biology*. Cambridge, MA: Harvard University Press.

Campbell, B. (1972). *Sexual Selection and the Descent of Man, 1871–1971*. Chicago, IL: Aldine.

Clark, Ronald W. (1968). *The Huxleys*. New York: McGraw-Hill.

Colp, R., Jr. (1977). *To Be an Invalid: The Illness of Charles Darwin*. Chicago, IL: University of Chicago Press.

Corsi, P. (1988). *The Age of Lamarck: Evolutionary Theories in France, 1790–1830*. Berkeley, CA: University of California Press.

Cronin, H. (1991). *The Ant and the Peacock: Altruism and Sexual Selection from Darwin to Today*. New York: Cambridge University Press.

Darwin, C. (1958). *The Autobiography of Charles Darwin 1809–1882*. Ed. by Nora Barlow. New York: Harcourt, Brace and Company.

Darwin, C.(1962). *The Voyage of the Beagle*. Garden City, NY: Doubleday.

Darwin, C. (1964). *On the Origin of Species*. Facsimile of the first edition. Cambridge, MA: Harvard University Press.

Darwin, C. (1965). *The Expression of the Emotions in Man and Animals*. With a Preface by Konrad Lorenz. Chicago, IL: University of Chicago Press.

Darwin, C. (1981). *The Descent of Man and Selection in Relation to Sex*. With a new introduction by J. T. Bonner and R. M. May. Princeton, NJ: Princeton University Press.

de Beer, Gavin, ed. (1983). *Charles Darwin. Thomas Henry Huxley. Autobiographies*. New York: Oxford University Press.

Degler, C. N. (1991). *In Search of Human Nature. The Decline and Revival of Darwinism in American Social Thought*. New York: Oxford University Press.

DiGregorio, M. A. (1984). *Thomas Henry Huxley's Place in Natural Science*. New Haven, CT: Yale University Press.

Eisley, L. (1958). *Darwin's Century: Evolution and the Men Who Discovered It*. Garden City, NJ: Doubleday.

Ghiselin, M. (1984). *The Triumph of the Darwinian Method*. Chicago, IL: University of Chicago Press.

Gillispie, C. C. (1951). *Genesis and Geology*. New York: Harper & Row.

Ginger, R. (1958). *Six Days or Forever? Tennessee v. John Thomas Scopes*. Boston, MA: Beacon Press.

Glick, Thomas F., ed. (1988). *The Comparative Reception of Darwinism*. Chicago, IL: University of Chicago Press.

Greene, J. C. (1959). *The Death of Adam. Evolution and its Impact on Western Thought*. Ames, IO: Iowa State University Press.

Greene, J. C. (1981). *Science, Ideology, and World View: Essays in the History of Evolutionary Ideas*. Berkeley, CA: University of California Press.

Gruber, H. E. (1974). *Darwin on Man. A Psychological Study of Scientific Creativity*. New York: E. P. Dutton & Co.

Hallam, A. (1983). *Great Geological Controversies*. New York: Oxford University Press.

Harrold, F. B., and Eve, R. A., eds. (1987). *Cult Archaeology and Creationism*. Iowa City, IA: University of Iowa Press.

Hull, D. L. (1973). *Darwin and His Critics*. Chicago, IL: University of Chicago Press.

Irvine, W. (1955). *Apes, Angels and Victorians: The Story of Darwin, Huxley, and Evolution*. New York: McGraw-Hill Book Company.

Jones, G. (1980). *Social Darwinism and English Thought*. Atlantic Highlands, NJ: Humanities Press.

Jordanova, L. J. (1984). *Lamarck*. Oxford: Oxford University Press.

Kelly, A. (1981). *The Descent of Darwin: The Popularization of Darwinism in Germany, 1860–1914*. Chapel Hill, NC: University of North Carolina Press.

King-Hele, D. (1968). *The Essential Writings of Erasmus Darwin*. London: MacGibbon and Kee.

Kropotkin, P. (1902). *Mutual Aid: A Factor in Evolution*. Boston, MA: Extending Horizon Books.

Lamarck, J. B. (1984). *Zoological Philosophy*. Chicago, IL: University of Chicago Press.

Larson, E. J. (1989). *Trial and Error: The American Controversy over Creation and Evolution*. New York: Oxford University Press.

Leakey, R. E., and Lewin, R. (1993). *Origins Reconsidered. In Search of What Makes Us Human*. New York: Doubleday.

Lovejoy, A. O. (1936). *The Great Chain of Being: A Study in the History of an Idea*. New York: Harper, reprint. 1960.

Lyell, C. (1863). *Geological Evidences of the Antiquity of Man: With Remarks on Theories of the Origin of Species by Variation*. New York: AMS Press, reprint 1973.

Mayr, E. (1982). *The Growth of Biological Thought: Diversity, Evolution, and Inheritance*. Cambridge, MA: Harvard University Press.

McNeil, M. (1987). *Under the Banner of Science: Erasmus Darwin and His Age*. Manchester: Manchester University Press.

Moore, J. R. (1979). *The Post-Darwinian Controversies*. New York: Cambridge University Press.

Morris, H. M. (1974). *Scientific Creationism*. San Diego, CA: Creation-Life Publishers.

Nelkin, D. (1982). *The Creation Controversy. Science or Scripture in the Schools*. New York: Norton.

Outram, D. (1984). *Georges Cuvier: Vocation, Science, and Authority in Post-Revolutionary France*. Manchester: Manchester University Press.

Pancaldi, G. (1991). *Darwin in Italy: Science Across Cultural Frontiers*. Bloomington, IN: Indiana University Press.

Paradis, J., and Williams, G. C. (1989). *Evolution & Ethics. T. H. Huxley's Evolution and Ethics with New Essays on Its Victorian and Sociobiological Context*. Princeton, NJ: Princeton University Press.

Playfair, J. (1956). *Illustrations of the Huttonian Theory of the Earth*. Urbana, IL: University of Illinois Press.

Pusey, J. R. (1983). *China and Charles Darwin*. Cambridge, MA: Harvard University Press.

Raven, C. E. (1986). *John Ray Naturalist: His Life and Works*. 2nd ed. New York: Cambridge University Press.

Richards, E. (1983). Darwin and the descent of woman. *The Wider Domain of Evolutionary Thought*. Ed. by David Oldroyd and Ian Langham. Dordrecht, Holland: Reidel, pp. 57–111.

Rudwick, M. J. S (1985). *The Meaning of Fossils: Episodes in the History of Palaeontology*. 2nd ed. Chicago, IL: University of Chicago Press.

Ruse, M. (1979). *The Darwinian Revolution: Science Red in Tooth and Claw*. Chicago, IL: University of Chicago Press.

Scopes, J. T., and Presley, J. (1967). *Center of the Storm, Memoirs of John T. Scopes*. New York: Holt, Rinehart and Winston.

Stauffer, R. C., ed. (1975). *Charles Darwin's Natural Selection*. New York: Cambridge University Press.

Tanner, N. M. (1981). *On Becoming Human*. New York: Cambridge University Press.

Vucinich, A. (1989). *Darwin in Russian Thought*. Berkeley, CA: University California Press.

Wallace, A. R. (1891). *Darwinism*. London: MacMillan.

Watanabe, M. (1991). *The Japanese and Western Science*. Philadelphia, PA: University of Pennsylvania Press.

Wilson, E. O. (1992). *The Diversity of Life*. Cambridge, MA: Harvard University Press.

Wilson, R. J., ed. (1989). *Darwinism and the American Intellectual: An Anthology*. Chicago: Dorsey.

Ziadat, A. A. (1986). *Western Science in the Arab World: The Impact of Darwin, 1860-1930*. New York: St. Martin's Press.

9

GENETICS

Just as the term *biology* was coined at the beginning of the nineteenth century to replace *natural philosophy*, so *genetics* was coined at the beginning of the twentieth century to separate new forms of scientific inquiry from previous studies of *generation*, *inheritance*, or *heredity*. In the 1920s genetics was still a new and fragile offshoot of natural history; within 50 years genetics could be called the new rootstock and unifying principle of the life sciences. Genetics can also be seen as a domain containing within its borders both the newest scientific developments and the oldest fragments of folklore.

The Bible reflects many ancient ideas about plant and animal breeding. Jacob, for example, put to good use the belief that offspring are influenced by what their mothers experience during pregnancy. While tending sheep for his uncle, Jacob was allowed to keep all the striped and spotted lambs. Such sports were usually very rare, but Jacob increased his allotment by peeling wooden rods into appropriate designs and showing them to the ewes. In addition to increasing the number of striped lambs, he improved their quality by showing the design only to the best of the ewes so that his own herd grew in vigor as well as in number. Twentieth-century advocates of "prenatal culture" continued to believe that maternal impression could influence the physical and mental traits of the offspring.

The relationship between the act of sexual intercourse and reproduction was obviously known in very ancient times and was exploited for practical purposes. Thus, animals and men were castrated to make them more tractable or useful, and animals with valuable traits were selected for breeding purposes. But the diversity of means of reproduction and differences in development were a source of constant confusion. Fertilization could be internal or external, the young could appear first as eggs, worm-like larvae, miniature versions of their parents, or totally different in form from the adults. Stories of strange hybrids and monstrous births are common themes in ancient myths and folk tales. Creatures such as centaurs, a cross between horses and humans, were certainly only products of the imagination, but a Sumerian proverb comments on the distance between the mule and its parents, and Greek philosophers discussed the sterility of this hardy hybrid.

Despite a general belief in ancient times that inbreeding was beneficial, eminent naturalists from Aristotle to Conrad Gesner (1516–1565) gave credence to the possibility of matings more bizarre than that of a jackass and a mare. Allegedly reliable sources reported that different species often met and mated at the water holes in Libya. This strange proclivity to cross-breeding gave rise to the saying "Something new is always coming from Libya." A cross between a camel and a leopard must have produced the giraffe, while a mating between a camel and a sparrow resulted in the ostrich. And was it not obvious that matings between humans and animals had produced such bizarre creatures as the minotaur, the manatee, and the great apes?

While many Greek myths contain examples of parthenogenesis or uniparental reproduction, such as the birth of Athena from the head of Zeus, Hippocrates argued that both parents contributed traits to their offspring. Aristotle rejected this theory and supported the idea that the male provided the *form* or blueprint for the embryo, while the female simply provided *matter* and served as a incubator. If the heat of the womb was insufficient, the paternal blueprint could not be fully executed and the unfortunate embryo developed into a female. When it came to the breeding of kings and warriors, the Spartans were, according to the story of King Archidamos II (fifth century B.C.), concerned with the physique of both parents. Archidamos was fined for marrying a short woman on the grounds that she would produce diminutive kinglets rather than robust kings.

Nineteenth-century scientists were faced with the task of reconciling ancient ideas and observations with new scientific theories concerning cell structure and function, embryology, and evolution. Rudolf Wagner (1805–1864), the prolific editor of the multivolume *Handbook of Physiology* (1842–1853), asserted that more than 300 different theories of procreation had already been published. None of them, he declared, had been satisfactory. Probably no scientist was ever more willing to pose hard questions about inheritance than Charles Darwin. In an attempt to accommodate heredity, variation, and evolution within a general framework, Darwin revived the Hippocratic theory of inheritance and renamed it the *provisional hypothesis of pangenesis*. Heredity, according to Darwin, reflected the direct transmission of qualities from parent to offspring, as influenced by external conditions. The prevailing view that inheritance involved the blending of characters from both parents created problems for Darwin's theory of natural selection, because matings between rare variants and ordinary members of the species would presumably produce offspring that were intermediate between the parental types. The favorable variation would, moreover, be even further diluted in succeeding generations. In defense of evolution, Darwin called attention to the astonishing array of odd and frivolous traits that had been selected and maintained by animal breeders and gardeners. Perhaps, Darwin suggested, domestication represented an extreme example of environmental influences that affected the reproductive system and actually stimulated variation.

According to a rather disingenuous account in his *Autobiography*, Darwin began to arrange his notes for the *Variation of Animals and Plants Under Domestication* just 2 months after the publication of the *Origin*. By Darwin's own reckoning the two volume *Variation* finally published in 1868, cost him 4 years and 2 months of hard labor. Towards the end of the second volume, Darwin presented the theory that he ruefully referred to as his "well-abused hypothesis of Pangenesis." Well-abused was an apt description, because even though Darwin's theory was (and is) regarded as highly speculative and wholly erroneous, it was one of the most widely discussed theories of inheritance in the late nineteenth century.

In preparation for attacks on his pet theory, Darwin began his exposition with a quotation from William Whewell (1794–1866), the British philosopher of science: "Hypotheses may often be of service to science, when they involve a certain portion of incompleteness, and even of error." Although Darwin probably began to formulate the provisional hypothesis in the early 1840s, he was always able to tell himself that his theory of natural selection did not require an explanation of exactly how variations arose or how they were transmitted. He did think, however, that the theory of pangenesis could explain major classes of generally accepted facts and observations concerning reproduction. Of course, some of the information about reproduction accepted without question as "facts" in the nineteenth century no longer enjoy that status. For example, Darwin assumed that the influence of the male on the female reproductive system was so profound that later offspring could be contaminated by the legacy of previous mates.

Darwin asserted that he had been "led, or rather forced" to believe that every organ, tissue, or cell of the whole body reproduced itself by giving off minute units that Darwin called *gemmules*. After circulating through the body, the gemmules were collected in the reproductive organs and incorporated into the gametes. Thus, all ovules and pollen grains, fertilized seeds, eggs, and buds consisted of multitudes of gemmules given off by "each separate atom of the organism." At conception, gemmules from both parents combined to form the embryo. In other words, the particles, or gemmules, came from all parts of the parental body and grew into the cells of the embryo and, subsequently, the tissues and organs of the new individual.

Pangenesis did not seem to throw much light on hybridism, but it was not incompatible with the possibility that novel gemmules might be formed in hybrids. Such gemmules could be transmitted to the offspring along with latent ancestral gemmules. In conclusion, Darwin summarized the complex facts subsumed by his general hypothesis with the assertion that: "The child, strictly speaking, does not grow into the man, but includes germs which slowly and successively become developed and form the man."

The provisional hypothesis of pangenesis gave Darwin a particulate theory of heredity, development, and growth. The characteristics of the new individual depended on whether the gemmules for particular traits came from the maternal or paternal line. But Darwin did not think of sexual reproduction as either the cause of variation or a means of providing diversity in a population. Usually, the characters of the parents blended, but in some cases one character exerted "prepotency," or dominance, over the other. Darwin accounted for the phenomenon known as *atavism* or reversion by postulating that some gemmules could be transmitted but not expressed. Latent ancestral gemmules might be transmitted for several generations before making their presence known. Reversion could, therefore, be attributed to dormant ancestral gemmules, which became active under certain conditions. Thus, metaphorically speaking, each animal and plant was like a garden full of seeds, some of which germinated, some of which remained dormant, and some of which perished. With appropriate manipulations, pangenesis could be used to account for the prepotency of some traits, the latency of others, regeneration, and malformations. If the limb of a salamander were cut off, limb gemmules circulating in the blood could go to the site and direct the formation of a new limb. Malformations might occur when the wrong gemmules were expressed at a specific site.

Ambiguous and amorphous as it was, the hypothesis that Julian Huxley later called Darwin's great "edifice of speculation" could not be ignored.

Skepticism about the reality of Darwins's gemmules prompted his cousin Francis Galton (1822–1911) to subject the provisional hypothesis to an experimental test. Unlike Darwin, Galton believed that the hereditary material was produced in the reproductive organs, with little if any contribution from other bodily tissues. If it were true that the gemmules were generated by body tissues and passed through the circulating fluids of the body to the reproductive organs, as Darwin suggested, a blood transfusion should transfer gemmules from donor to recipient. Rabbits that had received blood transfusion from rabbits of a different color were used in these experiments. Contrary to predictions drawn from Darwin's hypothesis, allowing the putative gemmules to be mixed by means of blood transfusions had no effect on the coat color of the offspring.

PLANT HYBRIDIZATION AND GENETICS

The controversial theory of evolution and Darwin's speculations about heredity attracted increasing attention to the problem of variation, but the work most closely associated with the development of modern genetics was the ancient and wholly respectable study of plant hybridization.

Plant hybridizers can be divided into two traditions: practical horticulturalists, who wanted to establish new and commercially useful varieties, and naturalists who were interested in basic questions, such as was it possible to create new species by cross-breeding existing types? Linnaeus, who had a profound influence on plant and animal breeders, suggested that God might have originally created one prototype for each plant genus. Modern species might then have formed as a result of variation and hybridization over the centuries. Hybridization provided evidence for Linnaeus's theory of sexual reproduction in plants. Experimental evidence for plant sexuality had been established in the 1690s by Rudolf Camerarius (1665–1721). Sexual reproduction in animals, in contrast, did not require specific scientific demonstrations.

Joseph Gottlieb Koelreuter (1733–1806) was one of the first botanists to systematically make and test hybrids. The scope of his work can best be appreciated by noting that Koelreuter carried out more than 500 different hybridization tests with 138 species and studied the shape, size and color of pollen grains from more than 1000 species. To avoid errors caused by inadvertent self-pollination, Koelreuter developed techniques for artificial hybridization: removing the anthers, pollinating by hand, and covering the flower to prevent contamination by extraneous pollen. These experiments and observations called into question widely accepted Linnaean ideas about hybrids. Linnaeus had assumed that any plant that showed characters intermediate between two known species must be a hybrid; it was not necessary, therefore, to conduct breeding tests to determine purity of type or reconstruct putative hybrids from the alleged parental lines.

According to Linnaeus, the leaves and the outer layers of the stem represented the paternal contribution, while the inner portions of the flower and the stem arose from the maternal line. Generally, Koelreuter's hybrids were intermediate between both parents, but in defiance of Linnaean theory, reciprocal crosses usually produced identical kinds of offspring. While hybrids produced from very different parents were often sterile, other hybrids exhibited greater vigor than the parents. Because Koelreuter accepted the idea that new species could not be produced, he assumed that there must be barriers to hybridization in nature. Given the fact that the initial hybrids were generally uniformly intermediate between the parents, while the next generation was extremely diverse, he assumed that the diversity of the second generation was an anomaly caused by his experimental interference with nature. After all, he had forcibly mated species that God had intended to keep separate and distinct.

Even though Koelreuter was sure he had unequivocally settled the question of plant sexuality, critics dismissed his conclusions on the grounds that he had mutilated plants by castrating the flowers, forcibly dusting them with foreign pollen, and growing them in pots rather than open fields. Surely these unnatural conditions must reduce fertility and produce monstrosities or degenerate forms rather than true hybrids. Despite such reservations, Koelreuter's work was carefully studied and extended by Carl Friedrich von Gaertner (1772–1850), a physician who turned to botanical research in order to complete the work of his father, the eminent botanist Joseph Gaertner (1732–1791). As Gregor Mendel politely noted, although Gaertner's *Experiments and Observations on Hybridization in the Plant Kingdom* (1849) was based on a formidable series of experiments with hundreds of species, it added little to the theoretical aspects of the debate about hybridization and heredity.

Many botanists observed the phenomena now recognized as dominance and segregation, but, because they were accustomed to treating the plant as a whole, they had little incentive to pay attention to the distribution of nonessential traits among the progeny of hybrids. Nevertheless, Charles Naudin (1815–1899) did call attention to the possibility of the operation of some mechanism leading to the segregation of traits in reproduction. Attempting to clarify taxonomic relationships in the potato and cucumber families, Naudin resorted to hybridization tests and became intrigued with the evolutionary significance of hybridization. Still, he continued to believe that hybrids were unnatural entities, abhorred by nature. When he discovered that the offspring of certain hybrids reverted to the parental types, he suggested that segregation of the two original species had occurred. That is, nature dissolved hybrid forms by separating the specific essences of the original species. From 1862 until 1882 Darwin and Naudin exchanged publications and letters, but Darwin did not think that Naudin's hypothesis concerning segregation in hybrids could explain reversion to distant ancestral traits. As all students of modern biology know, while Darwin was puzzling over Naudin's observations, a remarkably systematic experimental analysis of this phenomenon was being conducted in an Augustinian monastery garden by Gregor Mendel (1822–1884).

JOHANN GREGOR MENDEL

During the nineteenth century, Johann Mendel, the son of poor peasants, achieved success beyond his family's wildest expectations as a member of the Augustinian order and as the abbot of his monastery at Brno. During the twentieth century, Gregor Mendel was virtually canonized as the patron saint of modern genetics. The basic generalizations of genetics are known as Mendel's laws and the Mendelianum at the Moravian Museum in Brno has become a shrine to his memory. But in the 1960s, as scientists celebrated the Mendel Centennial, historians began the assault on the "Mendel myth" that eventuated in the charge that Mendel should not even be considered a Mendelian, much less the discoverer of the fundamental laws or theoretical framework of modern genetics. While such charges may be aimed at the wrong target, it is certainly true that the modern tendency to see Mendel's work as absolutely clear and compelling exacerbates the question of why it was ignored until about 1900 and obscures irreconcilable differences between science in the 1860s and the twentieth century. Perhaps all participants in the debate about Mendel's true goals and methods are fortunate that so few of his private papers have survived as a possible means of refuting mythmakers of all persuasions.

The story of Mendel's life has been told as both a mystery and a tragedy, because his work and ideas seemed to disappear for over 30 years. Johann Mendel was born in a village in an area that has been known as Austrian Silesia, Moravia, and Czechoslovakia. Because of Mendel's exemplary performance at the local primary school, the vicar urged his family to continue his education. Unfortunately after Mendel's father was injured and could no longer manage the farm, Mendel's financial situation became desperate. Stress, anxiety, and malnutrition led to a physical breakdown and forced him to abandon his studies until his sister Theresia generously gave him part of her dowry. With family finances still precarious, Mendel gratefully accepted the opportunity to join the Augustinian order in Brno. Entering the monastic community as Brother Gregor allowed Mendel to continue his studies and escape the bitter struggle for existence. The Brno monastery was well known as a center of learning; many of the monks taught at the local *Gymnasium* or Philosophical Institute. Several monks were involved in serious studies of natural science and experimental agriculture.

Mendel did not claim to have had a special call to enter the religious life, but he conscientiously studied theology in preparation for his new role, while also taking courses in agricultural science at the Philosophical Institute. After completing his religious instruction, Mendel assumed his parish duties, but he became depressed and ill as a result of his efforts to meet the spiritual needs of patients suffering and dying in the local hospital. Seeking an appropriate alternative, Abbot Napp dispatched Mendel to a grammar school as a substitute teacher. The schoolmaster was impressed with Mendel's work and recommended that he take the university examination that would qualify him for a regular teaching appointment in the natural sciences. At his first attempt to qualify, Mendel passed the physics and meteorology sections but failed geology and zoology. To prepare for another attempt, Mendel was sent to the University of Vienna to attend lectures in physics, chemistry, zoology, botany, plant physiology, paleontology, and mathematics.

When Mendel returned to Brno he was appointed substitute teacher of physics and natural history at Brno Technical School and became a member of the natural science section of the local Agricultural Society. He also began to study hybridization and worked out some of the problems of artificial pollination. In May 1856 Mendel made a second attempt to pass the examination for his regular teaching license. This time he seems to have experienced a psychological breakdown during the written examinations. Unable to cope with the stress, Mendel decided to forgo all further attempts at certification. He did, however, continue to pursue a variety of scientific investigations.

Mendel studied the pea weevil, grew many different strains of peas, established about 50 beehives, and attempted crosses between American, Egyptian, and European bees. Clearly, Mendel hoped to breed better bees, but instead his new strains tended to be excessively belligerent. He apparently conducted some preliminary genetic experiments on mice, but his superiors seem to have objected to this line of research. Mendel also served as Brno's correspondent for Austrian regional weather reports. Based on his daily records of temperature, humidity, rainfall and barometric pressure, he attempted to perform systematic statistical analyses of common meteorological phenomena, as well as sunspot activity and unusual storms. With Mendel's support, a novel system of weather forecasts was established to assist Moravian farmers. While Mendel's work in meteorology might be dismissed as a fruitless enterprise, his serious treatment of complex weather patterns was not unlike his approach to plant hybridization in that it was based on the careful collection of massive amounts of data.

The sacrifices Mendel and his family had made for the sake of his education were amply rewarded in 1868 when Brother Gregor was elected abbot of the monastery. Unfortunately for science, the burdens of administrative work drastically curtailed his research. Despite his new responsibilities, Mendel remained involved with local agriculture and science societies, including the Central Board of the Agricultural Society, the Brno Horticultural Society, and the Society of Apiculturists. From 1875 on he became embroiled in a dispute about taxes on monastery properties; as usual, conflict caused his health to deteriorate. His death was attributed to chronic inflammation of the kidneys and cardiac hypertrophy.

Failing to appreciate his scientific work, his successors at the monastery destroyed most of his private papers. According to his nephews, Mendel expected them to publish some of his papers posthumously, but when he died they received nothing from the monastery. Thus we have very little direct information about Mendel's preliminary ideas, sources, and plans. What little evidence remains has been collected at the Brno Mendel Museum. In any case Mendel never kept a diary and, given his position at the monastery, he had to be extremely cautious in expressing his philosophical views. He seems to have been very reserved in his relationships with his monastic colleagues, and his surviving letters reveal little of a personal nature. Some of his books survived and his annotations are rather suggestive. In his copy of Gaertner's treatise on plant hybridization, Mendel carefully marked all references to *Pisum* variants and the characteristics of pea hybrids. A copy of a German translation of the *Origin* with Mendel's own marginalia is preserved in the Mendelianum. While Mendel presumably accepted the fact of evolution and rejected the provisional hypothesis of pangenesis, his views on Darwin's attempt to prove that natural selection was the mechanism of evolution are uncertain. His nephew, Ferdinand Schindler, later recalled that Mendel had great esteem for English science and was glad to know that his nephew had learned the language of Darwin and Shakespeare.

Despite the existence of many accounts of Mendel and the origins of genetics, his inner life and sources of inspiration remain a mystery. Given the uncertainties about Mendel's ideas and the differences between modern genetics and the bare bones presented in Mendel's surviving writings, it has been possible for some historians to argue that Mendel himself was not a Mendelian and was not even interested in heredity. Perhaps it is better to simply acknowledge the difficulty of reading Mendel's work without imposing our own ideas on Gregor Mendel and the Mendelians of the early twentieth century. Other sources of information about the intellectual atmosphere of the area in which he lived and worked, the university he attended, and the books and journals available to him tell us that Mendel was not completely cut off from the scientific community of his time. Nevertheless, no complete and satisfactory explanation has emerged to tell us why and how the milieu that nourished and stimulated Mendel's work also caused it to be misunderstood and forgotten.

In the summer of 1854 Mendel began working with 34 different strains of peas; 22 kinds of peas were selected for further experiments. All the traits chosen for study were distinct and discontinuous and exhibited clear patterns of dominance and recessiveness. These preliminary experiments were apparently done to test various strains for constancy in the transmission of selected traits; they were followed by years of tedious and painstaking work. It is virtually impossible to believe that such a prodigious amount of effort would have been undertaken without a well-conceived hypothesis and a precise experimental program. From 1856 to 1863 Mendel grew and tested over 28,000 plants and analyzed seven pairs of traits. The putative hereditary particles, or *elements*, appeared to occur in pairs, with one member of each pair coming from each parent. The first generation of hybrids displayed the trait from only one of the parents, i.e. the *dominant trait*. By breeding these hybrids Mendel proved that the corresponding *recessive trait*, which had seemingly disappeared, had been dormant rather than diluted or destroyed. What is now called Mendel's first law, or the law of segregation, refers to Mendel's proof that recessive traits reappear in predictable patterns in subsequent generations.

In plants, the hereditary elements responsible for dominant or recessive traits are passed on to each generation by means of pollen grains and egg cells. For the sake of simplicity, the elements carried by the pollen grain and the egg cell may be represented by the symbols A and a. If the elements do not influence or contaminate each other, or blend during fertilization, the dominant trait (A) and the recessive trait (a), could meet at random and produce offspring of the following kinds: 1 AA, 2 Aa, and 1 aa. Mendel actually expressed the results of breeding tests using two different traits in the form A + 2Aa + a. Although there are three categories, because AA and

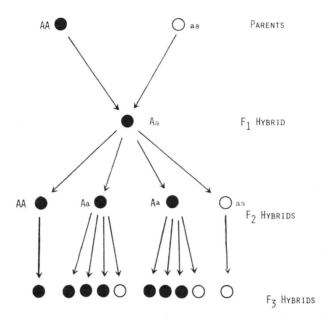

Mendel's hybridization experiments illustrating the 3:1 ratio. The diagram illustrates the results of crossing two strains of peas differing in a particular trait. The dominant trait appears in individuals with two A factors (genotype AA) or one factor A and one factor a (genotype Aa) (●). The recessive trait appears only in individuals with the genotype aa (○).

Aa are indistinguishable, the ratio of offspring would appear to be three A to one a.

In his simplest experiments, Mendel crossed strains of peas that differed in only one trait. This series of experiments demonstrated the famous 3:1 ratios, but he also studied the offspring of parental types that varied in two or three traits. Such bi- and trifactorial crosses were used to test predictions based on the hypothesis that independent nonblending factors were transmitted from generation to generation. When Mendel analyzed the offspring of such matings he found complex, but now predictable patterns of recombination. The law of independent assortment, also known as Mendel's second law, refers to this demonstration that the behavior of any pair of traits is independent of all other pairs of traits. However, Mendel did not explicitly formulate the law of independent assortment. Even after the rediscovery of Mendel's work, geneticists did not clearly explicate the second law until they

were forced to cope with embarrassing deviations from independent assortment.

The results of Mendel's experiments on peas were discussed in a paper
read to the Brno Society for Natural History in March 1865 and published
in the 1866 issue of the Society's *Proceedings*. In order to reach experts on
hybridization, including Karl von Nägeli (1817–1891), professor of botany at
Munich, Mendel acquired 40 reprints of his article "Research on Plant Hybrids." Some of these reprints were apparently still available as late as 1921
when copies were acquired by a Japanese scientist making a pilgrimage to
Brno. But in the 1860s, neither the members of the Brno Society for Natural History nor Karl Nägeli showed much interest in Mendel's ideas. Firmly
convinced that all hybrids produce variable offspring, Nägeli saw nothing of
interest in Mendel's data on the mathematical relationships among discrete
traits and he dismissed the analyses as "merely empirical, not rational." Obviously, Mendel and Nägeli differed in their understanding of the terms "rational" and "empirical." In his second letter to Nägeli (March 4, 1987),
Mendel called the expression $A + 2Aa + a$ the "empirical simple series for
two different traits." When this simple series was extended to any number
of differences between the two parental plants, Mendel felt that he had "entered the rational domain." He felt that this claim was justified because his
previous experiments proved that "the development of a pair of different
traits proceeds independently of any other differences."

Despite Nägeli's expressed reservations, Mendel believed that his results
provided an explanation for the development of hybrids. His contemporaries
and later critics, however, would say that Mendel's statistical approach to
plant hybridization did not confirm, establish, or generate a general theory
of inheritance, reproduction, development, and growth. Mendel had applied
the methods of physics and mathematical thinking to a fundamental biological problem. His method generated, predicted, and tested specific numerical
ratios, but the ratios were not of immediate interest to his contemporaries.
While Mendel's experiments clearly appear to disconfirm or falsify the
theory of blending heredity, Mendel apparently interpreted his results cautiously and did not make explicit claims about the existence of material particles in the germ cells. The reproducible pattern of ratios of types found in
the offspring of hybrids might be thought of as an "empirical regularity."
Further testing and confirmation could raise the status of such a "regularity"
to that of a new empirical law. Regular and predictable statistical patterns
would not automatically reveal the nature of the mechanism generating such
ratios, but they did support the concept that the inheritance of visible characters directly depended on the distribution of invisible factors.

Table 1 A Summary of Mendel's Results

Trait studied	Number of plants		Ratio of dominants/recessives
	Dominants	Recessives	
Length of stem: tall/short	787	277	2.84:1
Position of flower: axial/terminal	651	207	3.14:1
Shape of pod: inflated/constricted	882	299	2.95:1
Color of pod: green/yellow	428	152	2.82:1
Shape of seed: round/wrinkled	5474	1850	2.96:1
Color of cotyledons: yellow/green	6022	2001	3.01:1
Color of seed-coat: gray/white	705	224	3.01:1

Results are tabulated from Mendel's results for the F_2 generation in his seven major series of experiments. For each trait studied, the characteristic listed first is the dominant one.

Since Mendel planned further experiments with other species, Nägeli encouraged him to switch from peas to hawkweeds (*Hieracium*). Although known to Mendel for his studies of hybrids, Nägeli was irrevocably committed to his own theory of blending heredity, which involved the alleged fusion of the maternal and paternal *idioplasm* (genetic material). If Nägeli was right, Mendel's allegedly pure types in the F_2 generation were actually mixtures; further inbreeding should reveal the presence of the hidden characters. Unwilling to consider the possibility that he might be wrong, Nägeli gave Mendel little attention or encouragement and no recognition in his own publications. The scientist who had earnestly promised in his 1853 survey of botanical research that "every new method brings us new results, every new result is a source of new methods" did not mention Mendel, or his methods and results, when he published his major treatise on inheritance and evolution in 1884.

Sent down the wrong garden path by Nägeli, Mendel spent 5 years trying to hybridize *Hieracium*. Despite his valiant attempts to reproduce, verify, and validate the results he had obtained so neatly for peas, his hawkweeds refused to conform to the laws of segregation and independent assortment. By 1903 it became clear that Mendel had observed the paradigmatic case in peas; the aberrant results obtained with hawkweed were due to peculiarities in the reproductive pattern of a genus in which parthenogenesis (apomixis) is not uncommon. Discouraged by the lack of response from the scientific community, Mendel published only a few brief papers on hybridization after 1865. Those who knew him best described him as an extremely modest man, but, despite rejection and lack of recognition, he remained convinced of the fundamental value and universality of his work.

It is tempting to imagine that if Darwin and Mendel had met and discussed their research interests and problems, a grand synthesis could have been achieved by the 1870s. Creating such an imaginary grand synthesis, perhaps in a work of fiction or a "thought experiment," would, of necessity, require misinterpreting or misrepresenting complex issues embedded in nineteenth century ideas about generation, heredity, and evolution. The modern synthesis in which Darwinism and Mendelism seem to be inseparable soulmates tends to obscure the nineteenth-century intellectual context, which made it virtually impossible for Darwin and his contemporaries to see Mendelism as the solution to their problems. That is, evolutionary theory requires rethinking the problem of heredity: instead of keeping offspring like the parents, it requires some means of passing on differences and stabilizing some variations. Although there has always been great interest in the qualities of parents and offspring, human and animal, before the Mendelian revolution naturalists did not see the study of heredity as a separate area of investigation. Heredity or inheritance was part of the broad domain referred to as generation. Rather than ask how individual traits are transmitted, naturalists wanted to know how the new organism was constructed. On the other hand, although Darwin and Mendel proposed very different hypotheses about the nature of heredity, both seem to have been able to think about individual variations in terms of unit-characters rather than taking a "holistic" view of the individual as a reflection of the essence of the species.

Apparently Mendel had a clear goal in mind at the outset of his tests of garden peas for purity of type and suitability as an experimental system. Mendel worked with simple, separable traits in pure lines, and rather than looking at only the interesting cases, he counted all the progeny from his crosses. Indeed, the experimental design reported in Mendel's classic paper is so elegant that Sir Ronald A. Fisher (1890–1962), statistician and geneticist, suggested that the experiments were actually a confirmation, or demonstration, of a theory Mendel had previously formulated. Furthermore, Fisher claimed that Mendel's ratios are closer to the theoretical expectation than sampling theory would predict, and he insisted that such results could not be obtained without an "absolute miracle of chance." According to Fisher, the fit that Mendel reported would be expected only once in 30,000 repetitions.

Some skeptics accepted Mendel's results as proof of the power of pure abstract thought rather than the Machiavellian manipulation of bad data, while full-blown cynics assumed that Mendel's experiments were entirely fictitious. More generous reviewers suggested that Mendel might have simplified his presentation much as a teacher does in the classroom. Perhaps the good monk had been misunderstood by an assistant who deliberately "fudged" the results in order to simultaneously please Mendel and avoid some very tedious work. Alternatively, Mendel might have gotten his problematic ratios by unconsciously misclassifying ambiguous cases or by counting particular traits until the results looked good; when the ratios looked bad, he presumably kept classifying samples until they improved. In the 1980s, armed with more powerful statistical tools, a new breed of Mendel scholars reached the conclusion that Fisher's analysis had been faulty and that Mendel's data could be purged of the unwarranted aura of bias. In thinking about all these possibilities, perhaps we should remember where Mendel's garden was located and ask whether the prayers of this worthy monk were answered by ratios that came closer to theoretical expectations than secular statisticians would predict.

To the old charge that Mendel's ratios are too good to be true, skeptics added the argument that Mendel's independent assortment results are odd because some signs of linkage might be expected. That is, the pea has seven pairs of chromosomes, but the genes for the seven traits that Mendel selected do not lie on seven different chromosomes. Perhaps the choice of traits, arrangement of their genes, and the lack of complications due to linkage are all parts of the Mendel miracle.

MENDELIAN GENETICS: RECOGNITION OR REDISCOVERY?

Among the many attempts to explain the neglect of Mendel's work, the "obscure journal theory" might be considered the simplest. This theory is not, however, very compelling because the publications of the Brno Society were not sufficiently obscure; copies of the *Proceedings* could be found in the libraries of many universities and scientific societies throughout Europe and America. Given the numbers of scientific papers published in the 1860s, it could be said that no paper published in any reputable journal in Mendel's era was as obscure as most papers published today. Accounts of several citings and sightings of Mendel's paper have survived. For example, in

Plant Hybrids (1881), Wilhelm Focke, a German botanist, referred to Mendel's experiments on peas and the constant ratios found among his hybrid types. When George John Romanes, the Oxford biologist, asked Darwin for help with an article on hybridization, Darwin recommended Focke's treatise, which he considered unreadable because of its ponderous Germanic construction, but useful for its comprehensive bibliography. Another case, which ultimately led to a lifetime of devotion to Mendelian relics and documents, occurred in 1899 when a conscientious young man named Hugo Iltis began to prepare himself for graduate work in natural history by reading back issues of the Brno *Proceedings*. When Iltis discovered Mendel's paper, he excitedly showed it to his professor. Unfortunately, that learned man assured Iltis that Mendel's paper was of no importance, presenting as it did nothing but numbers and ratios. Later Iltis became a professor at the Brno Gymnasium and the author of a classic biography of Gregor Mendel.

Such examples suggest that it is more proper to say that Mendel's paper was misunderstood rather than unknown. Naturalists who encountered Mendel's paper apparently saw it as a collection of "mere facts," unconnected with a theoretical structure that could relate its numbers and ratios to what his contemporaries could call a *logical theory*. Another factor operating against Mendel was that the traits he studied, while admirable in an experimental system, were not the kinds of qualities that interested most biologists at the time. Scientists, as well as plant and animal breeders, considered properties like size, vigor, strength, milk and beef production in cattle, wool in sheep, speed in race horses, or intelligence in humans the kind of traits worthy of study. Such traits, however, generally do not show simple Mendelian genetics because they are governed by many genes and influenced by environmental factors.

In a practical sense, therefore, the discipline now known as classical genetics ostensibly began not with the publication of Mendel's papers in the 1860s but at the beginning of the twentieth century with the rediscovery of his laws of inheritance by three botanists, Hugo de Vries (1848–1935), Carl Correns (1864–1935), and Erik von Tschermak (1871–1962). During the intervening years, developments in the study of cell division, fertilization, and the behavior of subcellular structures had established a new framework capable of accommodating Mendel's "ratios and numbers." Yet, given the constraints of the intellectual, social, and professional milieu of the early twentieth century, it was just as difficult for the "rediscoverers" to see Mendel's work in terms of its 1860s context as it is for us to interpret Mendel's ideas, and theirs, without the distorting lens of contemporary theories. The revival of interest in discontinuous or saltative evolution was probably a key factor in the approach the rediscoverers brought to the study of heredity. That is, if the sudden appearance of new characters led to the formation of new species, studies of the transmission of clearly defined characters, i.e., Mendelian breeding experiments, might furnish the key to a new science of heredity and evolution.

Some historians argue that the original story of triple independent discovery is a myth created by geneticists who have exaggerated the sudden enlightenment that allegedly occurred at the moment of rediscovery. Certainly the rediscoverers hoped to emphasize their own creativity by claiming to have discovered the laws of inheritance before finding Mendel's paper. Having established the value of their own work, they could avoid a nasty priority battle with contemporaries by emphasizing Mendel's earlier work. Of this trio, de Vries had carried out the most extensive experimental program and was clearly the most disappointed at having credit for the discovery of the mathematical laws of inheritance ascribed to Mendel. A series of experiments begun during the 1870s was largely unsuccessful, but in 1892 de Vries resumed these experiments, using poppies with distinctive black and white markings, and soon obtained the 3:1 ratio among his F_2 plants. Self-pollination of the F_2 generation showed that the white poppies bred true, while the black ones were of two kinds: some bred true, but others produced both black and white offspring. Like Mendel, de Vries had observed the 3:1 ratio from his F_2 hybrids, the reappearance of the recessive trait, and had recognized that the F_2 plants that displayed the dominant trait could be separated into two groups distinguished by studies of their offspring. De Vries discussed these experiments in his lectures at the University of Amsterdam as early as 1896, but he chose to postpone publication until he had collected additional evidence; the delay was to prove very costly to his reputation. By 1900 he had demonstrated his law of segregation or disjunction in hybrids derived from some 15 different species.

At the turn of the century, the scientific world was ready to accept the concept of discrete hereditary factors and their independent segregation and recombination, but it was Gregor Mendel rather than Hugo de Vries who was honored as the founder of genetics. How unfair it all was, de Vries complained, to have prepared his case so carefully and then find all the credit awarded to a man who had studied peas in the 1860s. In a postscript to one of his hybridization papers, de Vries acknowledged Mendel's work as "quite good for its time," but he considered his own work more significant in terms of its scope and its linkage to his theory of intracellular pangenesis and August Weismann's (1834–1914) theory of the germplasm.

In *Intracellular Pangenesis* (1899) de Vries argued for a relationship between inherited characters and material particles or hereditary units in the cell nucleus. Emphasizing the need to study traits as separate units, de Vries proposed a theory of inheritance based on factors he called *pangenes*. Although de Vries's pangenes were just as obscure and invisible as gemmules or factors, unlike Darwin's gemmules, which supposedly circulated throughout the body, de Vries's hypothetical units remained intracellular. Relying on new developments in cell theory, de Vries proposed that pangenes were located in the nucleus of the cell, probably on the "chromatic threads." Presumably, pangenes multiplied within the cell nucleus during embryological development and passed into the cytoplasm where they became active. Microscopists had not yet observed such particles in the act of moving from the nucleus to the cytoplasm, but there was considerable evidence of the importance of the nucleus in cell function. When protozoa were cut into pieces, for example, only the part that contained the nucleus could regenerate into a new organism.

Unlike many of his contemporaries, de Vries focused his attention on a general theory of heredity and showed little interest in theories of embryological development. He was confident that a general theory of heredity would make it possible to explain the origin of species. In attempting to deal with ideas about evolution and the creation of new species, de Vries had been trying to transfer a character from one species to another by hybridization. The kinds of traits that intrigued de Vries were those that exhibited large or discontinuous variations. In *Mutation Theory* (1901, 1903) de Vries expounded his theory that evolution occurs not through gradual change, as Darwin believed, but through large discrete steps called *saltations* or *mutations*. When this two-volume treatise was translated into English, the section discussing Mendel's work was omitted. De Vries planned to publish a separate book on hybridization, but no such work appeared in English.

Experiments with mutant and normal varieties of the evening primrose (*Oenothera*) seemed to support de Vries's idea that segregation did not occur in the case of "progressive mutations," which resulted in the formation of new species. Research on mutability, de Vries predicted, would not only solve the great puzzle of the origin of species, it would lead to practical methods of inducing a species to mutate in particularly desirable directions. At that point there would be no limit to the power scientists could hope to gain over nature. Later work with the evening primrose revealed that de Vries was actually seeing the effects of gross chromosomal rearrangements, rather than simple mutations. *Oenothera* has been called a bizarre and unique genus. The "progressive mutations" that de Vries thought of as leaps towards the formation of new species might more properly be called "monster anomalies." Unlucky in his timing, de Vries was also unfortunate in his choice of an atypical model system.

For the public record, Carl Correns and Erich von Tschermak were more generous towards Mendel than de Vries; indeed they suggested the use of the terms *Mendelism* and *Mendel's laws*. But since they had been anticipated by de Vries, they did not have as much at stake in any potential priority war. Correns, who had studied with Nägeli, acknowledged that the task of discovering Mendel's laws in 1900 was much simpler than it had been in the 1860s. Growing hybrids of maize and peas for several generations and analyzing new developments in cytology apparently led Correns to think about the transmission of paired characters. Later he presented a rather dramatic account of how during a sleepless night the explanation of the 3:1 ratio came to him like lightning. Despite his familiarity with Nägeli's studies on *Hieracium*, from which he might have learned of Mendel, Correns claimed that it was Focke's section on peas that finally led him to Mendel's paper. Shaken by the discovery that Mendel had anticipated him by 35 years, the unhappy Correns soon encountered evidence that Hugo de Vries was about to lay claim to the same discovery. While Correns was eager to praise Mendel, his attitude towards de Vries has been described as malicious. Correns persistently implied that de Vries had merely imitated Mendel and had tried to suppress references to his predecessor.

Surprisingly, although Correns is chiefly remembered as a "rediscoverer" of Mendel, he eventually came to the conclusion that Mendelism was too limited in scope to provide a firm foundation for a general theory encompassing heredity, development, cytology, and evolution. His own experimental work on plants seemed to provide examples of non-Mendelian patterns of inheritance. Like many of his contemporaries, Correns was very interested in the possibility of the inheritance of acquired changes in plants and thought that some interaction between nuclear genes and the cytoplasm might be important in developmental processes.

Erich von Tschermak often complained that his contributions to genetics had been slighted. A botanist with a strong interest in practical aspects of plant breeding, Tschermak's training included work on commercial seed farms as well as advanced studies at the Universities at Ghent and Vienna. Tschermak's interest in Darwin's work on hybrid vigor led to a series of experiments on the growth promoting effects of foreign pollen. On analyzing the results of these experiments, Tschermak discovered the 3:1 ratio for hybrids between peas with yellow and green cotyledons and for smooth and wrinkled seeds. He also noted the 1:1 ratio for back-crosses of peas with green cotyledons and hybrid pollen from the second seed generation. At this point, Tschermak claimed, he was led to Mendel's paper by the citation in Focke's book. When he read Mendel's paper, Tschermak was shocked to find that Mendel had already carried out similar but more extensive experiments. Unlike Mendel's multigenerational studies, Tschermak had raised only two generations before he submitted his thesis for publication. In March 1900 Tschermak received another unwelcome surprise, a reprint of de Vries's paper "On the Law of Segregation of Hybrids." This bad news was followed by a copy of Corren's paper "Gregor Mendel's Law." Tschermak quickly prepared an abstract of his thesis and sent copies of his article on artificial hybridization to de Vries and Correns to establish himself as a participant in the rediscovery of Mendel's laws, but his claim has been regarded as marginal.

The apparently simultaneous discovery of Mendel's laws by de Vries, Correns, and Tschermak suggests that by 1900 "rediscovery" had become inevitable. Indeed, others, such as William Bateson (1861–1926), were working along similar lines. Before assuming the role of champion of Mendelian genetics, Bateson had been an evolutionary morphologist. Many aspects of Bateson's changing attitudes towards evolution, variation, embryology, genetics, and materialism are ambiguous, but it is clear that his early efforts to deal with fundamental issues in evolution and embryology had produced more frustration than enlightenment. After some unproductive attempts to use the study of embryological stages as a way to understand phylogenetic pathways in evolution, Bateson turned to the problem of heredity and variation, especially the role of discontinuous variants in evolutionary change.

During the summers of 1883 and 1884, Bateson studied the development of marine organisms under the direction of the American biologist William Keith Brooks (1848–1908), who apparently convinced Bateson that heredity was a subject worthy of intensive investigation. When Bateson returned to Cambridge as a Fellow of St. John's College, he took up the study of variation. In pursuit of relevant materials, Bateson traveled as far as Russia and Egypt. He was quite disappointed by his failure to find evidence of a causal relationship between changing environments and the appearance of variants.

In the 1890s Bateson essentially abandoned his field work and began to establish an experimental breeding program that would allow him to analyze discontinuous variations, the relationship between parental traits and those found in the offspring, and traits found in the offspring of experimental hybrids. Although disease and bad weather disrupted many of Bateson's experiments, these studies led to the publication of *Materials for the Study of Variation* (1894), a remarkable book calling for systematic investigation of the inheritance of discontinuous variations. Like de Vries, Bateson found the Darwinian concept of selection acting on small, continuous variations quite unsatisfactory. Convinced that Darwin had made a totally "gratuitous assumption," Bateson argued that discontinuous variation was an important and fundamental property of living beings, independent of natural selection.

By the time Bateson read Mendel's paper, he was already thinking along the lines of characters as discrete units. A reprint sent by de Vries led Bateson to the 1865 Brno *Proceedings* and a new mission in life as the passionate advocate and popularizer of Mendelian genetics. It should be noted, however, that Bateson did not believe that Mendelism required a belief in the existence of a specific genetic material. Long after Mendelism gave way to the modern chromosomal theory of genetics, Bateson continued to hope for the establishment of a general theory that would encompass both inheritance and embryological development. Despite some ambiguity about Bateson's personal beliefs concerning materialism, his struggle against the forces tending to separate embryology and genetics have been taken as evidence of his continued attachment to a more conservative and holistic philosophy of biology.

Many of the terms now used by geneticists were introduced by Bateson, including *genetics* (coined in 1905 from the Greek for descent), *allelomorph* (allele), *zygote, homozygote,* and *heterozygote.* In 1909 the Danish botanist Wilhelm L. Johannsen (1857–1927) introduced the term *gene* to replace older terms like factor, trait, and character. To clarify other aspects of the new science of genetics, Johannsen coined the terms *phenotype* and *genotype*, which are now used to indicate the appearance of the individual and its actual genetic makeup, respectively. Originally, however, phenotype and genotype were applied to discussions of populations rather than individuals.

Further experiments carried out by scientists in Europe and America extended Mendel's work and proved that Mendel's laws were applicable to animals as well as plants. Still, these studies did not resolve questions concerning the relationship between the inheritance of traits and evolutionary change. Indeed, orthodox Darwinists tended to view Bateson's approach as a direct challenge to the theory that evolution proceeded by the action of natural selection on continuous variations. Another assault on simplistic Mendelism was launched by scientists using a statistical approach to inheritance which was known as *biometrics.* Much ink was spilled in a wasteful war of words that raised doubts about the relative importance of continuous and discontinuous variations. Bateson counterattacked with a new book, *Mendel's Principles of Heredity* (1909). The text included a translation of Mendel's paper and Bateson's assertion that Mendel's laws would prove to be universally valid. While Bateson admitted that the biometricians might hope to contribute to the development of statistical theory, he warned readers that in the truly significant field of heredity their efforts "resulted only in the concealment of that order which it was ostensibly undertaken to reveal." Deluding themselves and others, he warned, the biometricians were delaying recognition of the value of Mendelism.

Charles Darwin's cousin Francis Galton can be called the founder of biometry, the application of statistics to biological problems. Like Darwin, Galton inherited enough money to abandon medical studies and the unpleasant prospect of working for a living. Discovering a talent for writing, he published some popular works on his travels in the Sudan, Syria, and Africa. Like Mendel, Galton devoted much of his energy to the study of weather conditions; he was instrumental in establishing the Meteorological Office and the National Physical Laboratory, but his primary interest was the inheritance of complex physical and mental characteristics in humans. This led to what he called anthropometric research, that is, quantitative and statistical studies of human populations, and the establishment of the Biometric Laboratory at University College, London. The use of fingerprints for identification was made possible by this approach.

Unlike Darwin, Galton was a talented mathematician who thought the problem of heredity could be solved through the statistical analysis of the distribution of traits in large populations. Like Francis Bacon, Galton believed in the virtues of collecting and analyzing data and hoped that the application of this methodology would eventually produce a rational society guided by a scientific priesthood. Apparently Galton believed that almost any question would yield to statistical treatment. For example, as a test of the efficacy of prayer, he analyzed mortality patterns among members of the royal family. Pious members of his family were most distressed by his statistical demonstration that the prayers of entire kingdoms had no effect on royal longevity. Similarly, he compared the rates of disasters for ships with and without missionaries. He also tried to design a quantitative scale for the measurement of beauty and love. Not surprisingly, the Galtonian approach has been criticized as a rather narrow-minded commitment to measuring whatever heritable traits are measurable and mismeasuring those that are not.

Glorying in the evidence of progress in science and industry that he saw all around him, Galton urged his countrymen to direct their own evolutionary path by applying the tools of science and mathematics to the study of human heredity and the origins of "natural ability." If human beings understood the laws of inheritance, Galton suggested, the "race of men" could be improved by getting rid of the "undesirables" and encouraging the multiplication of the "desirables." Statistical analyses of biographical encyclopedias supported his belief that distinguished individuals were most likely to be the offspring of distinguished families (not unlike the fruitful family tree that had produced Erasmus Darwin, Charles Darwin, and Francis Galton). Intellect, character, and talent, just as much as physical features, were, therefore, the products of heredity. In *Hereditary Genius* (1869) and other writings, Galton argued that genius was a conspicuous trait just as susceptible to measurement as height and eye color.

Statistical comparisons of eminent men in America and England were used to dismiss the possibility that the production of "men of genius" was related to extrinsic conditions like social advantage or disadvantage. Taking a hard position in the "nature vs. nurture" controversy, Galton contended that intelligence was determined by heredity and that attempts at education and other remedial methods could have little or no effect. He argued that human races appeared to remain stable even after migrations and adaptation to life under altered conditions. Children must, therefore, reflect ancestral traits rather than the conditions of their parents. His arguments were apparently quite persuasive, at least to men who shared his background, status, and social milieu, as indicated by Darwin's ready acceptance of Galton's belief that genius was inherited. Prior to assimilating Galton's work on hereditary genius, Darwin had assumed that all normal men were essentially the same in intellect, but varied greatly in diligence, patience, and dedication to hard work.

Galton saw variation and heredity as different aspects of a mechanism for passing on a collection of characteristics from a series of ancestors and combining them through sexual reproduction, a process that maintained variation in the population. The mean of a population, therefore, reflected "stability of type" from generation to generation. Ancestral characters, according to Galton, were transmitted by particles that could be divided and passed on in ever smaller but still effective units. Unable to explain just how this process might work, Galton focused on the statistical issues and attempted to deduce the laws of inheritance from studies of the distribution of deviations from the average in successive generations. Ancestral contributions diminished with each succeeding generation, but they were not lost. Just as certain poisons or drugs were very potent even in minute doses, tiny remnants of ancestral materials could make their presence known under the proper circumstances.

In 1883, long after he had worked out the principles of the mathematical treatment of human heredity, Galton coined the word *eugenics* from a Greek root meaning "noble in heredity" or "good in birth," but in the wake of the many twentieth-century cruelties committed in the name of eugenics, from American sterilization laws to the murderous racist doctrines of the Nazi regime, the term has assumed an irrevocably evil connotation. The goal of Galtonian eugenics was to improve the human race by applying the principles of scientific breeding to human beings. Ultimately, this was supposed to lead to the triumph of the "more suitable races" or blood lines over those disparaged as "less suitable." Perhaps, despite their love of statistics, Galton and his followers forgot that one half of the human race, when tested for any measurable trait, would fall below average. On a more intimate and immediate note, much to Galton's sorrow and regret his own eugenically propitious marriage failed to produce any evidence of the heritability of genius. Probably as the result of his own youthful indiscretions, Galton and his wife had no children.

As early as 1902 the mathematician Udny Yule (1873–1949) had suggested that the same mechanism could explain both simple Mendelian patterns and the complicated cases of blended inheritance if large numbers of factors were assumed to act together, but Mendelians and biometricians were not yet ready to compromise. Gradually it became apparent that the warring parties could be united by the *multiple gene hypothesis*. Sir Ronald A. Fisher expressed the union of the two schools of thought in 1918 in his paper "The Correlation Between Relatives on the Supposition of Mendelian Inheritance." His book *The Genetical Theory of Natural Selection* (1930) provides the classic introduction to population genetics, a discipline based on the statistical approach to patterns of inheritance. Slow, adaptive evolution, Fisher argued, could be explained in terms of the mathematics of shifting relative frequencies of genes in populations. Evolution by means of Darwinian natural selection could, therefore, be reconciled with Mendelian genetics. But, because the statistical methods used by early population geneticists ignored the implications of complex interactions among different genes and between genes and the environment, this approach has been disparagingly referred to as "beanbag genetics."

CYTOLOGY: THE STRUCTURAL BASIS OF MENDELISM

As late nineteenth-century cytologists attempted to link cell theory to theories of evolution and heredity, they became increasingly convinced that the cell nucleus was very different in form and function from the cytoplasm. New methods of preparing and staining biological preparations made it possible to see the nucleus and the special staining bodies known as chromosomes during cell division. Such studies supported the concept that the nucleus played a key role in growth, development, and heredity. Based on these observations, August Weismann (1834–1914) enunciated the guiding principles that provided a new theoretical framework for studies of cytology and heredity. The first was his theory of the continuity of the germplasm and the second was his prediction of, and rationalization for, the reduction division of the chromosomes during the formation of the germ cells. Bored with medical practice, Weismann turned to the study of zoology. In 1864, only 3 years after he joined the faculty of the University of Freiburg, Weismann experienced the first episode of a painful eye disorder that eventually forced him to abandon experimental work. It was, therefore, as a theoretician that he made his greatest, and remarkably diverse, contributions to science. Examining the major biological problems left in the wake of the Darwinian revolution, he realized that variation and inheritance were the subjects most in need of logical analysis and reassessment. Although Weismann was intensely interested in evolution, he realized that the stability of inheritance between generations was the most salient fact of heredity. Once that was understood, variation could be treated as a special case, or corollary. In other words, heredity must first be studied at the level of the cell and the individual, not in terms of evolving species or populations.

A critical examination of existing facts and theories concerning heredity, cell structure, reproduction, plant and animal breeding, and his own observations led to the suggestion that the precursors of the germ cells might be distinguishable from other cells at a very early stage in development. Moreover, in the course of development, maturation, and adult life, the descendants of the germ cells might remain separate and distinct from the other cells of the body. According to Weismann's theory of the continuity of the germplasm, there is a continuous line of descent from the germ cells of one generation to another. In other words, every organism is made up of two distinct parts: the *germplasm* and the *somatoplasm*. While the germplasm was continuously handed down to successive generations, the somatoplasm died away with each individual like the leaves of a tree in autumn. If the germ cells were not derived from the body of the parent, but from the ancestral germplasm, the concept of the inheritance of acquired characteristics could not play a role in heredity or evolution.

The germplasm was not presented as a Platonic idea to be imposed on mere matter, but as a specific component of the germ cell whose chemical and physical properties, especially its molecular structure, allowed the fertilized germ cell to become a new individual. The chemical nature of this putative cellular component was obviously unknown when Weismann published *The Germ-Plasm; a Theory of Heredity* (1893), but the theory of the continuity of the germplasm provided a framework for studies of many aspects of cell behavior, particularly the division of the nucleus and the chromosomes. Whereas *mitosis*, the division of the somatoplasmic cells, required keeping the chromosome number constant, Weismann predicted that a special *reduction division* must occur during the maturation of the germ cells in order to divide the number of chromosomes by half. Fertilization would restore the normal chromosome complement when the egg and sperm nuclei fused. By 1892 Weismann could feel secure in the conviction that his hypothesis had been vindicated by microscopic observations of the complicated steps involved in cell division. Indeed, Weismann insisted, his theory of heredity was not "theoretically superficial and cytologically impractical" like Darwin's, but was a logically based biochemical model. Ultimately, Weismann thought that the laws of inheritance would be explicable in terms of molecular movements, as indicated by his laudatory references to the doctrine of Hermann von Helmholtz (1821–1894), physicist, physiologist, and physician, that "all laws must be reduced in the last analysis to laws of motion."

By the turn of the century, some scientists had begun to suspect that there was a relationship between the classical factors tabulated in breeding experiments and the behavior of the chromosomes in mitosis and meiosis. The chromosomes were thought to fuse during fertilization in order to reconstruct the whole essence of the species. It was even possible, Weismann suggested, that each chromosome carried all the hereditary units that were needed to form a complete individual. It was known that chromosomes occurred in homologous pairs which conjugated and then separated during the formation of germ cells. Chromosome pairing was thought to provide the rejuvenating effect that many nineteenth century observers attributed to sex, but the maternal and paternal chromosome sets were thought to separate as groups. By 1910, some biologists were quite sure that the behavior of the chromosomes during gamete formation and the process of fertilization indicated that the paired factors could be on the paired chromosomes contributed by egg and sperm. This cytological approach was supported by the work of Walter S. Sutton (1877–1916) and Theodor Boveri (1862–1915), as well as Nettie M. Stevens's (1861–1912) studies of sex determination in insects and similar work carried out by Edmund B. Wilson (1856–1939).

Using sea urchin embryos with abnormal numbers of chromosomes to investigate what he called the "physiological properties" of the chromosomes in germ cell formation and embryological development, Boveri was able to test the idea that *each* chromosome carried *all* characters. Aberrations in the development of doubly fertilized eggs were traced to abnormal combinations of chromosomes in the experiments first briefly described in "Multipolar Mitosis as a Means of Analysis of the Cell Nucleus" (1902) and more fully elaborated in the next 5 years. Embryos with abnormal chromosome complements lacked certain parts and failed to develop normally. It was impossible for Boveri to identify the role of individual chromosomes, but it was clear that the chromosomes were not functionally equivalent.

Inspired by observations of the behavior of the sex chromosomes and Boveri's experiments on double-fertilized eggs, Sutton turned to the analysis of the chromosomes of grasshoppers for signs of constant and significant morphological differences among the chromosomes. Sutton had been a student of Clarence E. McClung (1870–1946), a pioneer in the study of the sex chromosomes. In 1902 McClung suggested that a special chromosome, known as the "accessory chromosome," was responsible for determining maleness or femaleness. The question of the individuality of the chromosomes was complicated by the possibility that between cell divisions, when the well-defined chromosomes were transformed into an apparently formless tangled mass of chromatin threads, their individuality might be lost. The reconstituted chromosomes that appeared at the next cell division might, therefore, be quite different individuals. Sutton concluded that the regular correspondence between the apparent size and shape of the "mother series" of chromosomes and the "daughter series" suggested the existence of morphologically distinct individuals. In 1903 Sutton concluded a paper "The Chromosomes in Heredity" with the prediction that further work would prove that the behavior of pairs of chromosomes, especially their separation during the reduction division in gamete formation, might constitute the "physical basis of the Mendelian law of heredity." Furthermore, Sutton suggested, the random assortment of different pairs of chromosomes could account for the independent segregation of pairs of genes. Sutton's papers were published while he was a graduate student with Edmund B. Wilson at Columbia University. Despite his evident skills as an investigator, Sutton never finished his graduate work; instead, he decided to become a physician.

Cytological work on the nature of the sex chromosomes was especially significant in providing evidence of the relationship between specific chromosomes and specific characters. Between 1900 and 1910 cytologists and embryologists puzzled over the role of the nucleus, the chromosomes, and the factors that established sexual differentiation. General explanations of sex determination included hypotheses that implicated external factors such as nutrition or temperature, those that postulated internal causes, and others that involved a combination of environmental and internal factors. Sex could be thought of as a trait that appeared in a 1:1 ratio, generation after generation, rather than the Mendelian 1:2:1. This could be explained if one sex was a heterozygote and the other a homozygote, but it did not explain the mystery of parthenogenesis in certain insects. Cytological studies of peculiar "unpaired" chromosomes suggested that the presence or absence of these "accessory" or "X" chromosomes might determine sex in various insects. This hypothesis was confirmed by Nettie Stevens and Wilson in 1905. Nettie Stevens is of special interest as an example of a very rare species, the early twentieth century American woman scientist. A native of Vermont and a graduate of Westfield State Normal School in Massachusetts, Stevens was able to earn both a B.A. (1889) and an M.A. (1890) from Stanford University. Her doctoral research was performed at Bryn Mawr. Although Bryn Mawr was a small woman's college, two of America's leading biologists, Edmund B. Wilson and Thomas Hunt Morgan, were associated with its faculty. Wilson went to Columbia in 1891, but he remained in close contact with Morgan, who taught at Bryn Mawr until 1904. After earning her Ph.D. (1903), Stevens won a fellowship that allowed her to study abroad and work with Boveri at the University of Würzburg. An award from the Carnegie Institution of Washington made it possible for her to conduct research on sex chromosomes at Bryn Mawr, free of teaching responsibilities, from 1904 to 1905.

Stevens saw her work as an investigation of the "histological side of the problems in heredity connected with Mendel's Law." Studies of the common mealworm demonstrated that the somatic cells of the females contained 20 large chromosomes, while those of the male contained 19 large chromosomes and one small one. Sex determination might, therefore, be caused by some unique property of this unusual pair of chromosomes. The male insects produced two kinds of spermatozoa: eggs fertilized by those containing 10 large chromosomes gave rise to females, while those fertilized by spermatozoa with the small chromosome developed into males. From these observations, Stevens concluded that there must be some intrinsic difference affecting sex in the chromatin contributed by the two kinds of spermatozoa. Conventionally, the large sex chromosome is referred to as X and the corresponding small chromosome is Y; females are then X,X and males are X,Y. Shortly after seeing Stevens's paper on the chromosomal nature of sex determination, Wilson reached a similar conclusion. His studies involved a species in which the male actually had one less chromosome than the female, rather than the more common case of the small Y chromosome. When Nettie Stevens's work is not totally ignored in modern textbooks, it is usually presented as if she had worked with Wilson or had confirmed his previous discovery. (Stevens did not even get an entry in the *Dictionary of Scientific Biography*, but she does appear in *Notable American Women*.)

The work of Boveri, Sutton, Stevens, and others suggested that the chromosomes were individuals in the morphological and the functional sense. Just how these individual entities differed and what unique hereditary qualities they carried could not be ascertained by cytological studies alone. But both Boveri and Sutton commented on the way in which the facts from cytology and Mendelian breeding studies complemented each other in supporting the hypothesis that the heredity Mendelian factors were actually constituents of the chromosomes. Sutton also predicted that each chromosome must carry many Mendelian factors because the number of characters that had already presented themselves for genetic analysis was much greater than the number of chromosomes in any cell nucleus. This meant that characters on the same chromosome should be inherited together, rather than independently. De Vries went further and predicted that during the formation of germ cells, some pangenes could be exchanged between pairs of homologous chromosomes, which he called "nuclear threads." It is unclear whether he thought that actual pieces of the chromosomes moved, or whether the chromosomes were simply vehicles, like buses or trains, that carried pangenes and remained intact when their passengers departed. At the time there was little evidence of linkage, but as more and more traits were closely studied by breeding tests, it became obvious that independent assortment was not a universal phenomenon. By 1905 there was good evidence that certain genes were always transmitted together. Although the Boveri-Sutton theory had many advantages and explained or predicted many aspects of both cytology and heredity, it did not gain immediate and universal approval. The importance of Wilson's enthusiastic endorsement and his experimental tests of the Sutton-Boveri hypothesis should not be overlooked. As the author of the *The Cell in Development and Heredity* and mentor of many prominent scientists, Wilson was instrumental in converting several skeptical scientists into advocates of the chromosome theory.

The experimental program that was most successful in exploiting correlations between breeding data and cytological observations in order to prove that genes are located on the chromosomes in a specific linear sequence was carried out by Thomas Hunt Morgan (1866–1945), and his research associates in the famous "fly room" at Columbia University (most notably Alfred H. Sturtevant, Calvin B. Bridges, Hermann J. Muller, Curt Stern), and the fruit fly *Drosophila melanogaster*. Morgan won the Nobel Prize in 1933 for his contributions to the chromosome theory of heredity, but he was a zoologist with broad interests, including cytology, descriptive and experimental embryology, and evolutionary theory. Originally trained as an embryologist, Morgan's text *The Development of the Frog's Egg; An Introduction to Experimental Embryology* (1897) was the first major book on the subject to be published in English. Indeed, his immersion in embryological concepts might account for his initial resistance to the chromosome theory. Many embryologists thought that the chromosome theory revived preformationism by assuming that adult characters were somehow preformed in the nucleus. The distinction between genotype and phenotype, which is usually attributed to Wilhelm Johannsen, but can be found in the work of August Weismann, helped to exorcise the preformationist demons.

Born in Lexington, Virginia, Morgan attended the State College of Kentucky and was awarded his B.S. in zoology in 1886. As a graduate student at Johns Hopkins, Morgan studied biology, anatomy, physiology, morphology, and embryology. After completing his doctoral research, he succeeded Wilson as head of the Biology Department at Bryn Mawr. In 1904 Wilson invited him to assume the chair of experimental zoology at Columbia. During this period Morgan made many visits to Europe and worked with founders of developmental mechanics, such as Hans Dreisch. He also remained involved in the work of former colleagues at Bryn Mawr, including Nettie Stevens and the physiologist Jacques Loeb (1858–1924). Morgan left Columbia in 1928 to establish the Division of Biological Sciences at the California Institute of Technology, where he remained until his death in 1945.

Until a visit to Hugo de Vries's garden aroused an interest in mutation theory, Morgan's primary research area was experimental embryology. Like Bateson and de Vries, Morgan was skeptical of the Darwinian doctrine of natural selection and the primacy of continuous variation as the raw material of evolution. This interest in mutations led to breeding experiments on various animals, including mice, rats, pigeons, lice, and the fruit fly. In *Drosophila melanogaster* Morgan discovered the ideal system for elucidating the complex relationships among genes, traits, chromosomes, and the statistics of recombination produced by breeding experiments. Indeed, Morgan exploited *Drosophila* so effectively it was later said that God had created this Lilliputian creature especially for him. Actually, other scientists, including W. E. Castle at Harvard, W. J. Moenkhaus at Indiana University, and Nettie Stevens at Bryn Mawr, had been breeding *Drosophila* in the laboratory some years before 1908 when Morgan began to work with it. In the 1980s, after being eclipsed by molds, bacteria, and bacteriophages, *Drosophila melanogaster* was again playing a leading role as experimental animal of choice in studies of molecular genetics, developmental biology, neurobiology, and other basic problems in metazoan biology.

Aristotle's description of a gnat produced from a larval form generated in the slime of vinegar appears to be the first scholarly reference to *Drosophila*. Originally placed in the genus *Oinopta* (wine drinker), the fly was later given the name "dew lover." The most commonly used species, *Dros-*

ophila melanogaster, which seems to have originated in southeastern Asia, probably arrived in the United States sometime before the 1870s as a stowaway in a bunch of bananas. The fruit fly has a life cycle of about 2 weeks, breeds prolifically, mates enthusiastically and indiscriminately, and is easily maintained in the laboratory. It is only about 1/8 inch in size, small enough so that thousands can be kept in a few bottles on inexpensive banana mash, but large enough to be easily studied with a low-powered microscope. A single pair of parents will quickly produce hundreds of offspring. More important, the fruit fly has only four pairs of chromosomes per nucleus and has scores of easily recognizable, even strikingly bizarre, heritable traits. Although *Drosophila* seemed a promising model of saltative evolution in the animal kingdom, the "fly group" discovered that even the most preternatural mutations did not establish a new species as de Vries had suggested.

Given Morgan's status as the champion of chromosome theory after 1910, it is not surprising that historians have been especially intrigued by his early objections to Mendelism and the chromosome theory. Geneticists, in contrast, are generally most interested in the research carried out by Morgan and his associates between 1911 and 1915. At first Morgan was quite skeptical about the universality of Mendel's work, especially after analyzing Lucian Claude Cuénot's (1866–1951) work on house mice. Later studies of coat color in mice revealed many complications, including the existence of multiple alleles, but initially Cuénot's finding of a non-Mendelian 2:1 ratio was most significant because it provided an example of what philosophers of science have called a "monster anomaly." The anomaly was resolved by postulating that certain genotypes were lethal during embryogenesis. In addition to providing exceptions to Mendelian ratios, these findings called into question the previously unexamined assumption that embryos of all possible genotypes were equally viable and would appear in the final count of phenotypes.

In 1910 Morgan sent a paper to the journal *American Naturalist* in which he argued that Mendelian factors could not possibly be carried by the chromosomes, because, if they did, characters on the same chromosome would have to "Mendelize" together. Even before this paper appeared in print, Morgan's own experiments on fruit flies had convinced him of the general validity of Mendel's laws as well as a powerful new way of demonstrating that apparent deviations from the law of independent assortment were due to *linkage*, that is, two characters being carried on the same chromosome. Linkage was first demonstrated for the sex-linked traits "white eye" and "rudimentary wings." Studies of double mutants revealed another important phenomenon, *recombination*, which could be explained in terms of crossing over, that is, the reciprocal exchange of genes between homologous paternal and maternal chromosomes. A cytological basis for reciprocal exchanges of bits of the chromosome had been provided by F. A. Janssens's (1863–1924) studies of "chiasma," that is, places were two homologous chromosomes seemed to twist around each other during the early phase of meiosis. In 1909 Janssens suggested the "chiasmatype hypothesis": chiasma might represent sites of chromosome breakage and healing. If the physical exchange of segments of chromosomal material actually occurred at such sites, the study of recombinants would represent another useful bridge between cytology and Mendelian genetics.

In fruit flies the white eye, yellow body, and rudimentary wing mutations were found almost exclusively in males because the traits are carried by the X chromosome. Because females (X,X) have two X chromosomes and males (X,Y) have only one X chromosome, reciprocal crosses involving these mutants did not produce identical results. X-linked mutations also provided a powerful test of the idea that the behavior of chromosomes establishes the physical basis of Mendelism. If genes are carried by the chromosomes in a linear array, then the degree of linkage established by mating tests should serve as a measure of the distance between genes on a chromosome. Based on the proposition that there must be a quantifiable relationship between the strength of linkage between genes and their linear sequence on the chromosome, Sturtevant constructed the first chromosome map. "The Linear Arrangement of Six Sex-Linked Factors in Drosophila, as Shown by Their Mode of Association" was published in the *Journal of Experimental Zoology* in 1913. By 1915 the Drosophila group had described four groups of linked factors which corresponded to the four pairs of chromosomes. In place of the "beanbag" image previously associated with Mendelian genetics, the chromosome theory of inheritance conjured up a picture of genes as beads on a string.

In 1926 Morgan published *The Theory of the Gene* as a summation of developments in genetics since the rediscovery of Mendel's law. But he also used the book as a means of defending his theory, refuting criticism, and marshaling evidence for the real existence of the gene. Other theories, especially those based on the belief that cytoplasmic, or extranuclear inheritance was at least as significant as the genetic material sequestered in the cell nucleus, continued to compete with so-called Mendel-Morganist genetics. By about 1930 Mendelian genetics had essentially become synonymous with orthodox or textbook genetics, but the alternative theories did not entirely vanish. Based on statistical studies of inheritance in *Drosophila,* Morgan assigned five principles to the gene: segregation, independent assortment, crossing over, linear order, and linkage groups. The chromosome theory, or as Morgan preferred to call it the theory of the gene, was not immediately accepted by all geneticists, but after visiting Morgan's laboratory in 1922, the redoubtable William Bateson professed his readiness to renounce his doubts about the chromosome theory of inheritance and pay respectful homage to the stars that had arisen in the West.

Morgan and his students saw the theory of the gene as a powerful and sophisticated generalization that explained the nature of the link between generations. Mendel's laws and the apparent exceptions to Mendel's laws could now be attributed to linkage groups, crossing over, multiple alleles, and so forth. Through his extensive writings and those of his associates in the "fly group," Morgan exerted a profound influence on the development of genetics and cytology. Sturtevant remembered the days in the fly room as a time of comraderie and cooperation, but historians have focused on signs of tensions: interpersonal friction, differences in scientific, political, and philosophical issues. Special attention is due to Hermann Joseph Muller's (1890–1967) work on induced mutations and the pressures that apparently caused him to feel that he had not been given sufficient credit for his work at Columbia.

Critics of Mendelism often pointed to the kinds of mutants studied by the "fly group" as evidence that mutations were basically "pathological" and could not, therefore, play an important role in evolution. In rebutting this argument, Morgan and his colleagues admitted that the kinds of traits studied in the laboratory were generally deleterious; this did not mean that all mutations were as striking as "rudimentary wings" or "white eyes." Presumably, many mutations produced character changes that were physiologically advantageous or deleterious but went undetected. Some geneticists continued to believe that the study of genetics and embryology could be harmoniously integrated, perhaps through studies of mutation and a new approach to "physiological genetics." Most notably, Richard Goldschmidt (1878–1958) introduced the idea that geneticists might discover mutations that affected the rates of major embryonic processes and thus gave rise to "hopeful monsters." The mutations investigated by Muller certainly did not support de Vries's hope that new species could be created in one step, but they did confirm his predictions about the unlimited horizons that would result from the experimental study of mutation.

When de Vries visited the United States in 1904, he was asked to give the inaugural lecture at the Station for Experimental Evolution at Cold Spring Harbor, New York. Emphasizing his mutation theory, de Vries suggested that the recently discovered roentgen and curie rays produced by radium might be used to induce mutations in plants and animals. Shortly afterwards, Morgan made some attempts to induce "Devriesian mutations" in various animals, including *Drosophila*, by subjecting them to radium, acids, alkalis, salts, sugars, and proteins. Failing to find macromutations or new species among the immediate offspring of the treated animals, Morgan decided that the results were not worth publishing. While further work demonstrated the value of mutants in genetic analysis, the natural rate of mutation was too low for practical quantitative studies of mutation as a process. To overcome this obstacle, Hermann Joseph Muller focused his energies on finding methods that significantly increased the rate of mutation in *Drosophila*.

While still a high school student in New York City, Muller became interested in the sciences, particularly evolution. Majoring in genetics at Columbia University, he came under the influence of E. B. Wilson and the chromosome theory of heredity. In 1910 Muller earned his B.A. and enrolled in Cornell Medical School, but he soon decided to join Morgan's *Drosophila* group at Columbia University. He earned his Ph.D. in 1915 for research on crossing-over. Known for his ingenuity in experimental design, Muller clarified obscure aspects of chromosome behavior and genetic mapping.

From 1921 to 1932, Muller, the quintessential New Yorker, was a member of the faculty of the University of Texas. By 1927 he had successfully demonstrated that ionizing radiation could be used to induce mutations in fruit flies. When Muller gave his report on X-ray–induced mutations at the Fifth International Congress of Genetics in Berlin in 1927, members of the audience immediately knew that they had been witnesses to a decisive advance in science. Geneticists believed that through Muller's discovery they had acquired the means of fulfilling de Vries's dream of deliberating transforming the hereditary material. Nevertheless, brooding over his feelings of isolation in Texas and his failure to be elected to the National Academy of Sciences (a notoriously arbitrary and capricious procedure), Muller accused his former mentor of sabotaging his professional advancement. His state of mind in 1932 can be judged by the suicide note found in his pocket when he was discovered alive but dazed after taking an overdose of sleeping pills. Perhaps this sign of emotional distress, as well as problems caused by his outspoken defense of radical ideas and socialist causes, expedited Muller's departure from Texas.

While working in Berlin in 1932, Muller and Nicolai Timofeeff-Ressovsky (1900–1981) attempted to use the induction of mutation as a means of gaining insight into the identity of the gene. If the gene is a real physical entity of a particular size and shape, bombarding it with radiation should provide some information about its nature. These rather frustrating preliminary experiments later led to a collaborative exposition of the "hit" or "target" theory of mutation by Timofeeff-Ressovsky and Max Delbrück (1935). Because of the growing threat of Hitler and Nazism, Muller left Germany and accepted Nikolai Ivanovitch Vavilov's (1887–1943) invitation to work at the Institute of Genetics, which was part of the Soviet Academy of Sciences. Muller and Vavilov, an eminent botanist, plant breeder, and geneticist, hoped to encourage the study of genetics and fight the growth of Lysenkoism, a pseudoscientific dogma that denied the very existence of the gene and the whole theoretical framework of "idealistic" Mendel-Morganist genetics. The neo-Larmackian doctrine promoted by Trofin Denisovich Lysenko (1898–1976) was chosen by societ dictator Joseph Stalin (1879–1953) and the Central Committee of the Russian Communist Party as true Marxist biology. Thoroughly disillusioned by the losing battle against Lysenkoism, after 6 tumultuous years Muller left Russia. Although he remained a socialist, he never became a Communist. At a meeting of the Lenin All-Union Academy of Agricultural Sciences in 1936, Muller compared the choice between Mendelian-Morganist genetics and Lysenkoism to a choice between medicine and shamanism, or astronomy and astrology. Of course, Muller would be very disappointed to know that more than 50 years later Russians and Americans were still likely to favor astrology over astronomy or creationism over evolution. In 1941 Vavilov was sent to prison; malnourished and ill, he died 2 years later. From 1937 to 1940 at the University of Edinburgh, Muller studied the chromosomal basis of embryonic death from radiation damage. World War II forced Muller to return to the United States. In 1945 he became professor of zoology at Indiana University, where he remained until his death.

When Muller began his systematic study of mutations, the term *mutation* was applied to a confusing collection of distinct phenomena which, from the genetic point of view, were totally unrelated. Mutation generally referred to the sudden appearance of a new genetic type, but Muller found that some so-called mutations were special cases of Mendelian recombination, some were due to abnormalities in chromosome distribution, and others were caused by changes in individual genes or hereditary units. Muller argued that, in the interests of scientific clarity, the term should be limited in usage and redefined as an alteration within an individual gene.

Studies of the evening primrose, which once formed the backbone of mutation theory, had to be set aside because the transformations involved were not due to actual alterations of specific genes, but were caused by abnormalities in the distribution of chromosomes. Beginning a full-scale attack on the problem of true mutations, now defined as heritable changes in specific genes, Muller tried to increase the frequency of mutations in *Drosophila* by means of various agents. When he embarked on this program, *Drosophila* geneticists had painstakingly identified about 100 spontaneous mutations. Muller demonstrated that it was possible to induce mutations in *Drosophila* by means of X-rays. His article on the "Artificial Transmutation of the Gene" established the new field of radiation genetics. Using X-rays, Muller produced several hundred mutants in a short time. Most of these induced mutations were stable over many generations and behaved like typical Mendelian factors when subjected to breeding tests.

The Nobel Prize in Physiology or Medicine was awarded to Muller in 1946 for his research on the effects of X-rays on mutation rates, but his interest in genetics went far beyond laboratory studies of the fruit fly. Muller exploited the publicity generated by his Nobel Prize in order to campaign against the medical, industrial, and military abuse of radiation. He remained a vocal critic of Lysenko and resigned from the Soviet Academy of Science in 1947 in protest. He was also interested in biology education, especially genetics and evolution, in the secondary schools. Ever since his undergraduate days, Muller had had a deep interest in evolution and human genetics, including a concern for the preservation and improvement of the human gene pool. Most mutations, he noted, were stable, deleterious, and recessive. Knowledge of mutation, therefore, had profound implications for eugenics and human reproduction. The eugenics movement in the United States was, however, a strange mélange of conservative and radical elements. On many occasions Muller publicly condemned the "mainline creed" of the eugenics movement as a perversion based on pseudoscience and racial prejudice which served the interests of fascists and reactionaries.

Galtonian eugenics held that since civilized values did not permit a return to nature where the unfit were simply eliminated by starvation, disease, and so forth, in order to maintain and elevate the human race society must limit the reproduction of the unfit and encourage the fit to procreate. Such ideas required replacing the gospel of *laissez-faire* with social management. Social scientists usually oppose biological determinism, while the rich generally favor the hereditary position. If the deficiencies associated with the "unfit" can be attributed to intrinsic biological factors, then attempts to improve their lot by serious remediation efforts would obviously be not only prohibitively expensive, but also inevitably futile. The hereditarian position supports the proposition that the only biologically sound solution is to keep the unfit from producing ever more unfit progeny.

In the early twentieth century interest in eugenics and intelligence testing coincided with the establishment of Mendelian genetics. Coincidence and sequence are not proof of causality, but some historians claim that the rising tide of hereditary attitudes in Western culture was, at least in part, responsible for the establishment of hereditarian theories in biology. Certainly it is true that hereditarian theories are part of the ancient nature-nurture debate, but scientists certainly will object to a history of science in which Mendelism and classical genetics are treated as the products of "imaginations disordered by ideological pressures" rather than the result of systematic and tedious experiments. Although some historians argue that the new Mendelian genetics necessarily dictated biological determination of human characteristics, with all the philosophical and social baggage such a theory implied, this is not a foregone conclusion. As the continuing controversy over evolutionary theories indicates, human beings are quite capable of seeing themselves as separate and distinct entities, above and outside the "animal" world.

Muller's commitment to eugenics was very strong, but his ideas were more subtle and complex than those who promoted the "mainline creed" based on race and class prejudice. For Muller the goal of the true science of eugenics was the admittedly distant one of using scientific understanding to consciously guide human biological evolution rather than the immediate goal of purging the world of the "unfit." As indicated by the "Geneticists' Manifesto" signed by Muller and about 20 other American and British scientists, the goal was to change attitudes towards sex and procreation so that newly advanced social and scientific attitudes would lead to a better genetic endowment for future generations. In various writings, Muller warned his audience that, in the absence of natural selection, undesirable genes would inevitably accumulate in the human gene pool until the germplasm became "riddled through with defects." His concern for the quality of the human gene pool was forcefully expressed in his 1949 presidential address to the American Society of Human Genetics on "Our Load of Mutations." Modern medicine and technology were such dysgenic forces, he warned, that our genetic load had already reduced evolutionary fitness; that is, the forces of natural selection that had previously purged the germ pool of harmful genes could not operate in the advanced nations. It was therefore necessary for geneticists to provide direction and guidance to establish voluntary but socially sanctioned eugenic reproductive controls. People with bad genes should refrain from reproducing as a social duty. Those blessed with a good genetic endowment should be encouraged to participate in positive eugenic programs, including "germinal choice." That is, suitable women should be artificially inseminated with the sperm donated by great men. In order to be sure that the men involved were truly worthy, their semen could be collected, frozen, and then used to create better babies after history had made its judgment. Open to the idea that nurture as well as nature affected human development, Muller attempted to determine the relative importance of environment and heredity, primarily by studies of twins. Nevertheless, he predicted that there would be a "complete and permanent collapse of the evolutionary process" unless humans beings, or nature, resorted to periodic inbreeding and selection.

With the success of mutation studies, Muller saw the question of the basic mechanism of evolution translated into the problem of the nature, frequency, and mechanism of mutation. His successful work with the fruit fly allowed him to serve as a gadfly to the scientific community and the eugenics movement. Because he saw eugenics as a special branch of evolutionary science, he was convinced that mutations must be of fundamental concern to eugenics. Certainly Muller raised many serious questions about human evolution and the human burden of mutations, and in doing so he challenged other scientists to question the wisdom and propriety of attempting to apply selective or eugenic measures to human beings.

SUGGESTED READINGS

Adams, M., ed. (1988). *New Perspectives in the History of Eugenics*. New York: Oxford University Press.

Allen, G. (1978). *Thomas Hunt Morgan. The Man and His Science*. Princeton, NJ: Princeton University Press.

Baltzer, F. (1967). *Theodor Boveri: Life and Work of a Great Biologist, 1862–1915*. Trans. by D. Rudnick. Berkeley, CA: University of California Press.

Bateson, W. (1902). *Mendel's Principles of Heredity—A Defense*. Cambridge, England: Cambridge University Press.

Bateson, W. (1928). *William Bateson, F.R.S., His Essays and Addresses, with a Memoir by Beatrice Bateson*. Cambridge, England: Cambridge University Press.

Bennett, J. H., ed. (1965). *Experiments in Plant Hybridization. Gregor Mendel*. Edinburgh: Oliver and Boyd.

Bowler, P. J. (1989) *The Mendelian Revolution: The Emergence of Hereditarian Concepts in Modern Science and Society*. Baltimore, MD: Johns Hopkins University Press.

Brink, A. P., and Styles, E. D., eds. (1967). *Heritage from Mendel: Proceedings of the Mendel Centennial Symposium*. Madison, WI: University Wisconsin Press.

Brush, S. (1989). Nettie M. Stevens and the discovery of sex determination by chromosomes. *ISIS 69:* 163–172.

Carlson, E. A. (1966). *The Gene: A Critical History*. Philadelphia: Saunders.

Carlson, E. A. (1981). *Genes, Radiation and Society: The Life and Work of H. J. Muller.* Ithaca, NY: Cornell University Press.

Corcos, A. F., and Monaghan, F. V. (1993). *Gregor Mendel's Experiments on Plant Hybrids. A Guided Study.* New Brunswick, NJ: Rutgers University Press.

Darden, L. (1991). *Theory Change in Science. Strategies from Mendelian Genetics.* New York: Oxford University Press.

Darwin, C. (1868). *The Variation of Plants and Animals Under Domestication.* 2 vols. New York: Orange Judd and Co.

Dobzhansky, T. (1970). *Genetics of the Evolutionary Process.* New York: Columbia University Press.

Dunn, L. C., ed. (1951). *Genetics in the Twentieth Century.* New York: MacMillan.

Fisher, R. A. (1929). *The Genetical Theory of Natural Selection.* New York: Dover.

Iltis, H. (1932). *Life of Mendel.* Trans. by E. and C. Paul. New York: W. W. Norton & Company, Inc.

Jacob, F. (1973). *The Logic of Life: A History of Heredity.* Trans. by B. E. Spillman. New York: Pantheon.

Kevles, D. J. (1986). *In the Name of Eugenics. Genetics and the Uses of Human Heredity.* Berkeley, CA: University of California Press.

Ludmerer, K. (1972). *Genetics and American Society: A Historical Appraisal.* Baltimore, MD: Johns Hopkins University Press.

Luria, S. (1973). *Life: The Unfinished Experiment.* New York: Scribner.

Mayr, E. (1982). *The Growth of Biological Thought.* Cambridge, MA: Harvard University Press.

McKusick, V. A. (1988). *Mendelian Inheritance in Man. Catalogs of Autosomal Dominant, Autosomal Recessive, and X-Linked Phenotypes.* 8th ed. Baltimore, MD: Johns Hopkins University Press.

Morgan, T. H. (1926). *The Theory of the Gene.* New Haven, CT: Yale University Press.

Morgan, T. H., Sturtevant, A. H., Muller, H. J., and Bridges, C. B. (1915). *The Mechanism of Mendelian Heredity.* New York: Henry Holt and Company.

Nardone, R. M., ed. (1968). *Mendel Centenary: Genetics, Development and Evolution.* Washington, DC: The Catholic University of America Press.

Olby, R. (1985). *Origins of Mendelism.* 2nd ed. Chicago: University of Chicago Press.

Orel, V. (1984). *Mendel.* Trans. by Stephen Finn. New York: Oxford University Press.

Orel, V., and Matalová, A., eds. (1983). *Gregor Mendel and the Foundation of Genetics.* Brno, Czechoslovakia: The Mendelianum of the Moravian Museum.

Provine, William B. (1971). *The Origins of Theoretical Population Genetics.* Chicago: University of Chicago Press.

Reilly, Philip R. (1991). *The Surgical Solution. A History of Involuntary Sterilization in the United States.* Baltimore, MD: Johns Hopkins University Press.

Riddell, N. N. (1903). *Heredity and Prenatal Culture Considered in the Light of the New Psychology*. Chicago, IL: Riddell.

Roberts, H. F. (1929). *Plant Hybridization before Mendel*. Princeton, NJ: Princeton University Press.

Russell, N. (1986). *Like Engend'ring Like: Heredity and Animal Breeding in Early Modern England*. New York: Cambridge University Press.

Sager, R. (1972). *Cytoplasmic Genes and Organelles*. New York: Academic Press.

Sapp, J. (1987). *Beyond the Gene. Cytoplasmic Inheritance and the Struggle for Authority in Genetics*. New York: Oxford University Press.

Shine, I., and Wrobel, S. (1976). *Thomas Hunt Morgan. Pioneer of Genetics*. Lexington, KY: The University Press of Kentucky.

Stern, C., and Sherwood, E., eds. (1966). *The Origin of Genetics: A Mendel Source Book*. San Francisco, CA: W. H. Freeman.

Stubbe, Hans (1972). *History of Genetics, From Prehistoric Times to the Rediscovery of Mendel's Laws*. 2nd ed. Trans. by T. R. W. Waters. Cambridge, MA: The MIT Press.

Sturtevant, A. H. (1965). *A History of Genetics*. New York: Harper & Row.

Vries, H. de. (1905). *Species and Varieties. Their Origin by Mutation*. Chicago, IL: Open Court

Weismann, A. (1893). *The Germ-Plasm; A Theory of Heredity*. Trans. by W. N. Parker and H. Ronnefeldt. New York: Schribner.

Weismann, August (1972). *The Evolution Theory*. 2 vols. New York: AMS Reprint.

Weismann, August (1989). *Essays upon Heredity and Kindred Biological Problems*. 2 vols. Chicago, IL: Open Court.

Wilson, Edmund B. (1925). *The Cell in Development and Inheritance*. 3rd ed. New York: Macmillan.

10

MOLECULAR BIOLOGY

Modern genetics can be seen as the result of the integration of three lines of investigation: statistical analysis of patterns of inheritance; microscopic studies of the intriguing behavior of subcellular entities; and the biochemical researches that elucidated the nature of various cellular components. Classical genetics could address the question of how the gene was transmitted, however, it could not answer the question of how the gene works. Until the mechanism of gene action could be analyzed, geneticists essentially used the gene as a symbol for analyzing Mendelian phenomena. Many geneticists regarded the gene as an abstract concept that was useful in organizing patterns detected by breeding experiments, rather than a material entity. Of course, in the early twentieth century many chemists felt the same way about the atom. But well before the nature of the gene had been clarified, Mendelian-Morganist genetics triumphed over alternative concepts such as soft heredity, cytoplasmic inheritance, and the inheritance of acquired characteristics. A blurring of focus concerning the actual nature of the gene allowed geneticists to think of chromosomal genes and Mendelian factors *as if* they were known to be the same entity before molecular biology transcended classical genetics and directly attacked the problem of what genes are and how they work.

The reception accorded to Mendelism, like Darwinism, varied in different countries in response to subtle social, intellectual and professional priorities and constraints. Some historians posit a vicious battle between those who thought in terms of the new nuclear genetics and those who maintained traditional, broader concerns about growth and development. This dichotomy does not seem to account for the long-term popularity and respect for Edmund B. Wilson's book *The Cell in Development and Heredity*, which first appeared in 1896. Subsequent editions provide valuable guides to prevailing views of the cell, before and after the rediscovery of Mendel's laws, as well as the state of cytology and genetics. At first the gene represented a necessary, but still shadowy and hypothetical, entity. That is, the new vocabulary of genetics helped clarify the distinction between the characters seen in breeding tests and the units transmitted through the germ cells. Early twentieth-century geneticists attempted to resolve questions about the difference between the nature of the "unit-factor" and that of the "unit-character." After T. H. Morgan established the chromosome theory, the language of biology shifted towards a new dialect of genes, alleles, genotypes, and phenotypes.

Similarly, at the turn of the century, cytologists were able to phrase questions about cell function in terms of the role of the nucleus, chromatin, and the chromosomes. Explorations of the behavior of the chromosomes during *mitosis* (somatic cell division) and in *meiosis* (the formation of germ cells) proved to be especially fruitful. Microscopists knew that the sperm and the egg contained an equal number of chromosomes, which was half the number of chromosomes found in body cells. During mitosis, the chromosomes appeared to divide longitudinally, while a special reduction division during meiosis halved the number of chromosomes distributed to the gametes. Microscopic studies had shown that fertilization involved the combination of male and female germ cells. The behavior of the chromosomes during the cell cycle and gamete formation was certainly intriguing, but all the chromosomes seemed to be essentially equivalent. Thus, the Mendelian segregation of "traits" or "characters" was generally explained in terms of the germ cells, rather than the chromosomes.

WHAT IS THE GENE?

The work of Morgan, Muller, and others on the genetics of *Drosophila* established the program for increasingly detailed studies of the mechanism of inheritance. More powerful approaches to the production and analysis of mutations, in particular, led to a quest for understanding the actual physical nature of the gene. A virtual obsession with mutations did not, however, obscure a fundamental feature of the gene, that is, its stability. Genes nor-

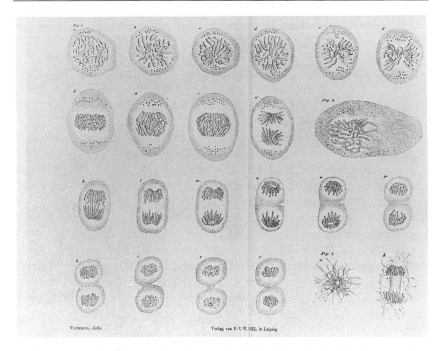

Dividing salamander larvae cells as depicted in Walter Flemming's *Cellsubstance, Kern und Zellthielung*, Leipzig, 1882

mally produce tens of thousands of accurate copies. Errors or mutations are rare, but once they occur these new forms of the gene are stable in turn. After a period in which concern for genetics overshadowed interest in evolution, the study of mutations was recognized as an integral part of the Darwinian puzzle; the mutational event was the source of variation. Natural selection acted ultimately to choose among the infrequent but critical mutations that affect the fitness of the individuals carrying them.

After Muller's discovery of the mutagenic effect of X-rays, other kinds of radiation, such as ultraviolet light, were found to have similar effects. Physicists interested in the mechanism of heredity surveyed these observations and framed new hypotheses about the nature of the gene; such speculations led to collaborations between geneticists and physicists. For example, in a paper entitled "On the Nature of Gene Mutation and Gene Structure" (1935), the geneticist Nicolai Timofeeff-Ressovsky, the physicist Karl Zimmer, and the theoretical physicist Max Delbrück drew attention to the applicability of hit and target theories of radiobiology to the mutational process. Much of the inspiration for Delbrück's formulation of a quantum-mechanical approach to the putative gene molecule was drawn from the essay "On Light and Life" (1933), by Niels Bohr (1885–1962), a founder of quantum theory. Erwin Schrödinger (1887–1962) later made Delbrück's speculations about the gene molecule and mutation better known through his book *What Is Life?* (1944), a work that has been called the "*Uncle Tom's Cabin* of molecular biology."

An understanding of the hereditary substance, Schrödinger admitted, would not soon come from physics, but from advances in biochemistry, acting in conjunction with physiology and genetics. This in itself was a remarkable admission for one of the founders of wave mechanics, but Schrödinger also confessed that he could only see one general conclusion coming from Delbrück's molecular model and that presenting this idea was his only motive for writing *What Is Life?* This general conclusion was the tantalizing possibility that while living matter might not actually elude the known laws of physics, an exploration of Delbrück's model of the hereditary substance might well reveal other, as yet unknown laws of physics. Although Schrödinger later became the almost legendary hero in the foundation myths of what had been dubbed the informational school of molecular biology, many biochemists found Schrödinger's attempts to fuse chemistry and biology superficial, naive, and misleading. Citing Schrödinger's uncritical use of analogies between the growth of crystals and cells, chemists noted that Schwann and Schleiden had been ridiculed for expressing similar ideas almost 100 years before. While several eminent molecular biologists later claimed that their move from physics to biology was influenced by Schrödinger's book, some biochemists were repulsed by the "temerity" of a physicist who wrote about the cell and the gene without an appreciation of the relevant chemical knowledge. Echoing Shakespeare, scientists who respected Schrödinger's contributions to physics attempted to warn readers of *What Is Life?* that there might well be "more things in chromosomes than are dreamt of even in wave mechanics."

Studies of cytology in the 1930s also began to focus on the question of the nature of the hereditary material. Various plants and animals provided promising materials for such studies, but just as *Drosophila*'s mating habits provided advantages for breeding experiments, its giant salivary gland chromosomes seemed specially designed for cytologists. Calvin Bridges (1889–1938) and others demonstrated a relationship between specific bands arrayed along the chromosomes and the linear sequence of genes on linkage maps. Correlations were found between certain peculiar traits and duplications of particular chromosomal sections. The effect of certain genes appeared to be related to their position on the chromosome. The recognition of chromosome banding and the discovery of means of resolving bands along mammalian chromosomes can be regarded as one of the key events in the development of modern cytogenetics. The excitement generated by Bridges's work on chromosome banding was comparable to that surrounding the methodology that has totally eclipsed it—genome sequencing. The painstaking investigations carried out during the golden age of chromosome banding provided many insights into chromosome organization, but with the historic transition from chromosome banding resolution to genome sequence analysis that marked the 1980s, it became obvious that many large questions about chromosome structure and organization have become unfashionable and unfundable and will, therefore, probably remain unresolved, even after the entire human genome has been sequenced.

The correlation between chromosomal structure and measurements of linkage became well established in the 1930s for several plant and animal species. During this time period, the development of ultraviolet microspectrophotometry and special staining techniques extended the range of cytological studies. Ingenious use of such techniques brought cytochemists tantalizingly close to understanding the chemical nature of the chromosomes. In retrospect it is possible to trace a direct path to the double helix; in practice, however, the complexity of the chromosomes and the incommensurable methodologies and assumptions that characterized chemistry and genetics created an apparently inescapable labyrinth of paradoxes and misconceptions. Indeed, some geneticists argued that it was *impossible* to imagine that any particle or molecule could constitute the genetic material. As William Bateson complained in his review of *The Mechanism of Mendelian Heredity* (1915) by Morgan and others, it was inconceivable that particles of chromatin or any other particular substance, no matter how complex, could possess the qualities appropriate to the gene. Even the most dedicated materialist, he argued, could not believe that particles of chromatin, which were apparently homogeneous and indistinguishable from each other by any known test, could "by their material nature" carry all the properties of life. Given the state of biochemical knowledge and methodology at the time, this rather pessimistic statement was not totally unfounded. The fundamental differences between particles of chromatin were not, at the time, susceptible to chemical investigation.

Nevertheless, other geneticists were sure that the gene must be a chemical entity with a specific, defined molecular arrangement. Both August Weismann and Hugo de Vries thought of the gene as a special chemical entity. The gene might be a minute particle, or a single large molecule with the property of self-duplication; the action of hereditary factors might be comparable to that of cellular ferments (i.e., catalytic proteins or enzymes). Genes might be proteins, or genes might make proteins, but the idea that the genetic material could be nucleic acid was, until the 1950s, generally considered quite unlikely. During the early decades of the twentieth century the nucleic acids, which are now known to be the material basis of heredity, were thought to be simple, repetitive, and rather uninteresting chemicals.

In many respects, the story of Johann Friedrich Miescher (1844–1895), the discoverer of the nucleic acids, is like that of Gregor Mendel. But Mendel was rediscovered, his work vindicated, his genius eulogized, and his name was attached to the fundamental laws of genetics, whereas Miescher remains a forgotten man. Indeed, the discovery of the nucleic acids is often confused with the elucidation of the double helical structure of DNA. Perhaps the lack of respect for chemistry that figures so prominently in many accounts of the discovery of the double helix and the neglect of experimental methodology, advances in instrumentation, and what practitioners of biochemistry honor as "craftsmanship" has been a major factor in the neglect of Miescher's work. That is, just as premodern physicians who used their heads rather than their hands were regarded as superior to surgeons, scientists who appear to be more interested in ideas and disciplinary boundaries than skills, instruments, methods, and empirical information receive more attention from historians.

After receiving his M.D. degree in 1868, Miescher studied physiological chemistry with Adolf Strecker (1822–1871) and Felix Hoppe-Seyler (1825–1895) at Tübingen. The project assigned to Miescher involved inves-

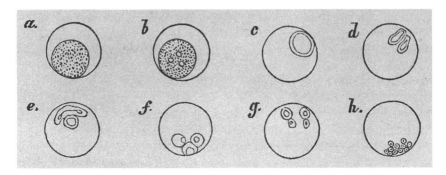

An engraving of pus cells showing the various stages of disintegration of the nucleus (Edmond Montgomery, *On the Formation of So-Called Cells in Animal Bodies*, London, 1867)

tigating the chemistry of pus cells, which were obtained by washing out bandages used in the surgical clinic. Given the high rate of postsurgical infection found in European hospitals at the time, a plethora of pus cells was always available. Attempting to prepare pure nuclei, Miescher extracted fatty materials from the pus cells with alcohol and then, to remove proteins, subjected them to an acid extract of pig gastric mucosa, which served as a crude preparation of the protein-digesting enzyme called pepsin. After this treatment, the remaining nuclei contained an organic acid, which had a remarkably high phosphorus content. The solubility properties of this organic acid, and its resistance to pepsin, suggested that it was a previously unknown cell constituent. This novel material was called nuclein. Working in Basel in the 1870s, Miescher was able to prepare nuclein from salmon sperm. Salmon were not only more pleasant to work with than pus cells, Miescher also realized that he had discovered a more productive source of nuclein, because sperm heads are essentially jet-propelled cell nuclei.

The difficulties involved in preparing nuclein should not be underestimated. Nuclein was unstable, and great care was needed to isolate it. Although Miescher recognized the problems inherent in his studies—especially the issue of the purity of his nuclein preparation—he was exasperated by what he considered excessively cruel attacks on his work. Indeed, he complained that he had struggled with biological substances so complex, under conditions so difficult, that "the real chemists shun them." Much of the criticism of Miescher's work was justifiable, but not very constructive. Some chemists claimed that nuclein was chemically undefined and might simply be an impure albuminous preparation. Critics charged that so-called nuclein was a crude, chemically undefined mixture or simply albumin contaminated with phosphate salts. Stung by harsh criticism, cautious, and reluctant to publish preliminary data, Miescher turned to safer studies of the physiology of the Rhine salmon. About 5 years before his death, Miescher resumed his chemical work on nuclein; these studies were published posthumously.

Just what physiological function Miescher ascribed to nuclein is not entirely clear. Before cytologists clarified the role of the sperm nucleus in fertilization, it did not seem possible for sperm to carry specific chemical substances that could act as fertilizing agents. But, Miescher argued, if we assume that a single substance, acting as a ferment, or through some other chemical mechanism, was the specific cause of fertilization, it would have to be nuclein. By the end of the nineteenth century, developments in histology made it possible to unequivocally establish the role of the nucleus in fertilization and heredity. Work with newly available dyes revealed the specific localization of cellular components. Walter Flemming (1843–1905), who suspected that nuclein was an important component of the cell nucleus, coined the term *chromatin* because the relationship between nuclein and the chromatic threads had not been clearly established. Miescher, however, considered his own chemical work superior to the studies of histologists, whom he contemptuously described as the "guild of dyers." Privately, Miescher speculated on the possible role of nuclein in the transmission of heritable traits. He thought its atoms could form *isomers*, or alternative spatial arrangements, which could account for variations. Still, it was not easy to see how enough variation could exist in such molecules to explain hereditary phenomena. Proteins were better known and understood and, because of their amazing variability, seemed the more logical vehicle to serve as the physical basis of heredity. Nevertheless, by the end of the nineteenth century suspicion was growing that chromatin was identical to nuclein, and that nuclein had some intimate relationship to the material basis of heredity. Oscar Hertwig (1849–1922) suggested in 1885 that nuclein was probably responsible for fertilization and the transmission of the hereditary characteristics. He argued that fertilization was a physical-chemical and morphological process, and ridiculed those who claimed that fertilization was a kind of fermentation process in which the sperm merely acted as a catalyst.

In 1895 Edmund B. Wilson pointed out that the chromosome complements contributed by the two sexes were precise equivalents and concluded that the two sexes play an equal role in heredity, even though different species vary in other aspects of reproduction and development. Because the physical basis of heredity must reside in something that the sexes contribute equally in all types of reproduction, it must reside in the chromatin. Since chromatin seemed to be essentially the same as Miescher's nuclein, this substance must be the genetic material. Studying nuclein from thymus and yeast, Albrecht Kossel (1853–1927) proved that there were two kinds of nucleic acids. Considered a pioneer of physiological chemistry, Kossel was awarded a Nobel Prize in 1910. The two types of nuclein prepared by Kossel, long known as *thymus nucleic acid* and *yeast nucleic acid*, are now designated deoxyribonucleic acid (DNA) and ribonucleic acid (RNA). Both nucleic acids contain the bases adenine, cytosine, and guanine, but in DNA the fourth base is thymine and in RNA it is uracil. DNA and RNA also differ in their sugar component, containing deoxyribose and ribose, respectively. RNA was initially thought of as the exclusive property of yeast cells, but better chemical techniques demonstrated its presence in all cells, including bacteria. Some scientists thought that RNA might serve as an energy reservoir, capable of supplying phosphorus atoms for cellular metabolism. Waves of confusion rather than enlightenment seemed to follow in the wake of these discoveries, but even as early as the 1830s the redoubtable Justus von Liebig had been driven to complain that the rapid pace of scientific progress could make one crazy.

Paradoxically, further chemical studies of the nucleic acids and microscopic studies of chromatin during the period from about 1910 to 1930 militated against the concept that nuclein or chromatin could serve as the hereditary material. Chromatin seemed to behave very peculiarly during the cell cycle; at times it apparently vanished. This did not seem consistent with the characteristics proper to the genetic material, particularly the need for stability from generation to generation. Finally, work carried out by Phoebus Aaron Levene (1869–1940) seemed to rob nuclein of complexity, another fundamental requisite of any chemical claiming to be the genetic material. After receiving his M.D. degree from St. Petersburg Imperial Medical Academy in 1891, Levene left Russia and emigrated to New York. While practicing medicine in New York City, he studied chemistry at Columbia University. Five years later, while recuperating from the effects of overwork and tuberculosis, he decided to abandon his medical practice. Totally dedicated to chemical research, Levene joined the newly formed Rockefeller Institute for Medical Research. During his association with that institution he investigated almost every important class of biological compound, from sugars and lipids to proteins and nucleic acids. His prodigious labors resulted in more than 700 papers in the *Journal of Biological Chemistry*.

Levene's analyses of the nucleic acids suggested that the four bases were present in equimolar amounts in nucleic acids obtained from many different sources. These data led to the *tetranucleotide interpretation*, an explanatory device that began as a working hypothesis but soon became the primary paradigm of nucleic acid chemistry. A minor revolution in biochemical concepts, and a great deal of hard chemical work, was needed to dismantle the multiple layers of explanation subsumed by the tetranucleotide theory. In essence, Levene's chemical analyses indicated that there were equal amounts of all four bases in DNAs from all sources. On a more sophisticated level of organization, the tetranucleotide interpretation claimed that all DNA polynucleotides were combinations of a fixed and unchangeable sequence of units, which were themselves a combination of four nucleotides. Ultimately, this led to the conclusion that DNA was a highly repetitious polymer, analogous to glycogen, and, therefore, incapable of generating the diversity that was essential in the genetic material.

Although several studies conducted in the 1940s suggested that DNA might be the hereditary material, this possibility was dismissed because nucleic acids were thought to be rather boring tetranucleotide polymers.

Erwin Chargaff in 1953

When scientists thought about the chemical complexity required by the genetic material, proteins seemed to be the only logical candidates, at least until the work of Erwin Chargaff and other chemists challenged prevailing ideas about DNA. But as Chargaff said, it was only in the post–Watson-Crick era that the nucleic acids rose from minor curiosities to the center of biology thinking. Chargaff's work can be seen as an example of the linkage between changing ideas and changing methodologies that facilitates the emergence of new experimental possibilities, if not immediate revolutions in theories. In addition, Chargaff's initial assumption that nucleic acids might be as complicated, highly polymerized, and chemically diverse as proteins reflects the fruitful intersection of very different research programs. According to Chargaff, it was Oswald Avery's 1944 paper on the transformation of pneumococcal types that led him to redirect his research to DNA. If different DNAs did indeed have different biological activities, as indicated by Avery's report on the transforming principle, they must differ chemically, and the tetranucleotide theory must be wrong. If DNA was as complex as protein, its biological activity might depend on specific structural patterns or sequences, but there were no chemical techniques capable of revealing the architecture of the nucleic acids. Nevertheless, Chargaff argued that in the 1940s it was possible to see the beginnings of a new chemistry of heredity that would eventually establish a new grammar of biology.

Sequencing nucleic acids was patently impossible in the 1940s when many fundamental aspects of the structure of the nucleic acids were still unknown. Using the methods that first became available after World War II—paper chromatography, ultraviolet spectrophotometry, and ion-exchange chromatography—Chargaff proved unequivocally that the four bases found in DNA were not necessarily present in equal amounts. Precise comparative studies of nucleic acids from different sources, such as beef thymus, spleen, liver, human sperm, yeast, and tubercle bacilli, disproved the assumption that all DNAs were identical to calf thymus DNA. However, the base composition of DNA from different organs of the same species was constant and characteristic of the species. Although at the time no theory could account for his observations, Chargaff noted that the molar ratio of total purines to total pyrimidines and the ratio of adenine to thymine and guanine to cytosine from different sources were always about one. Given the state of DNA chemistry in 1949, such "regularities" could only be described as "striking," even if they were "perhaps meaningless," not unlike Mendel's "numbers and ratios." Further evidence of these regularities suggested that, despite the lack of any recognizable periodicity in nucleotide sequence, some fundamental structural principles must be shared by DNAs from different sources. From Chargaff's work it was clear that, like the proteins, nucleic acids were complex, interesting, and distinct entities. Further research proved that DNA was a macromolecule, that is, a long-chain structure of surprisingly high molecular weight, as revealed by measurements of viscosity and the rates of diffusion and sedimentation of carefully prepared solutions of DNA.

All too often, the history of molecular biology is treated as a grand epic of charismatic and eccentric personalities and brilliant guesses and ideas, while the importance of new physical instruments, techniques, and experimental craftsmanship are neglected. Thus, it is important to remember that during this period the ability to think about biological problems at the molecular level was made possible by the availability of new instruments, such as the density gradient centrifuge, the scintillation counter, and the electron microscope. Clearly, the history of the nucleic acids from Miescher to Chargaff suggests the need to reconsider the role of such technical factors in expanding the scope and depth of investigations of major biological problems. Similarly, the history of viral genetics involves much more than just the history of the bacteriophages that are so closely associated with Max Delbrück, the origins of molecular biology, and the path to the double helix. It should be noted that the nucleic acids were also of interest to scientists involved in research on plant viruses. In 1935 Wendell Meredith Stanley (1904–1971) applied methods that had recently been used to crystallize proteins to purify tobacco mosaic virus (TMV), the pathogen largely responsible for the establishment of virology.

TMV had been intensively studied by plant pathologists; it was also a favored model system in the development of the techniques essential to modern biochemistry and molecular biology, including centrifugation, crystallography, X-ray diffraction, immunology, and electron microscopy. Stanley found that even after repeated recrystallizations, TMV retained its physical, chemical, and biological properties. One year later, the presence of nucleic acid in highly purified TMV was established, but until 1957 it was unclear whether TMV protein, nucleic acid, or nucleoprotein was the infectious component. Most of the plant viruses examined by the 1940s contained RNA rather than DNA. Crystallizing a virus made it the "right organism for the job," if the job involved physical and biochemical aspects of genetics, but made it more difficult to think about viruses as organisms, that is, *living systems*. Biochemists were able to come to terms with this paradox by thinking about or implicitly redefining "viral life" in terms of the possession of a genetic system in which mutation and recombination of characters could occur.

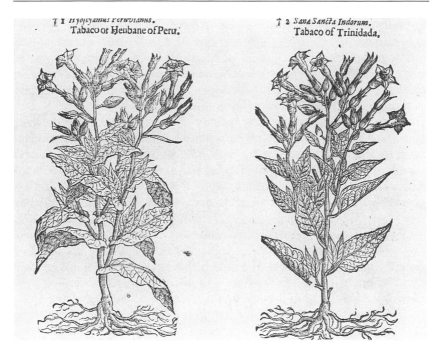

Tobacco plants as depicted in John Gerard's *The Herball* (London, 1633)

During the 1940s most chemists, physicists, and geneticists thought that the genetic material must be a protein, but another significant line of research on a phenomenon known as *transformation* suggested the nucleic acids played a fundamental role in inheritance. The first well-known series of experiments to challenge the assumption that genes must be proteins or nucleoproteins was carried out in the laboratory of Oswald T. Avery (1877–1955). In some respects, Avery's work was a refinement of an approach previously developed by Fred Griffith (1877–1941), a well-respected British bacteriologist. Various strains of *Diplococcus pneumoniae* were known to cause pneumonia in humans. These bacterial strains could be differentiated by measuring the ability of bacteria-free filtrates of young cultures to precipitate antibodies produced by rabbits that had been injected with known bacterial types. By 1920 several specific serological types of pneumococcus had been identified and differentiated; the difference between certain strains involved the polysaccharide capsule that envelops the bacterium. The encapsulated forms produced smooth, shiny, rounded colonies (S) when grown in vitro. Variants characterized by their lesser virulence were found to produce rough colonies of bacteria (R) that lacked the slimy polysaccharide capsule.

An intriguing but puzzling set of observations of pneumococci was reported by Griffith in 1928; these experiments suggested that various strains could undergo some kind of transmutation of type. From the blood of mice injected with living R pneumococci together with a heat-killed S type, Griffith isolated living virulent bacteria. This indicated that the living nonvirulent R form had acquired something from the dead S form that had caused transformation into deadly organisms with the S-type polysaccharide capsule. In retrospect, it can be said that Griffith had observed genetic transformation, but he probably did not realize that the phenomenon he had discovered involved the transfer of hereditary material. Sadly, Griffith died before this explanation emerged. In 1941, during the blitz of London, a German bomb blew up the dilapidated laboratory where Griffith and his colleague William Scott were working, killing them both. Several scientists confirmed Griffith's observations and proved that pneumococcal transformations could be effected in the test tube as well as in the animal body. The agent responsible was found to pass through a filter that could hold back bacteria. Intrigued by the possibility of finding some means of modifying the virulence of pathogenic bacteria, bacteriologists followed these experiments with more interest than geneticists. A paper published by Oswald Avery, Colin Macleod, and Maclyn McCarty in 1944, however, demonstrated that the phenomenon demanded the attention of geneticists and biochemists.

Having isolated the active transforming principle from S-type pneumococci, Avery and his colleagues identified it as a "highly polymerized and viscous form of sodium desoxyribonucleate." Tests for the presence of protein proved negative. Furthermore, the transforming principle was inactivated by crude preparations of an enzyme known to attack DNA, but it was unaffected by enzymes that broke down RNA and proteins. Although Avery noted that the transforming principle had been compared to genes and viruses, he was more cautious about publishing speculations about the genetic implications of these findings than his younger colleagues would have liked. In private he acknowledged that these studies of the transforming principle bristled with implications for the biochemistry of DNA, genetics, enzyme chemistry, cell metabolism, and so forth, but he thought it dangerous to publish speculations that might prove embarrassing later. The possibility that the biological activity of his preparations could be due to minute amounts of some other substance associated with DNA could not be excluded. Nevertheless, Avery concluded that the available evidence strongly suggested that DNA was the transforming principle. If further tests confirmed this hypothesis, scientists would have to think about DNA not merely as a structural component of the chromosomes, but as "functionally active in determining the biochemical activities and specific characteristics of pneumococcal cells." The chemistry of DNA would also, Avery noted, require further study in order to explain the biological specificity that must be inherent in the genetic material. That is, Avery and his associates realized that it was not possible to reconcile the biological activity of their DNA preparations with prevailing ideas about DNA chemistry. This enigma would not be solved until the acceptance of the double-helical structural model of DNA proposed by Francis Crick and James D. Watson in 1953. But between the publication of Avery's paper and that of Watson and Crick, many other observations were made that provided clues about the molecular basis of inheritance.

Work with bacterial viruses was especially significant in providing the basis for a direct assault on what Watson and Crick rather melodramatically referred to as the secret of life. The path that led to the Watson-Crick collaboration was largely determined by Watson's membership in the "phage group," followers of the German-born physicist Max Delbrück. Members of the group saw themselves as a special inner circle of scientists plugged into a privileged network of instant information that did not reach outsiders until many months later in the form of papers published in learned journals. Because of Delbrück's profound influence on the discipline that came to be known as molecular biology, he was sometimes referred to as the "pope" of the "phage church." Attitudes towards Delbrück among members of the phage group are closely linked to Schrödinger's alleged role in the foundation of molecular biology, a field that Chargaff defined as the practice of biochemistry without a license. Probably Delbrück's most important contribution was the organization in 1945 of a summer phage course at the Cold Spring Harbor Laboratory for Quantitative Biology; this led to the emergence of the American phage group, which regarded Delbrück as its intellectual leader.

Molecular biologists considered the bacteriophage a perfect model system for simple, rapid, and precise experiments probing the mechanism of reproduction. Most of the early phage studies were carried out by a small number of Americans working within the shared ideology and paradigms of Delbrück's phage group. Other scientists, such as André Lwoff, the French microbiologist, were also devoted to phage studies but adopted a different research program. Honored as a master craftsman rather than a would-be pope, Lwoff was especially interested in the role of growth factors in the life cycle of microorganisms and the phenomenon known as lysogeny. Lysogenic bacteria are those that quietly harbor viruses (prophage) within their own ge-

Max Delbrück, Nobel Laureate, 1969 (California Institute of Technology)

nome. Under appropriate circumstances the latent viruses can be induced to take over control of the host cell and produce large numbers of bacteriophages. Bacteriophages were also found to be capable of causing genetic transformation of bacteria by serving as the vehicles of new heritable traits. As a result of the work by Lwoff and his associates, geneticists could no longer dismiss bacteria as peculiar organisms with "no genes, nuclei, or sex." Although bacterial cells do not have a nucleus, structures analogous to the chromosomes of higher cells were discovered in bacteria. Joshua Lederberg, who was awarded the Nobel Prize in 1958, proved that genetic recombination occurs as a result of sexual matings of bacteria. Such research demonstrated that, in many ways, bacteria and bacteriophages were more favorable systems for an attack on the chemical nature of the gene than higher organisms. Indeed, viruses are often described as naked genes clad in protein coats. For many of the participants at the 1946 Cold Spring Harbor Symposium on Microbial Genetics, where Joshua Lederberg and Edward Tatum announced their discoveries concerning the sex life of bacteria, the meeting marked a new phase in the campaign to convince biologists that bacteria and viruses did indeed have mutable genes and chromosomes and that microbial genes undergo recombination and replication processes very much like the genes of plants and animals.

Participants later referred to the 1946 symposium as the landmark event at which the new discipline of molecular biology acquired "body and soul." Those who considered themselves outsiders noted that the members of the orthodox phage church had gathered around their three bishops: Max Delbrück, Alfred Hershey, and Salvador Luria, who shared the Nobel Prize in 1969. It was Delbrück, however, who defined the dogma of the orthodox faith. Ideas and results reported from outside the church were rejected by Delbrück and treated as heresy. This makes it rather surprising to realize that Delbrück and his disciples were among the many skeptics who resisted the idea that DNA could be Avery's transforming material. Because the possibility existed that minute amounts of contaminants in nucleic acid preparations might be the actual genetic material, a great deal of effort was devoted to purification of transforming factors. Even when highly purified DNA preparations proved to be effective transforming agents, skeptics could still argue that minute amounts of some other substance still attached to the nucleic acid might be the actual transforming principle. Not until the announcement of the results of the Hershey-Chase experiments of 1952 was there general agreement that DNA must be taken seriously.

Between 1951 and 1952 Alfred Hershey and his associate Martha Chase carried out experiments that were rapidly accepted by members of the phage group as proof that DNA served as the genetic material. Phage workers were beginning to think that bacteriophages might act like tiny hypodermic needles full of genetic material; empty virus coats might remain outside the host cell after the genetic material had been injected. To test this possibility, Hershey and Chase used radioactive sulfur to label phage proteins and radioactive phosphorus to label the DNA. After allowing phages to attack bacterial cells, the infected bacterial cultures were spun in a blender and centrifuged in order to separate intact bacteria from smaller particles. Hershey apparently began these experiments with considerable skepticism about the idea that nucleic acids would enter the bacterial cell. His colleague Thomas F. Anderson later recalled a day at Cold Spring Harbor Laboratory when he and Hershey discussed the "wildly comical possibility" that phage DNA could enter the bacterial cell and act like the transforming principle. Many scientists were surprised when the experiments indicated that the joke was "not only ridiculous but true."

Alfred D. Hershey, Nobel Laureate, 1969 (Director, Genetics Research Unit, Carnegie Institute of Washington)

On finding that most of the phage DNA remained with the bacterial cells while phage protein coats were released into the supernatant fluid, Hershey and Chase concluded that the protein served as a protective coat used by phages for adsorption to the bacteria and as an instrument for the injection of phage DNA into the host cell. The protein coat, however, had no function in the growth and multiplication of phages within the bacterial cell. Because their technique for separating labeled phage protein coats from infected bacterial cells introduced some uncertainty, Hershey and Chase could not claim that their data provided unequivocal proof that DNA was the genetic material. Somewhat tentatively, they concluded that further chemical work was needed to determine the identity of the genetic material, but that their experiments indicated that phage protein had no role in phage multiplication, while DNA apparently had some function.

Although the Hershey-Chase experiment did not in itself provide compelling proof that DNA was the genetic material, Hershey was a member in good standing of the phage group, and the Hershey-Chase experiment quickly transformed other members into true believers. Indeed, a junior member of the phage group named James Watson accepted the Hershey-Chase experiment as powerful proof that DNA was the primary genetic material. When the U.S. State Department refused Salvador Luria permission to attend the meeting of the Society for General Microbiology held at Oxford in April 1952, Watson went instead of his major professor. In presenting an account of the Hershey-Chase experiment, Watson defended the position that the data proved that DNA was the genetic material, probably much more vigorously than Hershey himself would have.

Thus, by 1953 phage workers had for the first time what Delbrück and Hershey called a workable party line. Although DNA was gaining acceptance in some quarters, because its structure was still obscure and misunderstood, it was impossible to establish a logical relationship between the chemical nature of the gene and the mechanism by which the gene acted as the vehicle of inheritance. This dilemma was resolved when James Watson and Francis Crick described a model of DNA structure which immediately suggested explanations for its biological activity. The success of the Watson-Crick model amply confirms one of the aphorisms attributed to Francis Crick: "If you can't study function, study structure."

WATSON-CRICK AND THE DNA DOUBLE HELIX

Rejecting any show of undue modesty, molecular biologists have hailed the elucidation of the three-dimensional structure of DNA as one of the greatest achievements of twentieth-century biology, comparable to the legacy of Darwin and Mendel. In brief, Watson and Crick based their model on a combination of guesswork and the inspection of Rosalind Franklin's (1920–1958) X-ray data. Several X-ray crystallographers had previously attempted to determine the three-dimensional structure of DNA. Outstanding for his pioneering work, and for his own estimate of his place in the history of science, William T. Astbury (1898–1961) referred to himself as the "alpha and omega, the beginning and end of the whole thing." The early, crude X-ray diffraction patterns of DNA suggested an arrangement of long polynucleotide chains with the rather flat nucleotides arranged perpendicularly to the long axis of the molecule. In 1947, about the time that Erwin Chargaff was disproving the tetranucleotide hypothesis, Astbury concluded that the degree of organization indicated by his X-ray findings supported Levene's theory of the regularity of DNA composition. Although his speculations about the regularity of the sequence of bases in DNA were in error, Astbury had demonstrated that DNA had a regular crystalline structure that might be clarified by the techniques of X-ray crystallography.

In Cambridge in 1951, James D. Watson, who seemed a typical enough American postdoctoral researcher, met Francis Crick, 12 years older but still a graduate student. Two years later both Watson and Crick were world famous. Both attributed their success to their special relationship: their ability to complement, criticize, and stimulate each other. Nonetheless, despite their intellectual compatibility, the contrasts between these two scientists were striking.

A former Quiz Kid, Watson entered the University of Chicago when only 15 years of age for an experimental early admissions program that allowed him to take special courses in the biological sciences. Although enthusiastic about ornithology, Watson did not seem especially interested in anything else. Teachers remembered Watson as a student who appeared to be completely indifferent to everything going on in the classroom but always came out at the top of the class. Formal training in genetics was negligible at the undergraduate level at the time, but Watson claimed that after reading Erwin Schrödinger's *What Is Life?*, finding the secret of the gene became his mission and obsession. Having been rejected by Harvard and Caltech, Watson decided to do his graduate work at Indiana University, which was a major center of genetic research in the 1940s. Distinguished members of the faculty included H. J. Muller, the *Drosophila* geneticist, Tracy Sonneborn, a pioneer in *Parmecium* genetics, and Salvador Luria, who was working with phage and bacteria. While Muller seemed the obvious choice as a thesis ad-

James D. Watson, Nobel Laureate, 1962

visor, Watson decided that *Drosophila*'s better days were over. The more progressive geneticists were now working with microorganisms, and sure that phage were preferable to paramecia, Watson decided to work for Luria, a friend of Max Delbrück, the hero of Schrödinger's *What Is Life?*. Thus, while still a graduate student, Watson became an acolyte of the phage church. Based largely on a paper Watson wrote for a course taught by Sonneborn, historians have suggested that the controversial German biologist Franz Moewus (1908–1959) might have had a greater influence on Watson than Schrödinger. Given the fact that Moewus was accused of having fabricated much of his data, Schrödinger would provide a more respectable inspirational figure. In his 1948 essay "The Genetics of *Chlamydomonas* with Special Regard to Sexuality," Watson suggested that while Moewus's ideas were certainly important, it was hard to believe that all of the experiments he reported had actually been carried out. The genetic analyses reported by Moewus, Watson concluded, seemed to represent "wishful thinking" rather than facts obtained by experimentation.

While Watson's remarkable memoir *The Double Helix* disingenuously suggests that he left Indiana University totally innocent of chemistry, physics, and the scientific literature, his transcripts and doctoral dissertation on the effects of X-rays on phage replication provide some evidence to the contrary. Having received his doctorate at 22, Watson was sent to Europe in hopes that further training in the Old World would produce a veneer of maturity and sophistication. After a year in Copenhagen, Watson transferred to Max Perutz's group at the Cavendish Laboratory to work on the molecular structure of nucleic acids extracted from plant viruses. At Cambridge, Watson found a kindred soul in Francis Crick, who shared his enthusiasm for genetics and its most essential part, the secret of the gene.

Francis Crick was born in 1916, in Northampton, England, where his father and uncle ran the shoe factory their father had founded. As a child Crick became interested in science and began to worry that everything of interest would have been discovered by the time he grew up. Rejecting his family's strict Congregationalist beliefs, Crick became a chronic skeptic and agnostic. Physics and chemistry, rather than vitalistic philosophies, would, he decided, provide the solution to problems thought to be utterly mysterious. After receiving a degree in physics from University College, London, in 1937, Crick began research on a rather dull problem concerning the viscosity of water under pressure. His research project was interrupted by World War II. As a member of the Admiralty Research Laboratory at Teddington, Crick carried out work on mine detection and development. When a mine blew up the apparatus he had constructed for his graduate work, Crick was able to put aside research on the viscosity of water. After the war, Crick hoped to do basic research in particle physics or physics applied to biology. Like several other physicists who made the transition to biology, Crick attributed his conviction that physicists could solve fundamental biological problems to Schrödinger's *What Is Life?*.

After several abortive research projects involving X-ray crystallography of proteins and nucleic acids, Crick joined Max Perutz and John Kendrew, who were planning to use X-ray diffraction to study the structure of proteins. Although colleagues acknowledged Crick's intelligence and intensity in argumentation, he seemed to find it impossible to settle down to one topic long enough to construct a doctoral thesis. Indeed, the paper that would lead to the 1962 Nobel Prize in Medicine or Physiology was published while Crick was still struggling to convert his "ragbag" of research topics into a presentable doctoral thesis.

In personality Crick and Watson may have been poles apart, but when they met both immediately felt they had found a collaborator who shared a special approach to biology. Watson was described as a loner, quiet, and introverted; in contrast, Crick was famous for his loud voice and laugh, an extrovert Watson described as the brightest person he had ever worked with and an ingenious theoretician who never had a "modest moment." Their enthusiastic collaboration illustrates Watson's talent for establishing useful scientific relationships, while Watson's rather calculated and devious relationship with Maurice Wilkins, an X-ray crystallographer working at King's College, London, provides another example of a peculiar but ultimately productive association. The Watson-Crick collaboration was a fortunate one, for Crick doubted that either he or Watson could have discovered the structure of DNA alone. In contrast, Wilkins and Rosalind Franklin (1920–1958), another X-ray crystallographer at King's College, were unable to collaborate effectively; indeed, Wilkins and Franklin were barely on speaking terms. According to Crick, the structure was there waiting to be discovered, and if he and Watson had not done so, the puzzle might have been solved by Rosalind Franklin, Maurice Wilkins, Linus Pauling, or by further refinements of biochemistry. Instead, working together, Watson and Crick arrived at the structure of the DNA double helix by a combination of guessing, model building, and the unacknowledged exploitation of Rosalind Franklin's X-ray crystallographic data.

In their first *Nature* paper of 1953, Watson and Crick described their "radically different structure for the salt of deoxyribose nucleic acid." Incorporating the usual chemical assumptions and evidence that the structure was a two-stranded double helix, the novel feature of the structure was "the manner in which the two chains are held together by the purine and pyrimidine bases." The planes of the bases in the proposed double helix were perpendicular to the fiber axis of the polynucleotide chain, and the bases on the two strands were joined by hydrogen bonds. The dimensions of the double helix indicated that a purine on one chain always paired with a pyrimidine on the other chain; specifically, adenine always paired with thymine and guanine with cytosine. Although any sequence of bases was possible on one

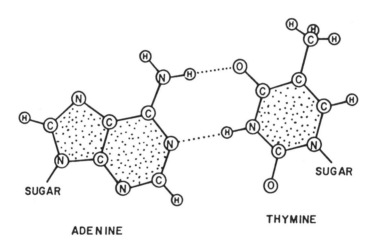

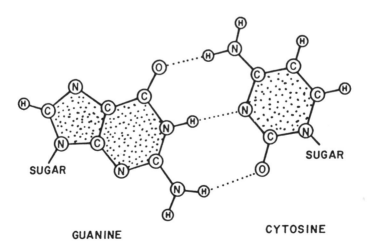

Diagram of the complementary base pairs found in the DNA double helix. Adenine pairs with thymine and guanine pairs with cytosine.

chain, the rules of base pairing automatically determined the sequence of bases on the other chain. Thus, the Watson-Crick double helix immediately explained Chargaff's data concerning the molar ratios of purines to pyrimidines and how DNA exhibits order and stability, as well as variety and mutability.

Despite the irresistible elegance of the double helical DNA model, many scientists felt that the Watson-Crick paper had a rather hollow tone. The X-ray data available in the literature at the time were, as the authors admitted, insufficient for a rigorous test of the hypothetical structure. More exact data, however, appeared in the two communications that followed the Watson-Crick proposal. One paper by Maurice Wilkins and his coworkers discussed the available X-ray data for calf thymus DNA, nucleoprotein preparations from sperm heads, and bacteriophages. A paper by Rosalind Franklin and R. G. Gosling presented the most refined X-ray diffraction patterns of DNA then available, and Franklin demonstrated that the double helix was consistent with the X-ray patterns for DNA.

Although Watson and Crick only alluded to the question of how the genetic material might function at the molecular level, the double helix immediately suggested means by which DNA molecules could be used to make copies exactly like the original. In a sentence described as one of the most coy statements in the scientific literature, Watson and Crick proclaimed: "It has not escaped our notice that the specific pairing we have postulated immediately suggests a possible copying mechanism for the genetic material." Crick, who insisted on some discussion of the genetic implications of the model, had settled for this "compromise" because of his coauthor's lingering uncertainty. Shortly after their first paper in *Nature*, Watson and Crick elaborated on the genetic implications of their model.

With the promulgation of the Watson-Crick model, researchers were soon engaged in transforming the *chromosome theory* into the *nucleic acid theory*. The gene was no longer an abstract *factor*, but an *information molecule*. Molecular genetics also provided the concepts of messages and feedback regulation. Studies of the mechanism of gene action and protein synthesis provided a way of dealing with the ancient paradox of cellular differentiation from identical cells. In his memoir *What Mad Pursuit* (1988), Crick assessed what he called the "classical period" of molecular biology, from the discovery of the DNA double helix in 1953 to about 1966 when the genetic code was elucidated. Crick, who began his scientific career as a physicist, argued that elegance and simplicity can serve as guides to researches in physics, a field in which basic laws can usually be expressed in exact mathematical terms, but the "laws" of biology are usually broad generalizations, reflecting the view of François Jacob that "evolution is a tinkerer."

Many scientists were left to wonder why Watson and Crick had been so successful in discovering a model for DNA when they began what has been called the race for the double helix so far behind their competitors Rosalind Franklin and Maurice Wilkins in London and the ingenious chemist Linus Pauling in California. Significantly, most of the explanations have focused on personal issues and interactions rather than specific scientific knowledge and technical opportunities. In retrospect, it is easy to see that the group at King's College was too split by personal incompatibilities and misunderstandings to collaborate effectively and that Linus Pauling's approach was based on obsolete and incomplete information. Pauling could not assess the new data available at King's College because he was unable to travel freely. The State Department had taken away his passport for allegedly subversive activities that primarily involved participation in the Ban the Bomb movement, for which Pauling won the 1962 Nobel Peace Prize. In contrast, having recognized the biological significance of DNA and the value of combining X-ray diffraction data with model building, Watson and Crick had the advantage of being at the heart of an active grapevine from which they were able to pick the ripe data of other members of the phage group and the progress reports of their competitors and even letters that Linus Pauling sent to his son Peter at Cambridge. Together and separately, Watson and Crick picked the brains of colleagues and visitors, such as Maurice Wilkins and Erwin Chargaff, for critical insights.

Despite their obvious advantages and intelligence, for two years Watson and Crick stumbled along the wrong path, making all possible incorrect choices for the structure of DNA. When they finally put together Chargaff's data on the base ratios and Franklin's study of the B-form of DNA with information about the proper forms of the bases, they were able to build a rational, elegant, and biologically interesting model of DNA in about 3 weeks. Despite the plethora of books and articles about the discovery of the structure of DNA, the most entertaining and controversial account of this remarkable burst of activity is still Watson's book *The Double Helix*. While Watson's account of the personalities involved in the story spared no one, his portrait of Rosalind Franklin was particularly harsh and, according to others who worked with her, totally fallacious. Franklin died of cancer in 1958 at only 38 years of age. She did not know that Watson, Crick, and Wilkins would share the 1962 Nobel Prize for the discovery of the DNA double helix, nor did she realize how her unpublished data had been used by Watson and Crick, because Watson's autobiography was not published until 10 years after her death.

It took about 5 years for experimental tests to prove the essential validity of the DNA replication scheme proposed in 1953. In 1958 Matthew Meselson, then a graduate student working with Linus Pauling, and Frank W. Stahl, a postdoctoral associate of Max Delbrück, traced the fate of parental strands of DNA labeled with radioactive nitrogen by the technique known as equilibrium density gradient sendimentation. They proved that DNA replication was semiconservative, that is, each old DNA strand became associated with a newly synthesized strand of DNA. Other important insights were provided by Arthur Kornberg, who in 1956 isolated an enzyme from *Escherichia coli* that catalyzed the synthesis of DNA. In 1959 Kornberg shared the Nobel Prize with his former mentor Severo Ochoa for their work on the biological synthesis of the nucleic acids.

Satisfied that the chemical nature of the gene had been discovered, scientists were able to link knowledge of what the gene was to studies of how the gene worked. Watson and Crick played an important role in formulating the general principles explaining how information stored in DNA is replicated and passed on to daughter molecules and how information was transferred from DNA to the metabolic activities of the cell. Even before the structure of DNA had been elucidated and confirmed, Watson wrote himself a message: "DNA → RNA → protein." The arrows in this slogan did not refer to actual chemical transformations, but represented the transfer of genetic information from the base sequences of DNA to the amino acid sequences in proteins. Watson's cryptic slogan about the unidirectional flow of information from DNA to RNA to protein has been called the central dogma of molecular biology. Already certain that DNA was the template for RNA, Watson predicted that RNA chains might serve as templates for protein synthesis. Eventually, scientists would discover that information could flow from RNA to DNA by means of an enzyme called reverse transcriptase; this insight has been critical to understanding a group of viruses called retroviruses, one of which is HIV (human immunodeficiency virus), the virus associated with AIDS.

Given the assumption that genes determined the amino acid sequence of proteins, the double helix structure suggested that instructions for building proteins must be encoded in the base sequence of DNA. Because DNA was found in the nucleus, whereas protein synthesis seemed to occur in the cytoplasm, some means of transmitting a copy of a gene from the nucleus to the cytoplasm was necessary. It was fairly easy to imagine that the hypothetical copy of the gene existed in the form of an RNA molecule made from DNA according to the rules of base pairing. The process of transferring information from a DNA template to an RNA copy is known as *transcription*; the information is still present in the language of nucleic acids. The really difficult conceptual problem was imagining how it could be possible to use any form of nucleic acid to establish the sequence of a protein. Transferring information from the language group of nucleic acids to that of proteins is known as *translation*; the difficulty factor in this step can be compared to translating ancient Egyptian hieroglyphics into modern English. In other words, how can about 20 amino acids, and a seemingly infinite number of proteins, be described by nucleic acid molecules containing only four bases? Presumably this could be done in a manner analogous to the way that Morse code uses dots and dashes to represent all the letters of the alphabet, an alphabet that can be used to generate an infinite number of meaningful (and meaningless) texts.

In the case of the genetic code, taking the bases three at a time provides $4 \times 4 \times 4 = 64$ triplet combinations, or *codons*. Speculations about the possibility that special "cavities" along the DNA molecule could serve as direct templates for the 20 amino acids led Crick to the adaptor hypothesis. Given the unlikelihood that strands of nucleic acid could serve as direct templates for the assembly of amino acids into polypeptide chains, Crick proposed the existence of 20 adaptors and 20 enzymes that could join an amino acid to its adaptor. The combination would then fit itself to a specific messenger RNA (mRNA) so that the amino acids could be linked together as proteins. By 1957 biochemists had confirmed the existence of adapter molecules, which were called *transfer RNAs* (tRNA); each amino acid had at least one specific tRNA. In 1965 Robert Holley and coworkers determined the complete sequence of the tRNA for alanine. Holley shared the 1968 Nobel Prize for Medicine or Physiology with Har Gobind Khorana and Marshall Nirenberg for their "interpretation of the genetic code and its function in protein synthesis."

Crick and other theoretical biologists expected to solve the genetic code by logic and guesswork, but Marshall Nirenberg, Heinrich Matthaei, Severo Ochoa, Har Gobind Khorana, and others showed that, with truly Herculean labors, the puzzle could be solved experimentally. When the codons were first worked out and studied in several different species, scientists exuberantly called it the *universal genetic code*. As exceptions and complications eventually emerged, the more modest term *standard code* came into use. Crick's frustrating attempts to demonstrate the relationship between a mutation in a gene and its altered protein literally brought tears to his eyes. Working with the British biochemist Vernon M. Ingram, Crick planned to look for alterations in lysozyme, an enzyme found in egg white and human tears. Each day Crick would bring a slice of raw onion to his eyes and collect the tears. Although Ingram could easily demonstrate the difference between chick and human lysozyme, efforts to find mutant forms of chicken lysozyme were unproductive. Ingram was successful, however, in demonstrating that the mutation responsible for the disease known as sickle cell anemia caused a change in one amino acid in the hemoglobin molecule.

With molecular biology ostensibly firmly established in the 1960s, Crick decided it was time to move on and revolutionize other backward fields, such as embryology and developmental biology, neurobiology, and biogenesis. Initially attracted to the study of development and differentiation, Crick found the obstacles to progress in this field unacceptably intractable. Eventually other problems, such as the nature of mind and the origin of life, claimed Crick's attention. In 1981, Crick published a book called *Life Itself: Its Origin and Nature*, based on the *panspermia* theory introduced by the physical chemist Svante Arrhenius (1859–1927). According to Crick's version of "directed panspermia," life on earth might have arisen, not by random biochemical reactions in the primordial soup, but from bacteria or "seeds" deliberately sent to earth via a rocket launched by a higher civilization that had evolved elsewhere in the universe. This willingness to speculate on untestable ideas has been lauded as evidence of Crick's eclectic curiosity, creativity, and boldness in pursuit of theories, even theories that appear to be closer to science fiction than to orthodox science. After spending a sabbatical year at the Salk Institute for Biological Studies in La Jolla, California, Crick was offered an endowed chair and persuaded to stay. He decided to turn to research on the brain and soon discovered that the state of the neurosciences in the 1980s was rather like that of molecular biology and embryology in the 1920s and 1930s: the major questions were unanswered and probably unanswerable without new techniques and new ideas.

When the double helix was announced it served as a guide for much research, but where and how protein synthesis occurred remained obscure. Protein synthesis seemed to require a messenger from nucleus to cytoplasm, as summarized in Watson's slogan: "DNA makes RNA makes protein." In the 1950s it was known that most of the cytoplasmic RNA was found in small particles called *ribosomes,* which consisted of RNA and proteins. Although some molecular biologists assumed that the RNA in the ribosomes was the messenger and that each ribosome was involved in synthesizing one protein, experimental evidence indicated that ribosomal RNA was very stable and seemed to come in two fixed lengths rather than the assortment that should characterize the messengers presumably associated with highly diversified proteins. The puzzle was solved by a series of elegant experiments, including the famous 1959 PaJaMo experiment. This experiment was performed at the Pasteur Institute by Arthur Pardee, François Jacob, and Jacques Monod. At the time, Monod was studying bacterial growth factors and their inducible enzymes. Jacob and Elie Wollman had been studying the mechanism of the transmission of genetic information during bacterial mating. The ingenious Waring blender experiment that Jacob and Wollman had devised to determine the sequence of genetic transfer between bacterial cells by violently interrupting the mating process had been called a case of "molecular coitus interruptus" by Crick. The "blender experiment" was rather reminiscent of Spallanzani's experiments on frogs involved in the act of mating.

Working in a laboratory between those of Jacques Lucien Monod (1910–1976) and André Lwoff, François Jacob realized that both of these scientists had been working on different phases of the same genetic mechanism; that is, Lwoff's studies of the induction of the prophage in lysogenic bacteria and Monod's induction of enzyme synthesis resulted from gene expression made possible by the removal of a repressor. The sudden insight that came to Jacob in 1958 led to a new theoretical framework known as the "operon theory." Jacob and Monod's operon theory provided an explanation for the induction of enzymes by means of generating the synthesis of new enzyme molecules, and genetic control by means of the *operon*, a complete unit of bacterial gene expression and regulation, which consists of structural genes, regulator genes, and control elements. According to the operon theory of gene regulation, genes were endowed with the dual functions of encoding and regulating proteins. If this concept applied to both unicellular and multicellular organisms, it essentially eliminated the intractable paradox of how identical genomes could exist in biochemically differentiated cells. Modification of cells during embryological development or adaptation to new environments could occur despite identical nuclear genomes; the induction of specific proteins would establish fundamental differences in the cytoplasm of differentiated cells. Jacob, Monod, and Lwoff shared the Nobel Prize in 1965 for their "discoveries concerning the genetic regulation of enzyme and virus synthesis."

The collaboration between Jacob and Monod was very different from the better known relationship between Watson and Crick; it was more complex, subtle, and productive. Together, alone, and with many other scientists, and through their elegant and eloquent writings, Jacob and Monod made many contributions to science and to the philosophy and history of science. Another major contribution by Monod dealt with the interaction between enzymes and the small molecules that interact with them as substrates, activators, and inhibitors. According to Monod's theory of allosteric regulation, the presence or absence of small effector molecules controls enzyme activity by causing changes in the shape of the protein that result in changing affinity for specific substrates.

Historians have argued that French biologists such as Jacob and Monod were able to bring a valuable perspective to studies of the mechanism of gene expression because their scientific community had not abandoned complex and subtle questions of growth, nutrition, and development when it adopted modern Mendelian-Morganist genetics. While this idea is undoubtedly important in calling attention to much neglected aspects of the history of the molecular biology of the gene, it does seem to ignore that fact that many of the pioneers of molecular biology were physicists with almost no understanding or sympathy for classical genetics, the "witchcraft" of biochemistry, or arcane theories of generation, growth, reproduction, and embryology. This view of French science also fails to accord with Jacob and Monod's own attitudes towards philosophical aspects of science and their condemnation of Lysenko and Moewus, prototypical icons in the history of scientific fraud, who have been treated with a remarkable degree of sympathy by many historians of science.

Scientists who worked with Monod believed that the key to the power of his theories was their elegance and simplicity. Jacob, who described Monod as a cross between a Roman emperor and a Hollywood movie star, saw Monod as a Cartesian who ascribed elegance to nature and preferred unique solutions to each scientific problem. In contrast to Jacob, who did not see the world as strict and rational, Monod was confident that "a beautiful model or theory may not be right; but an ugly one must be wrong."

As scientists came to accept the implications of the PaJaMo experiment, the question of the role of various forms of RNA and ribosomes was finally clarified. Geneticists realized that the RNA playing the role of the genetic messenger resembled the species of RNA found in the bacterial host after infection by bacteriophage; this RNA was found to have a base composition like the infecting phage, rather than that of the host DNA. For Crick this insight led to a sudden flash of enlightenment that cleared away the dense fog caused by incorrect ideas about ribosomal RNA (rRNA). It became obvious that rRNA was just an inert reading head rather than the messenger that carried information about the sequence of amino acids in proteins. Presumably, ribosomes could read along any message and take part in the synthesis of various proteins. When a conference was held in 1989 to consider the structure, function, and evolution of the ribosome, ribosome studies were old enough on the time scale of molecular biology to merit two historical chapters. Research in ribosomology, in keeping with the emphasis of molecular biology, tends to be RNA-centric; although RNA and protein function in the ribosome are intricately interconnected, ribosomal proteins and translation factors have received less attention than rRNA.

Because messenger RNA presumably turned over quickly, biochemists realized that it should be possible to create and test synthetic messengers and decipher the genetic code. Using an ingenious cell-free protein-synthesizing system prepared from bacteria, Marshall Nirenberg added an artificial message, a synthetic RNA containing only the base uracil, and produced a polypeptide containing only the amino acid phenylalanine. In this manner, the first word in the genetic code was deciphered: UUU was the codon for phenylalanine. Within 5 years, Nirenberg and others had determined all 64 triplets. In 1968 Nirenberg shared the Nobel Prize with Robert W. Holley and Har Gobind Khorana for their "interpretation of the genetic code and its function in protein synthesis." During the highly creative period following Nirenberg's determination of the first codon, scientists established the exact order of nucleotides in each triplet and proved that the code consisted of three-letter, nonoverlapping words, including many that were redundant, and some that served as punctuation marks for beginning and ending the synthesis of polypeptide chains. Building on these discoveries and the painstaking techniques that had made them possible, in 1970, Khorana and his associates were able to synthesize the first wholly artificial gene. The DNA sequence synthesized by Khorana's group coded for alanine transfer RNA; Robert Holley, one of the discoverers of transfer RNA, had reported the complete nucleotide sequence of alanine transfer RNA in 1965.

For Francis Crick, the Cold Spring Harbor annual meeting in 1966, which was devoted to the genetic code, marked the end of classical molecular biology. Having begun biological research in 1947, Crick felt that the questions about the gene that had most intrigued him had essentially been answered. Many years later he admitted that he had not appreciated that molecular biology would make another great and unforeseen leap forward as the consequence of three new techniques: recombinant DNA, rapid DNA sequencing, and monoclonal antibodies. Nor had he foreseen the emergence of a new category of RNA molecules known as ribozymes, remarkable entities with catalytic activity. While the central dogma originally placed RNA in a rather passive and subservient role between DNA and protein, in the 1980s Thomas Cech and others demonstrated that RNA could function as a biocatalyst as well as an information carrier. The discovery of ribozymes (RNA molecules with catalytic activity) prompted some RNA enthusiasts to speculate that RNA molecules, despite their rather hierarchical and predictable structure, might, under the appropriate circumstances, perform virtually all the functions previously ascribed to DNA and proteins.

One way to explore the maturation and transformation of molecular biology is to survey back issues of the *Journal of Molecular Biology* (*JMB*), which was founded by John Kendrew in 1959 when the subject was so new that, according to Sydney Brenner, few people were willing to call themselves molecular biologists in public. At the beginning of the twentieth century, scientists had just rediscovered Mendelian genetics and were beginning to transform the hypothetical unit-factors into genes that could be explained in physical-chemical terms. By 1975 scientists had been able to synthesize functional genes in the laboratory. After 9 years of work by 24 postdoctoral fellows under the direction of Har Gobind Khorana, the gene for tyrosine transfer RNA, which contains 86 nucleotides, was synthesized and shown to be biologically active. Shortly after Khorana's success, another team of researchers announced the synthesis of the gene for rabbit hemoglobin, which has about 650 nucleotide units. By the 1970s molecular biology had became so pervasive that almost all biological researchers could be called molecular biologists. Many of the most fundamental discoveries of molecular biology appeared in *JMB*, including the discovery of messenger RNA; the mechanism of protein biosynthesis; the structure of the genetic code; DNA replication; and the methodology that established nucleic acid sequencing, cloning, transfection, restriction enzymes, DNA mapping, site-directed mutagenesis, and intricate cellular control systems. The "classics" of molecular biology found in *JMB* contain complex and difficult arguments that culminated in ideas which later appeared to be simple and obvious. Technical aspects of the transformation of molecular biology are best grasped by comparing the heroic protocols of classical experiments with the advertisements for commercially available kits and instruments in later issues of the journal.

HOW GENES WORK

For many years, Watson's central dogma served as a stimulus and a framework guiding investigations of genetic mechanisms at the molecular level. Since the beginning of the twentieth century, however, the study of inherited metabolic disorders has made it possible to gain a glimpse of the way the gene works or fails to work. That is, certain inherited diseases are due to defective enzymes, which represent the altered gene products of mutant genes. This correlation was systematically exploited by George W. Beadle and Edward L. Tatum (1909–1975) in their studies of the genetic control of biochemical reactions in *Neurospora*, the red bread mold, but Archibald Edward Garrod (1857–1936) had previously called attention to the link between Mendelian genes and certain metabolic diseases found in the most refractory but most interesting of all biological systems—*Homo sapiens*.

It might be said that the study of biochemical genetics began in 1902 with Garrod's report on alcaptonuria. Individuals with alcaptonuria cannot completely metabolize the amino acid tyrosine and, therefore, excrete an intermediate of tyrosine metabolism known as homogentisic acid. The condition is more frightening to new parents than it is distressing to afflicted individuals, because the urine of babies with alcaptonuria darkens when exposed to air and causes peculiar stains in their diapers. From tests of alcaptonuric individuals, Garrod concluded that the disorder was not the manifestation of an acquired disease, but the result of some alteration in a metabolic process. Other human "chemical abnormalities," such as albinism, cystinuria, and pentosuria, seemed to be similar kinds of disorders attributable to "inborn errors of metabolism." Analyzing the family trees of afflicted individuals suggested that such disorders were inherited as Mendelian recessives. Indeed, as would be likely for recessive genes, a large proportion of the affected individuals were the children of first cousins. Eventually, Garrod traced the defect in alcaptonuria to the penultimate step in the metabolic breakdown of the aromatic amino acids. Individuals with congenital alcaptonuria, Garrod concluded, lacked a special enzyme that was normally involved in splitting the benzene ring of tyrosine and phenylalanine. Garrod proved that his patients could metabolize precursors of homogentisic acid, but that the metabolic pathway was interrupted at that point. Homogentisic acid, therefore, accumulated and was excreted in the urine. Genes were related to enzymes, as indicated by the fact that individuals who had inherited two recessive genes lacked the enzyme needed for specific metabolic reactions. The parents of the affected individuals carried the recessive gene but did not display the condition because they also had one normal gene for the enzyme in question.

Since the publication of Garrod's observations, many other human disorders have been shown to be inborn errors of metabolism, or genetic diseases. But, unlike Garrod, who was primarily a clinician, modern geneticists have generally turned to smaller and smaller creatures for their investigations. Human beings, with their extended childhood, peculiar mating customs, and small family size, hardly seem to represent likely experimental animals. Nevertheless, George Beadle and Edward Tatum, who shared the 1958 Nobel Prize in Medicine or Physiology for establishing the "one gene, one enzyme" theory, later said that their research would have been less circuitous if they had known of Garrod's studies before they began their researches on *Neurospora*. Thus, although Sir Archibald Garrod was once a rather obscure figure, he underwent a rediscovery frenzy rather like Mendel when Beadle suggested that Garrod was the true father of biochemical genetics. Actually, Garrod was never totally forgotten; William Bateson included Garrod's work in *Mendel's Principles of Heredity* (1909) and *Problems of Genetics* (1913). While scientists consider Garrod a pioneer of the theory of gene action for his work on inborn errors of metabolism, Mendelian inheritance, and defective enzymes, some historians have argued that this is just another precursor myth fabricated by scientists. There is, however, no doubt that Garrod was interested in applying Mendelian laws to the pattern of transmission of congenital defects, and his careful attention to the biochemical writings of his contemporaries is obvious in his detailed examination of the biochemical basis of the clinical disorders found in his patients.

Born on a small farm in Wahoo, Nebraska, Beadle earned his B.S. and M.S. from the College of Agriculture of the University of Nebraska in 1926 and 1927, respectively. At Cornell University, Beadle studied maize genetics with Rollins A. Emerson, head of the department of plant breeding and an early American advocate of Mendelian genetics. After graduating from Cornell in 1931, Beadle worked in T. H. Morgan's laboratory at the California Institute of Technology as a postdoctoral fellow. Before accepting an appointment to the faculty at Harvard University, Beadle spent a year doing research on the biochemical genetics of eye color mutations of *Drosophila* at Boris Ephrussi's laboratory in Paris. Ephrussi, who was trained in embryology and tissue culture, had come to Morgan's laboratory to learn about *Drosophila* genetics. Ephrussi and Beadle were especially interested in the chain of reactions between gene and character. Beadle and Ephrussi attempted to analyze the biochemical basis of heredity through studies of the genetic control of eye color pigment in fruit flies. They demonstrated that a larval eye taken from the mutant type vermillion fly developed into an adult structure with the wild-type eye color when it was implanted in the abdomen of a wild-type larva. Tests with various mutants proved that the steps involved in the sequential synthesis of eye pigment were controlled by several separate genes. Some of the biochemical intermediates in the biosynthetic pathway to pigment were chemically identified. In 1937 Beadle became professor of biology at Stanford. When Morgan died in 1945, Beadle was awarded the position of chairman of the division of biology at Caltech.

While at Stanford, Beadle hired Tatum, who was trained in biochemistry and bacteriology, as a research associate. Tatum later became professor and head of the department of biochemistry at Stanford. Frustrated with the *Drosophila* work, they turned to the bread mold *Neurospora crassa* as a means of unraveling the biochemistry of the gene–gene product relationship. Work on the genetics of the common bread mold had been carried out in Morgan's laboratories at Caltech. Moreover, Tatum had considerable experience in studying the biochemical pathways and nutritional requirements of bacteria and fungi. Beadle and Tatum decided to reverse the procedures generally used to identify specific genes with particular chemical reactions. Instead of taking a mutant as their starting point and then searching for the chemical reaction it controlled, they decided to begin with known chemical reactions and then look for the genes controlling them. They therefore needed an organism with well-known biochemical pathways. *Neurospora* was ideal; it had a fairly short life cycle, and techniques for genetic analysis had already been worked out. Furthermore, the wild-type mold could be grown on a simple synthetic medium containing only a carbon source, certain inorganic salts, and the vitamin biotin. Mutations were generated using X-rays; mutants that had lost the ability to synthesize certain organic substances were then deliberately selected. Biochemical and genetic analyses proved that the mutant strains were genetically different from the parental type. During the 1940s, Beadle and Tatum's work was regarded with considerable skepticism and even hostility. Later there were attacks on their originality and priority. Archibald Garrod and Franz Moewus were cited, respectively, for having proposed that genes controlled the action of enzymes and for using microorganisms to analyze biochemical genetics.

In the 1940s, many biologists called Moewus one of the outstanding scientists of the twentieth century, but later critics believed that he was actually the perpetrator of a remarkable fraud. Moewus began work on the genetics of microorganisms in the 1930s. His first paper on the biochemical genetics of sexuality in *Chlamydomonas*, a unicellular green algae, was published in 1938, 3 years before Beadle and Tatum's first *Neurospora* paper. According to statistical critiques, Moewus's results were too good to have been the products of real experiments; attempts to replicate Moewus's work failed. In spite of subsequent charges of fraud, some biologists credit Moewus with creating interest in studies of the genetics of microorganisms. The controversy surrounding Moewus's work led contemporaries to conclude that Moewus could be described as truly great, but it was difficult to determine whether he was a great faker or a great scientist.

In 1945 when Beadle enunciated the one gene, one enzyme hypothesis, the chemical identity of the gene was still unknown. After 1953 the question "What is the gene?" was generally framed in terms of explaining how information encoded in DNA could be used to direct the synthesis of proteins. Viruses, which could now be seen as nucleic acids, or genes, wrapped in protein coats, seemed to be the perfect models for exploring the flow of genetic information. Nevertheless, despite the prominence of bacteriophage studies in the evolution of molecular genetics during this period, it was the analysis of another human disease that led to the first significant demonstration of the relationship between a heritable gene mutation and a specific molecular alteration in a gene product. By analyzing the abnormal hemoglobin produced by patients with sickle cell anemia (SCA), Linus Pauling and Vernon M. Ingram established the existence of a specific molecular disease. Hemoglogin, the oxygen-carrying protein present in red blood cells, has been aptly described as a "molecular lung."

Sickle cell anemia is a severe, chronic, and crippling disorder. Under the microscope, the red blood cells of individuals with sickle cell anemia assume a peculiar crescent or sickle shape. This ancient genetic disorder was not rare in Africa and the southern Mediterranean, but the first fully documented report of SCA did not appear until 1910, when Dr. James B. Herrick of Chicago published his case study of Walter Clement Noel, a 20-year-old West Indian student studying dentistry at the Chicago College of Dental Surgery. Noel died at the age of 32, 9 years after returning to Grenada. Although the general idea that "sicklemia" was an inherited disease was accepted by the 1930s, the relationship between *sickle cell anemia*, the disease that afflicts individuals with two copies of a gene for the abnormal hemoglobin known as hemoglobin S, and *sickle cell trait* was not fully clarified until the 1950s. The red blood cells of individuals who have inherited only one gene for sickle cell hemoglobin display sickle cell trait; although these cells usually appear to be normal, they can undergo the reversible change in shape known as *sickling* in response to alterations in the partial pressure of oxygen. People with at least one copy of the sickle cell gene appear to be resistant to the most virulent form of the malarial parasite. Given estimates that at least 20% of the Africans brought to America possessed the sickle cell gene, it is not surprising that slave owners came to believe that blacks were less susceptible to malaria than whites. Sickle cell trait and sickle cell anemia offered some protection from malaria, but children weakened by sickle cell anemia were very likely to die of nonspecific infections long before reaching their twentieth birthday.

Research in Pauling's laboratory in 1949 proved that the sickling phenomenon was caused by an alteration in hemoglobin. Pauling argued that sickle cell anemia represented a specific molecular disease, traceable to an alteration in the chemical and physical properties of the hemoglobin molecule;

that is, normal and sickle cell hemoglobins acted differently because of a specific chemical change in the sickle cell hemoglobin molecule. When T. H. Morgan came to Caltech in 1929, Pauling became actively interested in a wide variety of biological problems representing what he called "molecular architecture and the processes of life." During the 1930s Pauling was involved in studies of the relationship between the structure and function of the hemoglobin molecule. Thinking about the physical properties of hemoglobin led to a long-term interest in the structure of proteins and the nature of antibody-antigen reactions. Pauling became interested in sickle cell anemia in 1945 when he heard Dr. William B. Castle discuss his studies of the red blood cells of patients with sickle cell anemia. Immediately thinking of the disease in molecular terms, Pauling assigned the problem to Harvey Itano, an M.D. working on a Ph.D. in chemistry. In 1949 Pauling and his associates published a paper entitled "Sickle-Cell Anemia, a Molecular Disease," which demonstrated that sickle cell hemoglobin differed from normal adult hemoglobin in terms of its physical properties. After applying his methods to many other biological and medical problems, Pauling came to think of "molecular medicine" as a potentially valuable branch of molecular biology.

Ranked along with Newton and Einstein as one of the 20 most important scientists of all time, Pauling's willingness to tackle a wide array of social, political, and medical questions, has also earned him a reputation as one of the twentieth century's most controversial scientists. He has the distinction of having been awarded both the Nobel Prize in Chemistry (1954) and the Nobel Peace Prize (1962). Whether or not analysts see Pauling as genius, maverick, troublemaker, or national treasure, there is no doubt that his text *The Nature of the Chemical Bond* (1939) is a classic and a model of clarity. Pauling has written many other books and papers dealing with the structure of crystals, the nature of the chemical bond, quantum mechanics, proteins and antibodies, biology and medicine, war and peace, the dangers of atmospheric testing of nuclear weapons, science and world affairs. The State Department revoked his passport because of antiwar activities; an investigative congressional subcommittee subjected him to intensive and hostile questioning in 1960; he was attacked in the media for allegedly pro-communist sympathies; he launched a series of lawsuits and has became an authority on litigation and libel law; he has been involved in the holistic or alternative health movement and is a leader in the battle for orthomolecular medicine and megadose vitamin C therapy and prophylaxis for colds, schizophrenia, cancer, etc. His controversial ideas about vitamin C and cancer led to the unprecedented action by the editorial board of the *Proceedings of the National Academy of Sciences* (*PNAS*) in rejecting a paper co-authored by

Pauling and Ewam Cameron, a Scottish surgeon and cancer specialist. The results of clinical trials of vitamin C as a preventive agent against the common cold and cancer remain controversial, with claims of bias on both sides. In the 1990s Pauling also assumed the role of honorary president of the International Campus of the Earth Environment University Roundtable and Academy, which is dedicated to "absolute peace" through the teachings of Hisatoki Komaki.

Although Pauling's work proved that sickle cell anemia was a molecular disease, the precise nature of the alteration in the hemoglobin molecule was unclear. In 1957, Vernon Ingram published an account of the specific chemical difference between normal and sickle cell hemoglobin. Basing his approach on the methods used by Frederick Sanger to determine the complete amino acid sequence of insulin, Ingram was able to determine and compare the amino acid sequences of normal and sickle cell hemoglobin. During the 1940s, Sanger had chosen insulin for investigation because this important hormone was one of the few proteins then available in a pure form. Insulin's relatively small size was significant in allowing Sanger to establish a complete sequence, but little was known about the relative sizes of proteins when he began this work. Moreover, insulin is a complicated protein, which consists of two different polypeptides joined by cross-linking disulfide bridges. Sickle cell hemoglobin was found to differ from the normal type by only one amino acid; in hemoglobin S a valine residue had been substituted for the normally occurring glutamic acid residue. Given the fact that the two forms of hemoglobin differed with respect to only one amino acid, Ingram postulated that a corresponding change, a true point mutation, had occurred in the gene coding for hemoglobin. Since geneticists had already shown that sickle cell anemia was transmitted as a simple Mendelian recessive factor, the Mendelian inheritance of a significant human disease, genetic mutation, and the precise alteration in a gene product had been linked together for the first time.

REVERSE TRANSCRIPTASE, RECOMBINANT DNA, AND THE NEW CENTRAL DOGMA

As they eventually went their separate ways, James Watson, Francis Crick, and other members of the phage group devoted much of their attention to determining the mechanism by which information encoded in DNA was incorporated into the amino acid sequence of proteins. In this search they were guided by the central dogma encoded in the cryptic note James Watson had taped above his desk: "DNA → RNA → protein."

In 1958 Crick published an elegant and influential article entitled "On Protein Synthesis," which addressed special aspects of theoretical biology. Primarily, Crick used this occasion to discuss the central dogma and analyze the validity of what he called the "sequence hypothesis" (the sequence of nucleic acids determines the sequence of amino acids in proteins). The paper provides an excellent example of Crick's conviction that theories are valuable guides, even if they are later found to be incorrect. Proteins, Crick remarked, are the macromolecules that do most of the jobs needed for life. Indeed, proteins can do almost anything; proteins serve as enzymes and have vital structural functions. Yet it is DNA that is most fascinating, because it serves as the material basis of inheritance and is ultimately responsible for the synthesis of proteins. In looking at the base sequence of DNA and the amino acid sequence of proteins, Crick made many predictions about how DNA could direct the synthesis of proteins, including his famous adaptor hypothesis.

Analyzing the logic of the colinearity between the sequence of bases in nucleic acids and the amino acids in proteins, Crick elaborated on the implications of the central dogma. According to Crick, the central dogma stated that information is transferred from nucleic acid to nucleic acid, or nucleic acid to protein, but not from protein to nucleic acid. In this context, *information* refers to the precise determination of sequence. When information passes into protein it cannot get out again. However, within 10 years it became clear that the central dogma, at least in its original form, had failed to predict the possibility that information could be transferred from RNA to DNA.

In 1975 the Nobel Prize in Physiology or Medicine was awarded to Howard Temin, Renato Dulbecco, and David Baltimore for discoveries concerning the interaction between tumor viruses and the genetic material of the host cell and the possible relationship between viruses and human cancers. As a graduate student at the California Institute of Technology, Temin planned to study experimental embryology but was drawn to Renato Dulbecco's work on animal virology. In 1958 Temin and Harry Rubin, a postdoctoral fellow, developed the first reproducible in vitro assay for a tumor virus. Continuing his work on Rous sarcoma virus (RSV) at the University of Wisconsin, Temin noted that the inhibition of DNA synthesis and DNA-dependent RNA synthesis blocked RSV infection. This observation was surprising because the RSV genome is single-stranded RNA, and it was assumed that the RNA served as the genetic material. To explain his observations, Temin suggested that a DNA intermediate must be involved in RSV infection.

Howard Temin, Nobel Laureate, 1975

David Baltimore, Nobel Laureate, 1975

According to Temin's *provirus hypothesis*, after an RNA virus enters a host cell, a DNA provirus is synthesized. The DNA provirus contains the genetic information of the RNA viral genome, and progeny viral RNA is synthesized from the DNA provirus. This hypothesis could also account for the integration of RSV genetic material into the host genome. The use of various inhibitors in these experiments suggested possible artifacts, and Temin's hypothesis was not readily accepted. The major impediment, however, seems to have been the way in which the provirus hypothesis conflicted with the central dogma of molecular genetics. New ways of thinking about the problem appeared in 1970, when Temin and Satoshi Mizutani proved that RSV particles contain an enzyme that transcribes single-stranded viral RNA into DNA. Soon after Temin and Mizutani reported the existence of reverse transcriptase, David Baltimore discovered the same enzyme in Rauscher murine leukemia virus. The new enzyme provided an obvious mechanism for the formation of a DNA intermediate in RNA viruses.

The RNA tumor viruses and reverse transcriptase quickly became basic tools for studying the molecular biology of mammalian cells. Crick has argued that reverse transcriptase does not really contradict the central dogma inasmuch as the transcription of information from one form of nucleic acid to another proceeds through essentially the same base pairing process. If the discovery of reverse transcriptase, ribozymes, transposable genetic elements, and other novel aspects of "postclassical" molecular biology do not overturn the most cautiously elaborated formal versions of the central dogma, perhaps they have dealt a major blow to a more informal version of the dogma which essentially held that what is true for a phage is true for an elephant. Indeed, the molecular biology of the gene in higher organisms, along with many aspects of cell growth, differentiation, and regulation, proved to be much more complicated than that of the microorganisms that served as the first successful model systems for molecular biologists. Recognition of this fundamental fact made it possible for scientists to pursue entirely new avenues of research and helped to explain previously misunderstood and obscure studies of the genetics of plants and animals, such as Barbara McClintock's pioneering experiments on the relationship between structural aspects of plant chromosomes and the effects of various genes. In the 1940s, McClintock's analysis of complex chromosomal rearrangements and the chromosomal basis of crossing over led to her discovery of transposable genetic elements. McClintock's transposable factors were generally ignored or humorously referred to as "jumping genes" and treated as a minor curiosity or aberration peculiar to maize genetics until similar factors were discovered in bacteria and higher animals in the 1960s. When McClintock, who was born in 1902, was finally awarded the Nobel Prize in 1983, she became the first geneticist since H. J. Muller to win such an unshared award.

In the 1970s, scientists learned how to cut and splice DNA to form recombinant DNA molecules, and, using reverse transcriptase and specific DNA probes as tools, scientists were able to detect viral genetic information in normal and malignant cells. The new techniques of molecular biology make it possible to create genetic information that might be used to treat genetic diseases and manipulate the genetic materials of plants, animals, and microorganisms. The benefits and dangers of possible genetic engineering were debated, however, even before more powerful, rapid, and simplified techniques became available. The public alarms and fears raised by the new biology, especially genetic engineering and recombinant DNA, led to efforts to evaluate alleged biohazards through an unprecedented voluntary research moratorium. The controversy about the use and potential abuse of recombinant DNA technology dates back to the early 1970s. Although some concerns remained even after two decades of experience, by 1990 the excitement generated by the prospects for therapeutic and diagnostic benefits clearly outweighed the burden of fear and uncertainty raised by the new techniques of molecular biology.

At a small but prestigious conference on nucleic acids held in 1973, scientists announced that pieces of DNA taken from different organisms could be linked together in the test tube and reinserted into a host organism where the newly created hereditary material could freely multiply. News of this novel technique spread rapidly through the scientific grapevine, and within months scientists were publicly calling for widespread and deliberate discussions of the possible risks involved in creating unprecedented forms of genetically engineered DNAs. In essence, the new technique developed out of attempts to determine whether animal tumor viruses could be used to probe the mechanism of gene expression in higher organisms in a manner analogous to the way in which bacterial viruses had been used to work out the mechanism of gene expression in bacterial systems.

Eventually, when scientists and historians look back at the debates that marked the debut of recombinant DNA technology, they may wonder why such preliminary findings should have generated expectations that Frankensteinian biological monsters would soon escape from every biological laboratory. It could be argued that recombinant DNA need not have been singled out from related areas as uniquely dangerous. Some observers argued that the controversy was merely a product of media attention during an era of generalized and exaggerated anxiety, exacerbated by widespread misunderstanding of the nature of science. After all, despite the universal fear of everything possibly related to cancer, there was little interest in the potential hazards of working with tumor viruses and pathogens before the prospect of genetic engineering was addressed. Scientists, however, expressed concern about the potential dangers of combining work on molecular biology, cancer, and tumor viruses as it became clear that the exciting new field was attracting researchers with little or no background in microbiology and pathology who were ready to work in or establish laboratories lacking the safety equipment and conventions traditionally used when dealing with known and potential pathogens.

Just as remarkable as the initial controversy has been the virtual disappearance of the near-hysteria that peaked during the 1970s and the subsequent euphoria that developed in the course of the next decade with the prospect of a cornucopia of molecular biology–based commercial developments. Entrepreneurs and scientists rushed to create biotechnology firms, attract venture capitalists, and trade shares on the stock exchange. The focus of media and public attention rapidly shifted from theoretical risks to the expectation of miraculous "breakthroughs" ranging from cures for cancer and genetic diseases to designer fruits and vegetables. New techniques make it possible to obtain large quantities of human gene products such as insulin and growth hormone. By the 1980s scientists, physicians, and health policy analysts were considering the possibility that developments in genetics would make it possible to manage disease at the level of the gene, rather than through the relief of symptoms. The techniques of molecular biology allow scientists to study the morbid anatomy of the chromosome and locate defective genes, much as Giovanni Morgagni's (1682–1771) postmortems located specific pathological lesions at a specific site; indeed, by the 1980s scientists had established the existence of some 3000 human genetic diseases. The new tools and methodologies made possible by molecular biology obviously hold great promise for promoting human welfare and the terrible potential for causing unprecedented problems. While there is little doubt about the eventual technical feasibility of treating many genetic diseases, genetic engineering raises social, ethical, moral, and perhaps even long-term evolutionary questions.

There is little doubt that the techniques of molecular biology will provide powerful new diagnostic tools, which might well lead to better understanding of disease processes and cell function at the gene level and more rational pharmacological approaches to treatment. Genetic engineering might, however, better be used to produce "designer drugs," hormones, interferons, interleukins, and so forth. In theory the ability to produce genes and gene products in the laboratory should allow "replacement therapy" for the hundreds of genetic diseases for which the defective gene and gene product are known, but human beings are not single-celled organisms where proteins can easily be directed to the appropriate site.

Despite all the immediate hypothetical obstacles, the potential of genetic engineering became real enough by the 1980s to interest biotechnology companies and regulatory agencies. Tentative successes in a small number of cases in which patients lacking specific enzymes were treated with genetically modified cells containing a functional gene for the missing enzyme were achieved. On the other hand, some researchers were reprimanded for carrying out human experimentation that their peers considered premature, uncertain, and inappropriate. Perhaps the most exciting outcome of the revolution in biology in the last quarter of the twentieth century has been the combination of molecular biology, immunology, and cancer research which led to techniques for the production of monoclonal antibodies. Even though the prospects for gene therapy in the 1990s remained a long way from "off-the-shelf" genes on demand, some biotechnology companies were hoping that genetically engineered products would be possible, practical, and profitable in the not too distant future.

The exciting possibilities and the heartbreaking reality of genetic diseases make patients and their families very vulnerable to appeals to volunteer for experiments that might well be dangerous and ineffective. Some scientists argue that the technical and ethical dilemmas posed by experimentation with human gene therapy are unacceptable. Complications must be expected, because it is now clear that, just as the atom is not a miniature billiard ball, the mammalian gene is not a simple sequence of beads on a string. Genetic engineering, however, could have a profound impact on human health on a global scale if attention were focused on improving useful plants and animals in order to increase food production and provide more nutritious foods with better quality proteins, lower levels of saturated fat and cholesterol, and higher vitamin content.

Genetic manipulation in plants may not sound as exciting as a direct attack on human disease, but it may ultimately have global effects on human health. Indeed, if it is true that general improvements in human health owe more to nutrition than to medical interventions, then on a worldwide scale, the genetic manipulation of plants, and the development of safe bioinsecticides, may have profound effects on patterns of morbidity and mortality. For example, novel varieties of plants might be made resistant to insects, parasites, drought, flooding, heat, cold, high salt concentration, and pollutants. Changes in the protein composition of major food crops could prevent much of the malnutrition that plagues developing countries. And, since plants have always been nature's apothecary shop, genetic manipulation of plants might provide a new generation of pharmacologically active and commercially valuable products. Plants might even serve as "factories" for the production of the genes and proteins of other species, including mammals.

Molecular biology has made it possible to plan and execute international research programs such as the Human Genome Project, which is dedicated to mapping and sequencing all of the estimated 100,000 genes and the interconnecting segments found in the 23 pairs of chromosomes that carry the human genetic legacy. The human genome initiative was officially launched in 1990, although the possibility of carrying out such a project had been raised in the 1970s when the first primitive maps of human chromosomes were established. At a workshop on the human genome project held in Los Alamos in March 1986, Walter Gilbert, a pioneer in DNA-sequencing techniques and entrepreneurial biotechnology, declared that sequencing the entire human genome was the ultimate response to the primary commandment: "Know thyself." Scientists were urged to join the great quest to seize and sequence the "holy grail" of human genetics. The prospect of locating and sequencing the estimated 100,000 genes (ignoring the presumably nonfunctional interconnecting segments) seemed overwhelming, but new techniques for rapid and automated sequencing emerged in the 1980s and continuously improved. According to some geneticists, only about 5% of the base pairs in human DNA actually code for genes; the remaining noncoding regions are called *introns*, or *junk DNA*. By estimating the cost of sequencing at $1 per base pair, Walter Gilbert predicted that a crash program could provide the completed sequence for $3 billion. Some scientists warned against mortgaging the future of biology to a mindless routine of big science sequencing, but entrepreneurs and biotechnology firms eagerly anticipated exploiting the huge market for DNA diagnostic kits and novel drugs that even a partial sequencing of the genome is likely to generate. Many landmarks in the quest for the holy grail of genetics occurred in 1988. James Watson agreed to serve as director of the Office for Human Genome Research established by the National Institutes of Health; an international council known as the Human Genome Organization (HUGO) was established to coordinate research and encourage the exchange of information as a sort of United Nations for the human genome; and the European Commission of the European Community established its own human genome project.

Perhaps broader claims that the Human Genome Project will provide the complete script from which scientists will read or decode the "nature" of

James D. Watson at a National Institutes of Health DNA Conference

man (and perhaps even the nature of woman) can be put into historic perspective by remembering that Vesalius once claimed that his anatomical studies constituted a true reading of the book of man. Just as the *Fabric of the Human Body* did not mark the end of the search for understanding the nature of life, only the most malignant sort of hubris could make anyone believe that the eventual publication of the complete human genome will be the end of the quest rather than the beginning of a new text. Reading the complete book of the humane genome will certainly involve a more complete grammar of biology and a new avalanche of knowledge, but it will not in itself establish a new way of understanding or a new wellspring of wisdom. To avoid future disappointment, disillusionment, and a great deal of serious mischief, if not the great apocalypse some have predicted, it might be useful to establish a new version of the central dogma, which will simply remind us that "Knowledge is not equivalent to Wisdom."

SUGGESTED READINGS

Abir-Am, P. G. (1992). The politics of macromolecules. Molecular biologists, biochemists, and rhetoric. *Osiris (2nd series) 7*: 164–191.

Allen, G. E. (1975). *Life Science in the Twentieth Century*. New York: Wiley.

Barrett, J. T., ed. (1986). *Contemporary Classics in the Life Sciences*. 2 vols. Philadelphia, PA: ISI Press.

Beadle, G. W. (1963). *Genetics and Modern Biology*. Philadelphia, PA: American Philosophical Society.

Bowler, P. J. (1989). *The Mendelian Revolution: The Emergence of Hereditarian Concepts in Modern Science and Society*. Baltimore, MD: Johns Hopkins University Press.

Brenner, S., compiler (1989). *Molecular Biology: A Selection of Papers*. London: Academic Press.

Cairns, J., Stent, G. S., and Watson, J. D., eds. (1992). *Phage and the Origins of Molecular Biology*. 2nd ed. New York: Cold Spring Harbor Laboratory of Quantitative Biology. 2nd ed. 1992.

Carmen, I. H. (1985). *Cloning and the Constitution. An Inquiry into Governmental Policymaking and Genetic Experimentation*. Madison, WI: University of Wisconsin Press.

Cavalieri, L. F. (1981). *The Double-Edged Helix: Science in the Real World*. New York: Columbia University Press.

Chargaff, E. (1963). *Essays on Nucleic Acids*. New York: Elsevier.

Chargaff, E. (1978). *Heraclitean Fire: Sketches from a Life before Nature*. New York: Rockefeller University Press.

Cherfas, J. (1983). *Man-Made Life: An Overview of the Science, Technology and Commerce of Genetic Engineering*. New York: Pantheon Books.

Crick, F.(1981). *Life Itself: Its Origins and Nature*. New York: Simon & Schuster.

Crick, F. (1988). *What Mad Pursuit. A Personal View of Scientific Discovery*. New York: Basic Books.

Dubos, R. J. (1976). *The Professor, the Institute, and DNA*. New York: The Rockefeller University Press.

Dulbecco, R. (1987). *The Design of Life*. New Haven, CT: Yale University Press.

Fischer, E. P., and Lipson, C. (1988). *Thinking About Science. Max Delbrück and the Origins of Molecular Biology*. New York: Norton.

Florkin, M., and Stotz, E. H., eds. (1972). *A History of Biochemistry. Comprehensive Biochemistry*. Vol. 30. New York: Elsevier.

Fruton, J. S. (1972). *Molecules and Life: Historical Essays on the Interplay of Chemistry and Biology*. New York: Wiley-Interscience.

Fruton, J. S. (1992). *A Skeptical Biochemist*. Cambridge, MA: Harvard University Press.

Garber, E. D., ed. (1985). *Genetic Perspectives in Biology and Medicine*. Chicago, IL: University of Chicago Press.

Garrod, A. E. (1963). *Inborn Errors of Metabolism*. London: Oxford University Press (reprint).

Gibbons, J. H. (1988). *Mapping Our Genes*. Washington, DC: Office of Technology Assessment.

Glover, D. M., ed. (1989). *Genes and Embryos. Frontiers in Molecular Biology*. New York: Oxford University Press.

Gore, A. (1985). *Biotechnology, Implications for Public Policy*. Washington, DC: Brookings Institution.

Grobstein, C. (1979). *A Double Image of the Double Helix*. San Francisco, CA: Freeman.

Hall, S. S. (1987). *Invisible Frontiers. The Race to Synthesize a Human Gene*. New York: Atlantic Monthly Press.

Huemer, R. P., ed. (1986). *The Roots of Molecular Medicine: A Tribute to Linus Pauling*. New York: W. H. Freeman.

Jackson, D. A., and Stich, S. P., eds. (1979). *The Recombinant DNA Debate*. Englewood Cliffs, NJ: Prentice Hall.

Jacob, F. (1976). *The Logic of Life. A History of Heredity*. New York: Vintage.

Jacob, F. (1988). *The Statue Within. An Autobiography*. Trans. by Franklin Philip. New York: Basic Books.

Judson, H. F. (1979). *The Eighth Day of Creation: The Makers of the Revolution in Biology*. New York: Simon and Schuster.

Kaplan, N. O., and Robinson, A., eds. (1983). *From Cyclotron to Cytochromes. Essays in Molecular Biology and Chemistry*. New York: Academic Press.

Kay, L. E. (1993). *The Molecular Vision of Life: Caltech, The Rockefeller Foundation, and the Rise of the New Biology*. New York: Oxford University Press.

Keller, E. F. (1983). *A Feeling for the Organism: The Life and Work of Barbara McClintock*. San Francisco, CA: W. H. Freeman.

Kendrew, J. S. (1968). *The Thread of Life: An Introduction to Molecular Biology*. Cambridge, MA: Harvard University Press.

Kenney, M. (1986). *Biotechnology: The University-Industrial Complex*. New Haven, CT: Yale University Press.

Kevles, D. J., and Hood, L., eds. (1992). *The Code of Codes. Scientific and Social Issues in the Human Genome Project*. Cambridge, MA: Harvard University Press.

Kleinkauf, H., Von Dohren, H., and Jaenicke, L., eds. (1987). *The Roots of Modern Biochemistry. Fritz Lipmann's Squiggle and Its Consequences*. Hawthorne, NY: De Gruyter.

Kohler, R. E. (1982). *From Medical Chemistry to Biochemistry*. Cambridge, England: Cambridge University Press.

Kornberg, A. (1989). *For the Love of Enzymes. The Odyssey of a Biochemist*. Cambridge, MA: Harvard University Press.

Kornberg, A., Horecker, B. L, Cornudella, L., and Oro, J., eds. (1976). *Reflections on Biochemistry in Honour of Severo Ochoa*. Oxford, England: Pergamon Press.

Krimsky, S. (1984). *Genetic Alchemy. The Social History of the Recombinant DNA Controversy*. Cambridge, MA: MIT Press.

Lederberg, J., compiler (1990). *The Excitement and Fascination of Science: Reflections by Eminent Scientists*. Vol. III. Palo Alto, CA: Annual Reviews.

Lederman, M., and Tolin, S. A. OVATOOMB: other viruses and the origins of molecular biology. *Journal of the History of Biology* 26: 239–254.

Lee, Thomas F. (1991). *The Human Genome Project. Cracking the Genetic Code of Life*. New York: Plenum.

Leicester, H. M. (1974). *Development of Biochemical Concepts from Ancient to Modern Times*. Cambridge, MA: Harvard University Press.

Levene, P. A., and Bass, L. W. (1931). *Nucleic Acids*. New York: Chemical Catalogue.

Luria, S. E. (1984). *A Slot Machine, A Broken Test Tube. An Autobiography*. New York: Harper & Row.

Lwoff, A., and Ullmann, A., eds. (1988). *Origins of Molecular Biology: A Tribute to Jacques Monod*. New York: Academic Press.

McCarty, M. (1985). *The Transforming Principle. Discovering That Genes Are Made of DNA*. New York: Norton.

McClintock, B. (1987). *The Discovery and Characterization of Transposable Elements: The Collected Papers (1938–1984) of Barbara McClintock*. New York: Garland Publishing.

Monod, J. (1971). *Chance and Necessity. An Essay on the Natural Philosophy of Modern Biology*. New York: Knopf.

Monod, J., and Borke, E., eds. (1971). *Of Microbes and Life*. New York: Columbia University Press.

National Academy of Sciences (1977). *Research with Recombinant DNA: An Academy Forum*. Washington, DC: National Academy of Sciences.

Nelkin, D., and Tancredi, L. (1989). *Dangerous Diagnostics: The Social Power of Biological Information*. New York: Basic Books.

Nichols, E. K. (1988). *Human Gene Therapy*. Cambridge, MA: Harvard University Press.

Olby, R. (1974). *The Path to the Double Helix*. Seattle, WA: University of Washington Press.

Pauling, L. (1970). Fifty years of progress in structural chemistry and molecular biology. *Daedalus 99*: 988–1014.

Perutz, M. (1992). *Is Science Necessary? Essays on Science and Scientists*. New York: Oxford University Press.

Portugal, F. H., and Cohen, J. S. (1977). *A Century of DNA*. Cambridge, MA: MIT Press.

Quagliariello, E., Bernardi, G., and Ullmann, A., eds. (1987). *From Enzyme Adaptation to Natural Philosophy: Heritage from Jacques Monod*. New York: Elsevier.

Richards, E. (1991). *Vitamin C and Cancer. Medicine or Politics?* New York: St. Martin's Press.

Sager, R., and Ryan, F. (1961). *Cell Heredity*. New York: Wiley.

Sapp, J. (1990). *Where the Truth Lies. Franz Moewus and the Origins of Molecular Biology*. New York: Cambridge University Press.

Savitt, T. L., and Goldberg, M. F. (1989). Herrick's 1910 case report of sickle cell anemia. *Journal of the American Medical Association 261*: 266–271.

Sayre, A. (1975). *Rosalind Franklin and DNA*. New York: Norton.

Schrödinger, E. (1992). *What Is Life?* New York: Cambridge University Press (reprint).

Scriver, C. R. (1989). *Garrod's Inborn Factors in Disease*. Facsimile of the 1931 edition with commentary. New York: Oxford University Press.

Serafini, P. (1989). *Linus Pauling: A Man and His Science*. New York: Paragon House.

Srinivasan, P. R., Fruton, J. S., and Edsall, J. T., eds. *The Origins of Modern Biochemistry: A Retrospect on Proteins*. New York: New York Academy of Sciences.

Stent, G. S., ed. (1980). *The Double Helix*. A Norton Critical Edition. New York: Norton.

Teich, M. and Needham, D. M. (1992). *A Documentary History of Biochemistry, 1770–1940*. Rutherford, NJ: Farleigh Dickinson University Press.

Tiley, N. A. (1983). *Discovering DNA: Meditations on Genetics and a History of the Science*. New York: Van Nostrand Reinhold.

Watson, E. L. (1991). *Houses for Science. A Pictorial History of Cold Spring Harbor Laboratory. With Landmarks in Twentieth Century Genetics, A Series of Essays by James D. Watson*. Plainview, NY: Cold Spring Harbor Laboratory Press.

Watson, J. D., and Tooze, J. (1981). *The DNA Story: A Documentary History of Gene Cloning*. San Francisco, CA: Freeman.

Wills, C. (1991). *Exons, Introns, and Talking Genes. The Science Behind the Human Genome Project*. New York: Basic Books.

Wilson, E. B. (1925). *The Cell in Development and Heredity*. 3rd ed. New York: MacMillan.

Zilinskas, R. A., and Zimmerman, Burke K., eds. (1986). *The Gene-Splicing Wars: Reflections on the Recombinant DNA Controversy*. New York: Macmillan.

INDEX

Abbe, Ernst, 164
academy, 132–134
Academy (Plato's), 30–31, 33, 36, 73, 132
Academy of Sciences (France), 247, 248, 252, 266, 272, 331
Accademia dei Lincei, 133, 149
Accademia del Cimento, 133–134, 157, 231, 267
Accademia Secretorium Naturae, 133
achromatic lens, 163
achromatic microscopes, 193
acquired characteristics, 334–335
active host immunity, 297, 301
Adam, 177, 192, 326
Adams, John, 144
adaptor hypothesis, 456
agar-agar, 288
Agnodice, 52
Agricola, Georg Bauer, 340
AIDS, 289, 293, 294, 455
Albertus Magnus, 76–77

albinism, 462
alcaptonuria, 462–463
alchemy, 4, 79, 81, 102–107, 238, 241, 242
Alcmaeon, 17, 172
Alcock, Rutherford, 373
Alexander the Great, 33, 47, 56
Alexandrian Era, 47–56
Alhazen, 147
Al-Kindi, 79
allele, 409
allelomorph, 409
Althoff, Friedrich, 306
altruism, 372
American Academy of Arts and Sciences, 144
American Association for the Advancement of Science, 145
American Council of Learned Societies, 144
American Council on Education, 144
American Philosophical Society, 144